Aristotle and the Animals

With a novel approach to Aristotle's zoology, this study looks at animals as creatures of nature (*physis*) and reveals a scientific discourse that, in response to his predecessors, exiles logos as reason and pursues the logos intrinsic to animals' bodies, empowering them to sense the world and live.

The volume explores Aristotle's conception of animals through a discussion of his ad hoc methodology to study them, including the pertinence of the soul to such a study, and the rise of zoology as a branch of natural philosophy. For Aristotle, animal life stems from the body in the space of existence and revolves around sensation, which is entwined with pleasure, pain, and desire. Lack of human reason is irrelevant to an understanding of the richness of animal life and cognition. In sum, the reader will acquire knowledge of the "animal as such," which lay at the core of Aristotle's agenda and required a study of its own, separate from plants and the elements.

This book is intended for students of the history of science, ancient biology, and philosophy and all those who, from different fields, are interested in animal studies and the human–animal relation.

Claudia Zatta (PhD, Johns Hopkins University, USA) is the author of *Interconnectedness: The Living World of the Early Greek Philosophers* (2019, second edition) and numerous articles on different aspects of the classics. She currently teaches at the American College of Greece in Athens.

Routledge Monographs in Classical Studies

Recent titles include:

Travel, Geography, and Empire in Latin Poetry
Edited by Micah Young Myers and Erika Zimmermann Damer

Ancient History from Below
Subaltern Experiences and Actions in Context
Edited by Cyril Courrier and Julio Cesar Magalhães de Oliveira

Ideal Themes in the Greek and Roman Novel
Jean Alvares

Thornton Wilder, Classical Reception, and American Literature
Stephen J. Rojcewicz, Jr.

Married Life in Greco-Roman Antiquity
Edited by Claude-Emmanuelle Centlivres Challet

Future Thinking in Roman Culture
New Approaches to History, Memory, and Cognition
Edited by Maggie L. Popkin and Diana Y. Ng

Aristotle and the Animals
The Logos of Life Itself
Claudia Zatta

The *Aeneid* and the Modern World
Interdisciplinary Perspectives on Vergil's Epic in the 20th and 21st Centuries
Edited by J.R. O'Neill and Adam Rigoni

The Province of Achaea in the 2nd Century CE
The Past Present
Edited by Anna Kouremenos

For more information on this series, visit: www.routledge.com/Routledge
-Monographs-in-Classical-Studies/book-series/RMCS

Aristotle and the Animals

The Logos of Life Itself

Claudia Zatta

Routledge
Taylor & Francis Group

LONDON AND NEW YORK

First published 2022
by Routledge
2 Park Square, Milton Park, Abingdon, Oxon OX14 4RN

and by Routledge
605 Third Avenue, New York, NY 10158

Routledge is an imprint of the Taylor & Francis Group, an informa business

© 2022 Claudia Zatta

The right of Claudia Zatta to be identified as author of this work has been asserted in accordance with sections 77 and 78 of the Copyright, Designs and Patents Act 1988.

British Library Cataloguing-in-Publication Data
A catalogue record for this book is available from the British Library

Library of Congress Cataloging-in-Publication Data
Names: Zatta, Claudia, author.
Title: Aristotle and the animals: the logos of life itself/Claudia Zatta.
Description: Milton Park, Abingdon, Oxon; New York, NY: Routledge, 2022.|
Series: Routledge monographs in classical studies | Includes bibliographical references and index. |
Identifiers: LCCN 2021040054 (print) | LCCN 2021040055 (ebook) | ISBN 9780367409494 (hardback) | ISBN 9781032197425 (paperback) | ISBN 9780367816001 (ebook)
Subjects: LCSH: Aristotle–Influence. | Aristotle–Knowledge and learning. | Zoology–History–To 1500. | Zoology–History–Philosophy. | Human-animal relationships–History.
Classification: LCC QL15 .Z38 2022 (print) | LCC QL15 (ebook) | DDC 590.9–dc23
LC record available at https://lccn.loc.gov/2021040054
LC ebook record available at https://lccn.loc.gov/2021040055

ISBN: 978-0-367-40949-4 (hbk)
ISBN: 978-1-032-19742-5 (pbk)
ISBN: 978-0-367-81600-1 (ebk)

DOI: 10.4324/9780367816001

Typeset in Times New Roman
by Deanta Global Publishing Services, Chennai, India

**To my mother
and in memory of
Florio and Gilda Zatta**

Contents

Acknowledgement ix

Introduction 1
Notes 9

1   Aristotle, animal boundaries, and the logos of nature 11
   1.1 Away from the stars: Animals' common nature 11
   1.2 The centrality of sensation, reason, and the articulation of the
      common 15
   1.3 A new beginning 23
   1.4 Animals, tykhē, and the logos of nature 25
      1.4.1 Animals' logos from speech to body and life 29
      1.4.2 On the birth of zoology and animals' equality (and not) 34
   Notes 36

2   From reason to life: Aristotle on soul division 51
   2.1 Understanding ensouled bodies: Soul partition and
      homogeneity 51
   2.2 Problematic divisions and attributions: The bipartition and
      tripartition of the soul 54
      2.2.1 Under the rule of logos: From Plato's Republic to
         Aristotle's Ethics 55
   2.3 A new model: The geometry of the soul 63
   Notes 67

3   Animals and nature: At the core of Aristotle's zoocentrism 78
   3.1 Animality and the living body 78
   3.2 Nature, bodies, movement, and life 80
      3.2.1 From the coincidence of causes to the definition
         of growth 82
      3.2.2 Animal growth, nutrition, and the soul 86

3.2.3 Growth, movement, and the origin of animals' life 89
3.3 Nutrition, reproduction, and the desire for immortality 91
Notes 97

4   The sentient animal   109
4.1 Setting the problem 109
4.2 From the dialectics of sensation to a new form of
    alteration 110
4.3 Sensation and logos 114
    4.3.1 On the inability to sense 120
4.4 Relating to the world: Sensorial architectures and animal
    awareness 121
Notes 127

5   Animal pleasure: From sensation to imagination and beyond   137
5.1 The questions about pleasure 137
5.2 Pleasure and pain within and beyond morality 139
    5.2.1 From virtue to the naturalness of pleasure 140
    5.2.2 Life and pleasure 148
5.3 Animals' desire, phantasia, locomotion, and
    communication 150
    5.3.1 Dreams, memory, and the physiology of phantasia 154
    5.3.2 Body, sensation, and knowledge: In response to the
        Presocratics 159
Notes 162

6   The lives of animals   175
6.1 The History of Animals *in Aristotle's zoology 175*
    6.1.1 The articulation of differences and sameness 177
6.2 Body constitution, habitats, and life 179
    6.2.1 Diet, pleasure, and the fight for survival 184
6.3 Animals' characters and learning 187
    6.3.1 Between psychology and ethological physiology 190
6.4 The nonhuman paradox: Being political in Aristotle's
    zoology 196
    6.4.1 The plasticity of the political animals 198
Notes 202

Conclusion   215
Notes 219

Bibliography   221
Index   233

Acknowledgement

I started working on this book almost a decade ago, while teaching at Northwestern University. The project originally encompassed a study of Presocratic theories on living beings and life, which became a monograph of its own, titled *Interconnectedness. The Living World of the Early Greek Philosophers* (2019, second revised edition). I am very grateful to the Loeb Library Foundation for a fellowship that enabled me to carry out part of this project. My warm thanks go also to Richard Kraut and Giulia Sissa for their encouragement and support in the early phase of this project and to Aileen Ajootian for her long-lasting friendship and help. I'm grateful to Franco Giorgianni for the invitation to participate to the Seminar "Forme, parole, e metafore della physis" at the University of Palermo in May 2018 and to Arnaud Macé, Alessandro Buccheri and Leon Walsh for inviting me to the first Colloquium *PHYSIS KAI PHYTA* in Paris, organized by The University of Chicago and ANHIMA in June 2019. Both events gave me the opportunity to work on, and refine, some aspects of this book. I also thank Hussein Abrama, June Allison, Francesco Ciabattoni, Lowell Edmunds, Mauricio Fuks, Mali Ishtewi, Seth Jaffe, Mike Levine, Alexis Malliaris, Jenifer Presto, James Redfield, Livio Rossetti, Iannis Stamatellos, and Rosemary Wright, cherished friends and colleagues. The Blegen Library of the American School of Classical Studies has provided me with invaluable resources and I want to thank its staff and, particularly, Andrea Guizzetti. In the prepublication stage, I used the Philosophy library of the University of Padua, whose staff I wish to thank for their kindness and help. My final thanks are for the anonymous readers of the manuscript and the scholars, past and present, without whose studies I could not have written this one.

Feltre, 26 May 2021

Introduction

Since we have completed stating the way things appear to us about the divine things, it remains to speak about animal nature, omitting nothing in our power, whether of lesser or greater esteem ... For in all natural things there is something marvelous.

(Aristotle, *Parts of Animals* 1 645a4–18)

Hence when someone said that there is Mind in nature, just as in animals, and that this is the cause of all order and arrangement, he seemed like a sane man in contrast with the haphazard statements of his predecessors.

(Aristotle, *Metaphysics* 1 984b15–8)

This, then, is the principle through which all living things have life, but the first characteristic of an animal is sensation; for even those which do not move or change their place, but have sensation, we call living creatures, and do not merely say that they live.

(Aristotle, *On the Soul* 2 413b2–4)

All animals have, in addition, some measure of knowledge of a sort (some have more, some less, some very little indeed), because they have sense-perception, and sense-perception is, of course, a sort of knowledge. The value we attach to this knowledge varies greatly according as we judge it by the standard of human intelligence or the class of lifeless objects. Compared with the intelligence possessed by man, it seems as nothing to possess the two senses of touch and taste only; but compared with entire absence of sensibility it seems a very fine thing indeed.

(Aristotle, *On the Generation of Animals* 1 731a32–731b5)

For Aristotle, among all animals, only man has and lives by *logos*, or reason.[1] This exclusive possession enables him to form political communities and to be the most political living being. With the force of a spell, this claim from *Politics* has shaped interpretations of Aristotle for centuries and has, at the same time, laid the foundations for the human/animal difference in Western thought, from stoicism to St. Augustine to Thomas Aquinas and Hobbes, to mention a few.

DOI: 10 4324/9780367816001-1

Aristotle, the first to engage with a study of animals in a systematic way, is also often considered the originator of a crisis:[2] his verdict set an unbridgeable divide between living beings, marking nonhuman animals with a long-lived stigma of lack and inadequacy. In *Metaphysics*, however, Aristotle bursts into a eulogy of Anaxagoras as the only sane man—the only one who, in contrast with his predecessors, "said that there is mind (*nous*) in nature (*physis*), just as in animals (*zōa*) and that this is the cause of all order (*kosmos*) and arrangement (*taxis*)."[3] Further, notoriously, Aristotle calls other animals "political"—bees, ants, wasps, and cranes. How is it possible? How can animals have mind (*nous*) if they do not have reason? How can they be political without *logos*? Ultimately, what did Aristotle really think about animals?

Aristotle's so-called biological treatises have been the focus of several important studies in the last 40 years. Since Balme's pioneering commentary,[4] scholarly research has explored the connections between Aristotle's biology and philosophical issues such as his theory of science,[5] the notion of explanation,[6] the role of teleology,[7] and the issue of classification.[8] But these studies have used Aristotle's lessons of biology mainly to understand his philosophical thought, leaving the very protagonists of his treatises, the animals, in the background. On the other hand, in a parallel interpretive line spanning across the philosophy of mind and history of science, animals have received attention but only insofar as the opposite pole of the human/animal binary. Whether dwarf-like creatures bent under a god-like humanity,[9] or deprived of reason (*logos*) but with expanded emotions,[10] or credited with a spark of practical intelligence (*phronēsis*),[11] Aristotle's animals have been defined by a negative difference vis-à-vis their human fellows, often carrying in the treatment of their bodies, capacities, and/or souls the manifesto of Aristotle's anthropocentrism. Much remains to be done to understand the conception of animals qua animals in Aristotle's study, and the terms of their exclusion from *logos*, particularly in relation to the method through which Aristotle inquires about them.

In the passage from *Generation of Animals* quoted above Aristotle reveals a double perspective from which one can look at animals and which he himself had to confront when discussing them. One can look at animals from the perspective of the human being possessing intelligence (*phronēsis*), to be taken as a manifestation of that *logos* which realizes human nature and animals do not possess, or one can look at animals from the perspective of the lifeless kind of beings (*apsykhōn genos*), i.e. rocks and stones. In this case—he says—even the possession of the bare senses of touch and taste is a very fine thing compared to the lack of sensibility. In his study of animals Aristotle chooses this second standpoint. He defines animals by the possession of sensation (*aisthēsis*)[12] and appreciates them for the knowledge sensation in its different forms, patterns, and, so to speak, conversions that afford them to live. Considered in these terms, the compass of Aristotle's discourse on animals does not point to the human/animal alterity because it is the lack of sensation that makes the relevant difference. The framework is grander and the real alterity lies between the living beings endowed with sensation and the lifeless ones, which are without it. For inasmuch as human beings are sentient they too belong to the animal club. Hence *logos* as reason and its lack is marginalized in Aristotle's approach to and ultimate conception of animals.

Aristotle defines animals by the possession of sensation and devotes to them a treatment of their own. Their study is inscribed in his overall discussion of nature (*physis*) whose ample domain he reviews in *Meteorologics* 1, object after object, field after field, from the principles of natural beings and a theory of natural movement to the basic components of matter and their transmutation to the celestial movements and the sublunar phenomena to animals and plants. It is, then, the scope of this book to understand animals' life and capacities in terms of nature, to which they belong.[13] How does Aristotle use the principles adopted to discuss nature in general for the study of animals? How does sensation happen and what does it empower them to do? We have seen earlier that even the bare senses of touch and taste provide animals with knowledge. What is the cognitive apparatus that sensation in its different forms affords them to have and the ensuing relation they entertain, so equipped, with the world? Animals are creatures of nature (along with plants and the elements) because they originate their movement and rest, but as to the types of movement they originate they are proper of their own. For, as we will see, Aristotle adjusts the natural movements so as to account for phenomena that are exclusive to animals. Their growth (*auxēsis*), for instance, is not the same as that of the elements. Animals share it with plants, but in their case growth is more complex and leads at any rate to an articulate, functional body with a definite shape which is foreign to both elements and plants and for which, in his discourse of method in *Parts of Animals*, Aristotle mobilizes the notion of *logos* indicating their artful structure. Sensation on the other hand is a special form of alteration (*alloiōsis*), which only animals are able to experience by tuning into the objects of sense in their living environment. This operation is possible on account of a unique body malleability and, so to speak, integrity[14] that no other natural being possesses, for animals take on the form of the proper sensible objects (*ta aisthēta*) without matter.

In tune with his systematic vision of nature, in the so-called biological treatises, Aristotle never speaks of biology. He speaks instead of animals as creatures of nature (*physis*), hence inheriting the inclusive vision of the Presocratics—for whom the cosmos culminated with the rise of *ta zōa*, humankind included—and building his study by reframing the terms of his predecessors' inquiry. This book investigates Aristotle's departure from the Presocratics, and Plato, but also shows the points of continuity with them: a zoocentric focus and a multipartite notion of the soul, respectively. Only by analyzing Aristotle's study of animals as natural beings in terms of his critical approach to his predecessors,[15] I believe, can we truly grasp the novelty and import of his discourse, which on the one side explores the common (*to koinon*) among animals and, on the other, values animals' perceptual dimensions as coessential with their lives, for the exile of reason from his study of animals corresponds in that context to the marginalization of man. Overall, this book sets out to show that in approaching animals as objects of study

Aristotle's perspective is zoocentric, not anthropocentric, and that such a perspective is supported by, and in turn corroborates, a notion of animal egalitarianism.

To this end, *Parts of Animals* 1 and *On the Soul* are fundamental texts.[16] In the first, Aristotle reacts to the Presocratics' conceptions of animals and deploys a series of epistemological boundaries (*horoi*), which are emblematic of his philosophical position on animals. Among them Aristotle advocates the study of the soul but not of all its parts insofar as only those that have to do with animals as creatures of nature have a legitimate place. He excludes the rational soul because his intent is that of pursuing "the animal as such" (*toiouto to zōon*), a crucial expression which defines his zoological project and which in this book will be referred to with the term "animality."[17] Uninvolved with natural movement, reason does not pertain to it. As for *On the Soul*, while reacting to the Presocratics' materialistic, homogeneous psychological views Aristotle also responds to Plato and the Academics who narrowly focused on the human soul manifesting, we may add, the intrinsic bias toward reason such focus could carry. Aristotle comes up instead with a novel psychological partition that has the advantage of illuminating the study of nature inasmuch as the soul is the principle of animals (*arkhē tōn zoōn*).[18] This partition, whose terms will be discussed in depth in the course of the book, allows Aristotle to extricate the fundamental functions of life and to distinguish living beings from those not living but also to create a rupture between animals and plants, based on the crucial capacity to sense. For Aristotle only animals, the *zōa*, sense, plants do not. After him, plants will be long considered simple and senseless *phyta*, a position which, only recently, new botanical studies have proven utterly untenable.[19] But the clear-cut distinction between living and not living, and within the living world of sensing animals and senseless plants are not the only effects of Aristotle's soul partition. For, with a blunt strike against Plato, Aristotle's partition makes the life of reason build on the life of the senses, turning sensation (and the body) into the basis of that type of knowledge, which stems from the possession of *logos* and is exclusive to humans.[20] The beginning of *Metaphysics* spells out this relation clearly by highlighting a path to art (*tekhnē*) and knowledge (*epistēmē*) that starts from sensation and unfolds through the obliged stages of memory (*mnēmē*) and experience (*empeiria*).[21] As for *On the Soul*, Aristotle avoids charged, axiological terms and recurs instead to a neutral geometrical figure.[22] Sensation is indeed a very fine thing, the building block and essence of animals' life, which in this way is made to consist first of all in the awareness of the surrounding reality, whether only of tangibles and tastes or also colors, sounds, and scents. Hence Plato's bias toward reason is tamed in a discourse that, true, does acknowledge the immortality of the rational soul, its separation, and subsistence regardless of the other parts, but that in animals, human or not, roots the rational soul together with the other parts. That is, for the living beings of the sublunar world there are no rational soul or functions without the sensitive soul and the senses. These come first and are adequate enough to make animals' life full and complete.

To grapple with Aristotle's adjustments of the natural movements (*auxēsis, alloiōsis*, and *kinēsis kata topon*) and understand animals' living phenomena and their sensation-based cognitive apparatus, including *phantasia* and memory, along with what such apparatus enables them to do (i.e. the actual actions in the real world), the discussion will extend to other treatises besides *On Parts of Animals* and *On the Soul*. We will look at *On Generation and Corruption, On the Movement of Animals, Parva Naturalia, History of Animals*, and also *Politics* and *Nicomachean Ethics* for what pertains respectively to Aristotle's claim that some of the nonhuman animals are political and his conception of animal pleasure. It is indeed one of the aims of this book to address the obscure association among *aisthēsis*, pleasure, pain, and desire, which Aristotle presents early on in his positive treatment of the soul[23] but never explains. Such an association has determined the preferred choice to translate *aisthēsis* as sensation rather than simply perception, on the belief, to be argued for, that for Aristotle pleasure and pain also inhere in the actualizations of potentialities, or, said otherwise, in the fulfilment of natural functions, and, particularly, for living beings of sensing.[24] In fact, in dealing with these subjects too Aristotle is continuing a Presocratic thread: Empedocles considered sensation intrinsically connected with pleasure, Anaxagoras with pain.[25] Now, it may come as a surprise but, despite its human concerns, *Nicomachean Ethics* still provides us with significant evidence on Aristotle's conception of animal pleasure along with the pleasure that is proper to man *qua* rational living being.

The order in which Aristotle's treatises are discussed reflects the overall focus of this book, the effort being that of presenting his conception of animals in a systematic way, from his zoological principles to animals' living capacities to their actual existence.[26] The appeal of such an order lies also in its faithfulness to Aristotle's consideration of animals as *living* beings, whose bodies are teleologically conceived for the complex purpose that is life. The *ergon* argument, as it has been called, is more than the account of the striking functionality of animals' bodies and their parts,[27] its meaning not being simply exhausted with the organic outlook and the artfulness of nature, for, I believe, it ultimately points to animals' capacity to move and live, fulfilling their most recondite and essential effort to continue to do so. In the end, Aristotle's study of living beings and life converges on a philosophy of life. For his appreciation of living functions and activities, with the emphasis on sensation, stems from, and further corroborates, a consideration of the relationship of the living being to its environment, what I will call the "space of existence," and to the life the living being itself lives. Significantly, in Aristotle's treatment, the animal *qua* object of study powerfully emerges as the subject of its own life.[28]

The book consists of six chapters, plus an introduction and conclusion. Chapters are organized thematically, moving from Aristotle's discourse on method (Chapter 1) to his discussion of the soul (Chapter 2) to his treatment of nutrition and sensation (respectively, Chapters 3 and 4) followed by that of desire, pleasure, and *phantasia* (Chapter 5) to end with an analysis of his ecological and ethological theories, that is, animals' relations to their habitats and their behaviors in their space of existence (Chapter 6). Chapter 1 is dedicated to Aristotle's zoological project and looks at the way he subtracts animals from the cosmological discourse of the Presocratics, and Plato's *Timaeus*, to consider them in their autonomy as living beings that carry in themselves the origin (*arkhē*) of their movement and rest.[29] The focus of this chapter is *Parts of Animals* 1, whose methodological boundaries sustain the birth of zoology and are discussed in relation to the Presocratics' positions and as indicators of Aristotle's own conception of animals. In response to Democritus, who devoted attention to the inert shapes of animals' bodies and their parts, Aristotle constructs a discourse that centers on life and sensation and argues that the study of animals as creatures of nature should include those parts of the soul that are relevant to their existence—the nutritive, sensitive, and locomotive. In excluding the rational soul, Aristotle marginalizes the human being from a discourse ultimately directed to explore the commonality (*to koinon*) among living beings, which he articulates in terms of comparative anatomy and the differences that relate animals' parts by degree and analogy. Further, against matter, which in his reading has been at the core of the Presocratics' inquiry, Aristotle assesses the priority of the final and formal causes, and, therefore, of *logos* versus *tykhē*. In moving intrinsically and systematically toward an end, which is their *logos* and essence, animals are emblematic of nature itself; their development is seen as a *diakosmēsis*, a coming into order. Considered, thus, at the intersection of teleology and ontology, animals are full of a *logos* that constitutes, at once, both the rationality immanent in nature and ruling animals' physiological processes and living activities, and their perfectly constituted form qua living beings in the world. In appropriating animals so conceived as objects of study, Aristotle's zoology is claimed to be, with due qualifications, egalitarian. Part of this chapter is forthcoming in an essay titled "A New Beginning: Aristotle and the Birth of Zoology" in A. Vergados, A. Walter (eds.), Ἀρχή *and Origo: The Power of Origins Conference Proceedings*.

Chapter 2 examines *On the Soul* and discusses Aristotle's novel model of soul partition in light of the Presocratic monolithic, life-giving conception of the soul and Plato's ideological partite model centering on the human. It situates Aristotle's problematization of soul partition within his aim to understand embodied life (i.e. which natural bodies are ensouled) and turns to *On the Soul* 3.9, where Aristotle expresses dissatisfaction with current models of soul partition (both tripartite and partite). This chapter pinpoints such models in the Platonic versions of the soul in *Republic* and *Timaeus* and Aristotle's own psychological model in *Nicomachean Ethics*, arguing that both tripartition and bipartition centered on reason (and the human) to the detriment of the body and related capacities. Aristotle's new partition encompassing nutritive, sensitive, locomotive, and rational souls[30] is, then, shown to break the dominion of reason and to advocate a positive, essential, and integrated role for nutrition and reproduction, sensation, and locomotion. All living beings are grounded on the nutritive soul, which equally confers humans, animals, and plants the capacity to grow, nourish, and reproduce themselves. But while plants are *zōnta* (living things) and possess only the nutritive soul (a radical point of departure from the Presocratics), all animals (*zōa*), humans included, possess the sensitive soul, which seals their shared nature qua living beings by enabling them all to sense the world. In sum, it is argued, Aristotle's perspective is zoocentric rather than anthropocentric: his goal is to understand all forms of life in terms of continuity and a "shared core," with sensation constituting the cement of animal life.

Chapter 3 returns to *Parts of Animals* 1 and grounds Aristotle's reasons (methodological and physical) for excluding the rational soul from his study of animals on the conception of animality. Aristotle aims at comprehending the "animal as such" (*toiouto to zōon*), that is, at pursuing the relation animals entertain with movement and matter *qua* bodies exposed to the physicality of the world. In being disconnected from the body and movement, mind, its activity, and objects do not contribute to such an aim. After discussing Aristotle's adjustment of the natural movements (*auxēsis*, *alloiōsis*, and *kinēsis kata topon*) to understand animals' modes of life as linked to their complex bodies and due to the soul, this chapter focuses on *auxēsis* (growth) in light of *On Generation and Corruption* and *On the Soul*. Crucial for comprehending the animal as such, growth catalyzes the influence of the final cause in animal formation and the centrality of form/*telos* versus matter, and hence of *logos* as structure bound to limit (*peras*) and internal proportion (*logos*). *On Generation of Animals*, it will be added, complements this account of *auxēsis* by pursuing the transmission of form upon matter with regard to the beginning of (a new) life and the articulation and growth of the embryo. The chapter ends with a discussion of the other functions of the nutritive soul (i.e. nutrition and reproduction), their relation, and the status of reproduction as the most natural activity of all. Intended for the sake of the living being itself and fulfilling its intrinsic desire to live forever, reproduction will be shown to establish for Aristotle the autonomy of existence for every living being and to support his zoocentric view in contrast with the anthropocentric notion of finality presented in *Politics* for which all nonhuman forms of life are ultimately for the sake of the human being.[31]

Chapters 4 and 5 are complementary and address the cognitive apparatus of the living being qua sentient (*zōon aisthētikon*), a definition that stands at the core of Aristotle's thought on animals, humans included, and of central concern in his treatment of the soul as a principle of life in *On the Soul*. Chapter 4 discusses the general treatments of *aisthēsis* in DA 2.5 and DA 2.12. It pursues the co-extensiveness of life and sensation in light of Aristotle's re-conceptualization of *alloiōsis* pinpointing in sensation a preservative change (rather than destructive) and proceeds to analyze the sophisticated operation that allows the sense to receive the simple sensibles. This capacity, it will be argued, consists in a unique sharedness of the qualities of the world and depends on animals' body composition (*synthesis*) and hence on the *logos* that is exclusive to them as creatures of nature, discussed in Chapter 1, and extraneous to both plants and elements, which in their specific passivity are unable to sense. Next, this chapter discusses the different categories of sensibles (proper, common, and incidental) and their architectures along with the awareness of sensing and the simultaneous perception of different special sensibles as due to the common sense (*koinē aisthēsis*) and available to all animals, even the simplest ones. In the end, it is argued, Aristotle's pursuit of the animal *qua* sentient corresponds to a comprehension of the living being in the immediacy of its existence as resulting from being in touch with the environment it lives in and ensuing awareness. Chapter 5 explores the obscure association of sensation with pain and pleasure, in addition to desire, and the psychological capacities that derive from sensation and transcend its immediacy such as imagination (*phantasia*) and memory (*mnēmē*). If the desire for pleasure is conventionally (and rightly) intended as instrumental to nutrition and triggering locomotion, this chapter argues that since animal pleasure is an actualization of potentiality it is also enmeshed with sensation *tout court* beyond the bare circumscription to nutrition and the sense of touch advanced by a logocentric perspective. Evidence for this position is found in *Nicomachean Ethics* 7 and 10 which, despite specific differences and the human-centered focus of the treatise, predicate the naturalness of pleasure. Defined in relation to activity (*energeia*), from these texts pleasure emerges as intimately connected to life and arguably residing in animals' fulfillment of their sentient being, activities, and dispositions. Its universality expresses for both animals and humans the same desire to live and hence points to an egalitarian trend in Aristotle's zoology. Complementing *On the Soul* with *On the Movement of Animals* and *Parva Naturalia* this chapter continues with a discussion of the role of *phantasia* in animals' practical life, the acquisition of their objects of desire, and hence pleasure. It elaborates on the synesthetic nature of *phantasia*, its relation to different categories of sensibles, in particular the incidental, and its involvement in animal communication, dream-production, and memory. Shown to be depending on body physiology and movement, Aristotle's *phantasia* is finally reappraised in terms of a response to the Presocratics' identification of sensation and thought and in line with his pursuit of the animal as such.

Chapter 6 turns to *History of Animals* and discusses it in relation to Aristotle's other zoological treatises highlighting also in *HA* the articulation of animals' sameness by differences of analogy and degree which supported *PA*. It then proceeds to focus on animals' lives in their space of existence, moving from Aristotle's conception of animals' habitats, nutrition, and interpersonal relations to their character, learning (*mathēsis*), emotions, and intelligence (*phronesis*). Here too it will be possible to relate these aspects to animals' *logos* (i.e. body structure), underscoring, when pertinent, Aristotle's originality (or debt) in respect to the Preocratic tradition. Last, this chapter grapples with the paradox of the *politika zōa* —nonhuman animals which are political despite their lack of *logos* (in the sense of reason) and to which a long interpretive tradition has denied political life, considering them merely social. However, in describing the lives of the *politika zōa*—bees, ants, wasps, and cranes—*History of Animals* makes clear that for Aristotle certain animals are truly political (so they were, incidentally, for Hobbes reading Aristotle), and not only on the basis of their working toward a common goal but also on account of a hierarchical, or conversely anarchical, structure, and, further, on account of their relation to space as a community, a feature inherent in the conception of the "political" which Aristotle addresses in *Politics*.[32] A relation to space is in fact crucial to all political animals that are stationary and gravitate around an inhabited settlement—like wasps, bees, and ants—but also to cranes and, by extension, to all the other travelling birds and fish, which are able to maintain social cohesion and focus on the common goal, i.e. migration, while moving across territories. Thus, the *logos* which animals possess qua natural ensouled beings and which, "informing" their bodies, allows them to sense, move, nourish, reproduce, and pursue their desire for a never-ending life, also enables them to live politically where politicality too is instrumental to life. Part of this chapter constitutes a revision of an article originally published as "The Non-Human Paradox: Being Political in Aristotle's Zoology," in A. Avramidou and D. Demetriou, *Approaching the Ancient Artifact: Representation, Narrative, and Function. Festschrift in Honor of Alan Shapiro.*[33]

Notes

1 Aristot. *Pol.* 1 1253a9–18.
2 On this "crisis," see Sorabji, 1993.
3 *Met.* 1 3 984b15–8.
4 In 1972 Balme published a commentary on Aristotle's *Parts of Animals 1* and *On the Generation of Animals* 1, discussing major issues of Aristotle's philosophy of biology (1992 [1972]).
5 Gotthelf, 1987; Balme, 1987.
6 Lennox, 1987, 1991; Bolton, 1987; Lloyd, 1996.
7 Johnson, 2005; Leunissen, 2010.
8 Pellegrin, 1986.
9 Lloyd, 1983.
10 Sorabji, 1993.
11 Labarrière, 1990, 2005.
12 See the passage from Aristotle's *On the Soul* (2 413b2–4), quoted above.
13 In proceeding so, this book undertakes a new path in the interpretation of Aristotle's zoological treatises because it pursues Aristotle's study of animals in terms of his study of nature (i.e. physics) rather than his metaphysics and philosophy of science as it has become standard to do in the past decades. More specifically, this book calls attention to the natural philosophy that is applied to, and further developed in, the study of animals as natural objects distinct from other natural objects such as the elements and plants in order to comprehend Aristotle's conception of animals. In this respect, I believe, it is relevant to underscore the philosophical closeness of *Physics* 2 and *Parts of Animals* 1, which is not only an indication of the chronological proximity of these two texts (Balme, 1987b, 17) but also of the inscription and membership of the study of animals into that of nature.
14 I speak of integrity here because, as it will be seen in Chapter 4, for Aristotle animals' senses and by extension bodies hold a plasticity that allows them to grasp the *logoi* of the (proper) sensibles without being irremediably changed and impaired (which they are, however, in the event the sensibles result too strong).
15 It may be useful to recall at this point that in *Metaphysics* itself Aristotle remarks on the cooperative nature of scientific inquiry (*Met.* 2 993a30–34).
16 On the relevance of *On the Soul* for Aristotle's zoological studies, of which it provides a general framework and constitutes a general introduction, see respectively Lloyd (1992, 147–67) and Pellegrin (1996, 466–9); cf. Gotthelf and Lennox, 1987, 5.
17 I take "animality" to be a concept that encompasses all the aspects that constitute animals with regard to other natural beings, living and not (i.e. plants and elements), and that Aristotle reduces to the array of movements due to the soul and defining fundamental functions of life. For the variable meaning of this concept in contemporary philosophy, see Wyckoff, 2015; for the delineation of a set of theses that defines it, Cimatti, 2015.
18 *DA* 1 402a7–8.

19 See, for instance, Baluŝka, Mancuso, Volkmann, 2006; Baluŝka, Mancuso, 2009.
20 Cf. Pellegrin, 1996, 472, 474 on the "derivative" nature of human thought as "elaboration of sensation."
21 *Met.* 1 980a21–981b20.
22 *DA* 2 414b28–415a12, discussed in Chapter 2.
23 *DA* 2 413b23–5, 414b4–6.
24 I take sensation to have a richer meaning than perception in that it implies the feeling of pleasure and pain that inhere in perceiving as fulfillment of the living being's definitory potentiality to sense.
25 For Empedocles, see DK 31 B 107 and DK 31 A 86/Thph *de Sens.* 9; for Anaxagoras, DK 31 A 92/Thph. *de Sens.* 27.
26 Basically, the discussion moves from *Parts of Animals* (in which Aristotle presents his methodological pillars for the study of animals) and *On the Soul* (in which he gives an introduction to animal psychology and living functions) to *Parva Naturalia* to *History of Animals* with cross-referencing and integrations along the way from *Physics*, *On the Generation and Corruption*, *Nicomachean Ethics*, and *Politics*. Regarding the chronology of Aristotle's treatises dealing with (animal) life, see Balme, 1987.
27 See Chapter 1.
28 This outcome, which here synthetizes Aristotle's view on the finality of reproduction in *On the Soul* (see Chapter 3) and intersects Aristotle's discussion in *On the Generation of Animals*, may well be ultimately rooted in a metaphysical problem, as Balme points out (1987, 10–12), namely to illustrate "how can the subsistent individual animal—the Socrates—be formally definable when it contains matter," but will be pursued, consistently with the framework and goal of this book, in terms of the context in which it appears, i.e. the study of animals within the domain of nature, and for the philosophical position it represents with regard to Aristotle's conception of animals in his *ad hoc* study.
29 *Phys.* 2, 1.
30 *DA* 2.3–4.
31 1 1256a20–1256b27.
32 3 1276a18–9.
33 2014.

1 Aristotle, animal boundaries, and the logos of nature

1.1 Away from the stars: Animals' common nature

A distinctive, unannounced feature of the first book of Aristotle's *Parts of Animals* is the grand apology[1] that follows the presentation of the methodological criteria (*horoi*) to be reviewed in this chapter. Rather than a mere rhetorical tangent,[2] this apology has a foundational intent: it shows that Aristotle felt compelled to justify his interest in animals and the apparent new context of his inquiry. Why devise a method to study animals? To what end? And, at any rate, why not instead devote attention to other sublime specimens of beings? This apology is crucial for understanding not only Aristotle's general view on the animal world and its diversity but also his awareness of the project's novelty.

> Among the substantial beings constituted by nature, some are ungenerated and imperishable throughout all eternity, while others partake of generation and perishing. Yet it has turned out that our studies of the former, though they are valuable and divine, are fewer (for as regards both those things on the basis of which one would examine them and those things about them which we long to know, the perceptual phenomena are altogether few). We are, however, much better provided in relation to knowledge about the perishable plants and animals, because we live among them (*dia to syntrophon*). For anyone wishing to labour sufficiently can grasp many things about each kind. Each study has its attractions … Perishable things, however, take the prize in respect of understanding because we know more of them and we know them more fully. Further because they are nearer to us and more akin to our nature (*oikeiotera physis*), they provide a certain compensation compared with the philosophy concerned with divine things. Since we have completed stating the way things appear to us about the divine things, it remains to speak about animal nature, omitting nothing in our power, whether of lesser or greater esteem. For even in the study of animals disagreeable to perception, the nature that crafted them likewise provides extraordinary pleasures to those who are able to know their causes and are by nature philosophers. Surely it would be unreasonable, even absurd, for us to enjoy studying likenesses of animals—on the ground that we are at the same time studying the art,

DOI: 10 4324/9780367816001-2

such as painting or sculpture, that made them—while *not* prizing even more the study of things constituted by nature, at least when we can behold their causes. For this reason we should not be childishly disgusted at the examination of the less valuable animals. For in all natural things there is something marvelous. Even as Heraclitus is said to have spoken to those strangers who wished to meet him but stopped as they were approaching when they saw him warming himself by the oven—he bade them to enter without fear, 'for there are gods here too'—so too one should approach research about each of the animals without disgust, since in every one there is something natural and beautiful.[3]

In considering animals on a large basis that encompasses in anonymity all living beings (*ta zōa*), here Aristotle is adopting the perspective of the earlier physicists: no special position is assigned to humans; they merely blend into the larger category of the living. At the same time, however, as in Plato's cosmological account and in line with a belief held in the Academy,[4] stars and planets are considered members of the animal kingdom and endowed with a privileged status and biology.[5] The celestial animals are immortal and divine. In Plato's *Timaeus*, the body of the cosmos is spherical and self-sufficient: its movement is limited to rotation, and it does not need to eat to exist. It is a perfect body, undifferentiated, without members or organs, and its activity coincides with that of the rational soul. And so are the bodies of the celestial animals.[6] In Aristotle's *On the Heavens*, likewise, the sky, planets, and stars are assigned divine bodies (*theia sōmata*) and are said to enjoy a happy life— to hold the best disposition (*aristē diathesis*), unblemished by the *dyskhereia*, "difficulty," that afflicts the mortals.[7] In their immortality and constitution, the celestial creatures are opposed to the animals of this earth that come into life, grow and perish, and are equipped with a body that is suitable and functional to these biological processes.[8] Thus the animal kingdom is articulated through a series of oppositions—high/low, heavenly/terrestrial, immortal/mortal, and therefore divine/nondivine, worthy (*timios*)/unworthy (*atimos*)—where value is directly proportional to immortality and the degrees of approximation to it along with the respective animals' bodies and activities.

In the passage from *Parts of Animals* quoted above, however, while Aristotle maintains a hierarchy between higher and lower animals, he defines a field of inquiry that is circumscribed to the "less worthy," terrestrial animals, as opposed to the celestial ones, and, by extension, as we will consider in the next section, as hived off from the cosmological discourse that interested the Presocratics. Thus Aristotle reserved the discussion of the planets and stars for *On the Heaven* and deals separately with the other animals, striving, in Moreau's words, "to transfer to the study of the living beings the feelings of admiration and the religious emotions which for his contemporaries were attached to the contemplations of the stars."[9] Rescued from an evaluative comparison with higher beings,[10] mortal animals now become relevant in themselves: their bodies and the lives they sustain morph into objects worthy of contemplation. This is possible because Aristotle introduces a new scale of value, that of knowability.[11] The animals of this world, recognizes our philosopher, can be known better than the celestial ones[12] and hence provide inconceivable joys to those who are able to know the causes of things and are philosophical by nature.[13]

Animals are (better) knowable on account of two distinct, interrelated factors: on the one hand, their physical proximity, and, on the other, their nature. We can study animals more fruitfully "on account of living together" (*dia to syntrophon*),[14] Aristotle states at one point, adopting an expression (*syntrophos*), which in the ancient sources applies to animals, whether from the same species or not, to indicate a "shared life." And along with coexistence, this expression implies a sense of familiarity and recognition. In *History of Animals*, for instance, while describing the character of the lion, Aristotle remarks on the playfulness and affection the animal shows to "those reared with (*syntropha*) and familiar" to it.[15] But what makes animals knowable is also, and importantly, the possession of a nature that is more akin (*oikeiotera physis*) to that of humans.[16] This point is often neglected in commentaries on Aristotle, but it is interesting and is of the greatest importance, since, as we will soon see, it provides a legitimate basis for the study of animals in a comparative framework that forms the core of Aristotle's biological project. Animals are not only better known because they are accessible to direct observation, which made Lennox remark that "this is one of the strongest assertions of the centrality of extensive perceptual experience in developing scientific understanding of nature."[17] Importantly, animals are also better known because they share with us their perceptual experience. In their mortal bodies—and by extension in the activities and lives sustained by their bodies—animals are more closely related to us than the celestial entities, which are isolated in their complete rationality, senseless bodies, and perfect movements in far regions around the sky. And this notion of kinship (*oikēiosis*), which Aristotle interjects here and which tacitly sustains his comparative study of animals, will be forcefully developed by Theophrastus, who, departing from his teacher, will turn it into a foundational reason to oppose animal sacrifice and meat-eating.

Despite its application to an ethical discourse, which in fact is irrelevant to Aristotle's zoology, Theophrastus' testimony—a fragment belonging to the treatise *On Piety* and reported by Porphyry in his *On Abstinence*—is worth reporting in full. For in concisely discussing the human/animal relatedness it helps us define retroactively the focus, framework, and methodological questions of Aristotle's biological inquiry while also supporting the integral role that the apology plays in *Part of Animals*.

> In this way, too, we class all humans as related (*syngeneis*) both to each other and to all animals. For their bodily origins (*arkhai*) are by nature the same. By this, I do not mean to refer to the primary elements, since plants are also made of these, but for example skin, flesh and the type of fluids that are natural to animals. And much more are they related through their souls being no different in nature, I mean in their appetites (*epithymiai*), anger (*orgai*), and again in their reasonings (*logismoi*) and above all in their senses (*aisthēseis*).

But as with their bodies, some animals have souls more finely tuned, others less so, but they still all by nature have the same origins. And this is shown by their passions (*pathē*) being akin (*oikeiotēs*).[18]

An interplay of resonances connects Theophrastus' statement on the kinship between humans and the other animals with Aristotle's overall discussion in *Parts of Animals* and helps to illuminate the basis for animals' "more familiar nature" (*oikeiotera physis*). First, there is an emphasis on parts and their composition. Like Aristotle, Theophrastus acknowledges the presence of different orders of compositions (*synthesis*), forming animal bodies from the basic elements (*arkhai*) to the different organic parts, and he further alludes to the peculiar status of plants, which are composed of the same elements as the animals but have less differentiated bodies than animals do: plants too are considered "living beings" but ones deprived of sensation.[19] We will return to these aspects of Aristotle's treatment later in the chapter. Second, and more importantly, in Theophrastus' passage kinship between humans and animals is predicated on the basis of their sharing the same soul, intended as the source of desires (*epithymiai*) and of a diverse range of affections. Theophrastus mentions anger (*orgai*) and in general passions (*pathē*), but emphasizes, above all, sensations (*aisthēsis*). Humans are kin to the animals of this earth because unlike plants—or, we could add, the celestial animals—they are living beings that feel. They are made of the same basic flesh-stuff and share "in the perceptive powers to which this flesh is heir."[20] Thus, approaching the apology of *Parts of Animals*, and the treatise at large, from this particular angle, it is plausible to understand in Aristotle's acknowledgment of animals' *oikeiotera physis* his very definition of animals as sentient beings. In other words, the animals of this planet have a nature more akin to us than the celestial animals do because they are made of the same stuff and, like us, they feel.[21] And if the capacity to feel, in one or multiple ways, is what all animals share[22] and what, in fact, defines them and grants legitimacy to a study of them in their plurality, then at the bottom it is imperative to consider Aristotle's project in *Parts of Animals*, along with the other zoological treatises—despite their intrinsic specificities and individual goals—an exploration of the common nature of animals, humans included, and a forceful affirmation of the unity of all forms of (sentient) life.[23] Significantly, this aspect of Aristotle's inquiry finds a theoretical counterpart in *On the Soul*, which will be discussed extensively in Chapter 2. Here let it suffice to mention that in this treatise Aristotle proposes to find the most common (*koinotatos*) definition of the soul,[24] namely one that can be applied to all living things and beings (*zōnta/ zōa*). To this effect he devises a tiered model of the soul, whose basic part—the nutritive one—is shared by all forms all life, enabling them to perform the same essential functions such as nutrition, growth, and reproduction.[25] Indeed it is by sharing the same soul that trees, oxen, and humans are all able to be alive, to reproduce and nourish themselves, to grow and decay. In contrast to plants, however, the animals (*zōa*), humans included, also possess besides the nutritive soul the sensitive one, which enables them all to perceive despite the specificity of their physical conformation.[26] Thus those activities that animals share with plants via the nutritive soul, namely nutrition and reproduction, are accompanied by sensation, desire, pleasure, and pain, in a way that distinguishes them fundamentally from the vegetal world.

1.2 The centrality of sensation, reason, and the articulation of the common

Animals for Aristotle are sentient beings, all of them, with no exclusion. This definition returns over and over in *Parts of Animals*, almost with the force of an axiom, and is anchored in the cardio-centric model—or its organic equivalent in bloodless animals—that sustains animal physiology.[27] "An animal is defined," writes Aristotle, "by the fact that it possesses sensation: and the part of the body to have sensation first is the part that has blood in it first—in other words, the heart, which is the source of the blood and the first part to have it."[28] The heart is the first sentient part of animals' bodies. In fact, it is an animal in itself, and the site of the soul.[29] In *On the Generation of Animals* it appears as the part that gives life to the embryo and makes it grow,[30] from which we understand its overall primacy in animal physiology. But, while present first of all in the heart, sensation permeates animals' bodies. It takes place in the uniform parts that constitute them and that Aristotle distinguishes, in a first, basic examination of animal constitution, from the nonuniform ones, which are, by contrast, instrumental.[31] All animals share at least one sense, that of touch, which for the blooded animals resides in the flesh and for the bloodless ones in its equivalent.[32] Thus, on account of their constitutions, animals are intrinsically sentient, and sensation is coextensive with the very fabric of their bodies and, ultimately, with their lives.[33]

The centrality of sensation in the definition of animals has an ideological/methodological consequence, which Aristotle addresses extensively in the course of his discourse on method in *Parts of Animals* and which contributes to the new basis of his study vis-à-vis that of his predecessors. Aristotle situates himself against Democritus, who claimed that man and by extension all other animals could be defined by their shapes (*skhēma*), consisting ultimately in the arrangements of the bodily parts, and by their colors (*khrōma*).[34] Aristotle criticizes this position by advocating strongly an identity of bodily parts, and ultimately the body itself, on the one side, and bodily parts' functions, and the body's function, on the other, so that—as he claims in respect to the human animal—a hand made of bronze, and therefore incapable to "act," is only nominally a hand; it is not a real one.[35] And so it goes for the other animals as well: they cannot be identified merely in terms of their inert shapes, or parts, which must, instead, be considered in relation to their specific activities (*erga*), therefore in movement and in the context of life. This "*ergon* argument," as scholars have called it, is central[36] in Aristotle's study of animals, and it is significant not only as a pillar of Aristotle's teleological system and explanatory intent,[37] but also because it constructs a discourse on animals that centers on their being sentient and alive—each one of them in the specificity of its constitution and shape—and, by extension, on their actions in the external world. So in *Parts of Animals*, one after the other, the parts of animals' bodies are cast into their "vital" activities and reviewed systematically from the inside to the outside of the body, from the high and the low along a vertical axis provided by the body of man, but also intersecting, at the same time, the nutritive apparatus that is essential for any living being to exist.[38] "Everything that grows must of necessity take food," states Aristotle.[39] Whether intrinsic to each animal species or adapted, working under conditional necessity or not,[40] each animal part sustains and contributes to the life of a living being, from the crucial business of the heart—or its equivalent in bloodless animals—as the prime mover, the blood and heat producer, and the original site for sensation to the brain, whose cooling, counterbalancing effect provides "safety" (*sōteria*) to the entire body.[41] In Aristotle's biology, teleology, as Johnson has argued, is immanent,[42] and each living being is conceived in the tension and effort of being alive.[43]

That animals must be regarded as living, sentient beings calls for a fundamental feature of Aristotle's zoology, the inclusion of the study of the soul. One of the aporetic alternatives leading to the establishment of the methodological boundaries of Aristotle's inquiry on animals is in fact whether the natural philosopher should consider the soul in its entirety or only some parts of it.

> In view of what was said just now, one might puzzle over whether it is up to natural science to speak about *all* soul, or some part ... However, it is not the case that all soul is an origin of change, *kinēsis*, nor all its parts; rather, of growth, *auxēsis*, the origin is the part which is present even in plants; of alteration, *alloiōsis*, the perceptive part, *to aesthētikon*, and of locomotion, *phora*, some other part, and not the rational, *to noētikon*; for locomotion is present in other animals too, but thought, *dianoia*, in none. So it is clear that one should not speak of all soul; for not all of the soul is a nature, but some part of it, one part or even more.[44]

Although it is still a principle of life and, as such, shared by all living beings, just as it was for the Presocratics, the soul lacks the homogeneity for Aristotle that it had for his predecessors.[45] Instead, in line with the Platonic version, which is expounded in the *Timaeus* and other dialogues, the soul is a composite entity[46] whose discrete parts coincide with specific faculties of the living being: growth, sensation, locomotion, and thought. The soul is the source (*arkhē*) of these activities, which, in turn, constitute psychological powers (*dynameis*).[47] Yet not all parts of the soul are relevant to the study of animals, but only those that make creatures "move," where movement (*kinēsis*) is intended as the key phenomenon of nature[48] and in an array of manifestations. Indeed, movement (*kinēsis*) for Aristotle encompasses a range of changes that systematizes his predecessors' reflections on animals, subsuming under the same metaphenomenon a diversity of affections, from physical growth, or conversely, decay,[49] to the bodily alteration that accompanies the phenomenon of sensations to the specific ability to move from one place to another.[50] Related to the parts of the soul, which are responsible for them, these changes call for a consideration of the nutritive (*threptikon*), sensitive (*aisthētikon*), and locomotive soul. By contrast, Aristotle excludes from the study of animals the soul's intellectual part (*noētikon*), whose function is thought (*dianoia*)[51] and unrelated to physical, bodily movement.[52] Indeed, *dianoia* is distinctive of only one animal species, the human.[53] The rational soul, with *dianoia* and its other faculties (such as the exercise of reason (*logos*) through speech, and calculation (*logismos*)), is exiled from a project, and a field, that aims at discussing animals in terms of their filiation from nature and their intrinsic power to move in whatever form they possess it (growth and decay, sensation, and locomotion). True, as we will soon see, in this project reason may still appear as an attribute of the human animal,[54] but it does not constitute for Aristotle a proper object of zoology.

At the same time, while Aristotle decentralizes and exiles reason, he also marginalizes the human being, thereby departing from Plato too. For the exile of reason from his method releases reason from being the absolute standard against which to account for animals as Plato does in *Timaeus*.[55] Aristotle's animals are not seen—at least not from his methodological point of view as presented in *PA* 1—in terms of their progressively degrading distance from the sublimity of the human being, the only creature endowed with reason, nor do animals carry on their bodily shapes the visible marks of that distance, from birds to fishes through quadrupeds, each kind more and more removed from mortal perfection. It is true that the long excursus on animals' kinds in book 4 of *Parts of Animals* echoes *Timaeus*.[56] In the wake of Plato, Aristotle presents the human being as the only one among the animals who stands upright on account of his divine nature and essence (*physis kai ousia theia*) along with the power of reason these features equip him with. Indeed, Aristotle claims, the "work" of what is divine is to think and be intelligent (*noein kai phronein*); the human being is the only living being empowered to do so and a downward posture would hinder him in this capacity.[57] Likewise, in the same passage, Aristotle proceeds to compare blooded (nonhuman) animals to man, defining them notoriously as "dwarf-like" (*nanōdes*), and ultimately connects bodily structure and locomotion to intelligence.[58]

> The bird and the fish kind, and every blooded kind are, as has been said, dwarf-like, *nanōdes*. And because of this all animals are less intelligent, *aphronestera*, than human beings.[59]

But the terms of Aristotle's discussion are descriptive, not evaluative.[60] The attribute "dwarf-like," as he explains, indicates a specific proportion of a body in which the upper part is bigger than the lower part.[61] In human adults (alone) the two are proportionate. While *Timaeus* makes the failing of reason in men's lives the key for understanding animals' degenerate embodiments and progressively witless nature, Aristotle explains animals' body (at first that of quadrupeds, and subsequently that of the other land animals, including plants) in terms of "physical" constitution.[62] That is, the cause of animals' posture and difference in limb number (compared to the upright, biped human being) lies in the synergy of the heat and earthy material that composes their bodies and not in their lack of reason.[63] And it is because of their weighed-down body structure that all blooded animals are indeed "less intelligent" than human beings. Thus, in contrast to Plato's account, for Aristotle animals' growth (conceived of in terms of "spatial" orientation) and capacity to move across space take center stage, while their removal from (or lack of) reason becomes an attribute referred to, and explained through, their body constitution.[64] In line with Plato, reason instead remains central in explaining teleologically the upright position of the human being (and its bipedness),[65] but emblematically loses the centrality it had in Plato's consideration of the nonhuman animals.[66] In sum, in this excursus on animals' kinds influenced by *Timaeus*, Aristotle remains faithful to his discourse on method in book 1, where he voices the plurality of the soul and claims to privilege in his study those parts which make animals "move,"[67] to the exclusion of the rational part.[68] True, mind and intelligence still enter Aristotle's discussion, as in *PA* 4, where they characterize the human being and are "negative attributes" of nonhuman animals. But, as Aristotle has prescribed in his discourse of method, they are not the proper object of his study and remain rather marginal and tangential to a treatment that focuses on animals as creatures of nature that have in themselves the origin of movement and rest and that are defined by the power of sensation and their empowerment to live.

The human body may well offer a roadmap to follow in the treatment of animal parts,[69] namely an analytical itinerary that goes from the head down, but, as Lennox has shown, this does not happen because man is taken as a model. Of the three factors that Aristotle invokes to explain this methodological preference—human organic complexity and consequent good life (*eu zēn*), familiarity, and the matching of the order of the parts in the human body with the natural order[70]—one, in particular, stands out: the familiarity of the human body. Indeed, this will be the only reason adduced in *History of Animals* to justify a treatment of the animals' bodily parts according to the vertical axis provided by the human body.[71]

> And first, we should consider the parts of the human body. Every community reckons its currency in relation to that most familiar to itself, and we must do the same in other situations—and mankind is of necessity the most familiar of animals to us.
>
> (*HA* 1 491a20–23, transl. by A.L. Peck, with slight modifications)

This discourse, which equates the use of the human body to that of a community currency, betrays on Aristotle's part the awareness of the relativity of a method of inquiry that cannot avoid being partly self-referential in order to acquire knowledge of other life-forms that are less known because they are less familiar to us. And that the goal of this method lies outside the human animal itself, even though it is the point of departure, becomes evident if we consider the outset of *Physics*, where Aristotle anchors the inquiry into nature to the search of the first principles (*arkhai*) and recommends an inquisitive path that starts "from the things which are more knowable and obvious to us toward those that are clearer and more knowable by nature."[72] Used as a heuristic tool, the human animal with its body does not take a central stage in Aristotle's discussion and only appears in relation to its idiosyncratic uniqueness. For instance, in *Parts of Animals* the human animal is the one that blinks more often than the other animals with two eyelids, and it is the only one that has eyelashes on the two lids; it also possesses the hairiest heads and the softest tongues. The examples could continue[73] showing the coherence of a methodological principle that Lennox has forcefully argued for when discussing Aristotle's philosophy of biology. Lennox writes:

> Parts should be predicated of the widest kind that possesses them, and explained at the same level: and the corollary of this principle, features should be explained at a narrower level of extension only when they are distinctive of kinds of narrower extension. Thus humans are mentioned specifically only when they have a feature which distinguishes them from the various other kinds that Aristotle is discussing.[74]

In sum, the human animal is mentioned for what distinguishes it in terms of its body, and eventually physiology.[75] And if the reference to its exclusive intellectual faculties (*to noein* and *to phronein*) explains teleologically its body structure in Plato-style,[76] it is legitimated by the narrower level of extension of a part of its body, for the human animal is the only one among living beings to possess hands.[77]

While their presumed deficiency of reason is not relevant in Aristotle's overall study, animals are seen in their full, "kind-specific" aptness to live.[78] Those parts of the soul that for Plato explained animals' degradation in the absence and failing of reason are now invoked, in a different formulation and order,[79] to explain different forms of life, in a sequence of living beings, distinct from the scale addressed in the discussion above,[80] and going, in this case, from plants to stationary animals to the other animals that are endowed with "traveling movement" (*kinēsis poreuomenē*), humans included. "An animal can neither exist nor grow without food," states Aristotle in *Parts of Animals*, letting us soon after understand that, inasmuch as they take in food and grow, plants are also living beings (*zōnta*), but not fully animals (*zōa*).[81] Without a sensitive part, the nutritive soul alone is insufficient to grant a living being the status of animal.[82] Here too, in denying that plants are animals, Aristotle moves away from the Presocratic conception.[83]

Further, the inclusion of the consideration of the locomotive soul in his discourse on method seems directed toward comprehending the animal world by differentiating those living beings that can feel and move, first, from those that merely feel but are stationary (*monima*), and then further defining—albeit "horizontally"—the scale of nature.[84] For when considered in relation to the conformation of animal bodies, the power of locomotion explains animals' specific ways of moving across space, from flying to walking and crawling to swimming, and allows the identification of distinct animal types whose difference is articulated around modes of movement that are intrinsic to their being.

With all sharing the nutritive and sensitive souls, and with some also possessing the locomotive one, animals are for Aristotle ensouled bodies. On these premises, his method is based on inclusiveness rather than exclusiveness, of which the comparative anatomy exposed in *Parts of Animals* is only the most visible result. The first aporetic dilemma with which Aristotle starts his discourse on method, and the dilemma whose resolution sustains the plan not only of *Parts of Animals* itself but also, tacitly, of *History of Animals, On the Soul*, and *Parva Naturalia*,[85] is whether to address animals species by species or by those phenomena that happen in common (*ta koinēi symbebēkota*) to different groups of animals. This is a problematic point to which Aristotle returns several times in the course of his treatise on method, refining it and making it the first, fundamental boundary (*horos*) of his inquiry.

> Should we take each single species severally by turn (such as man or Lion, or Ox, or whatever it may be) and define what we have to say about it, in and by itself? Or should we first establish as our basis the attributes that are common to all of them because of some common character, *ta koinēi symbebēkota*, which they possess?—there being many attributes which are identical though they occur in many groups which differ among themselves, e.g. sleep, respiration, growth, decay, death, together with those other remaining affections, *pathē*, and conditions, *diatheseis*, which are of a similar kind.[86]

The solution to this dilemma arrives later in the treatise, once Aristotle has laid out another methodological boundary, that which gravitates around division (*diairesis*)[87] and which consists in a strategy to identify animal species based on the simultaneous embodiment of multiple differences against Plato's adoption of simple dichotomy.[88] In the end, Aristotle opts for the common attributes (*ta koinēi symbebēkota*).[89] As Stenzel pointed out, Speusippus also had adopted a method based on "*koinēi*" and considered both animals and plants in terms of similarities, an aspect of the inquiry which Aristotle accepts,[90] although from *Posterior Analytics* it is clear that he distances himself from Speusippus' epistemological holism,[91] which was probably also at the core of his discussion of life's forms in the treatise *Likes*. In *Parts of Animals* Aristotle articulates the common traits in terms of degree (*kata genos*) and by analogy (*kat' analogian*), thereby fathoming animals' commonality (*to koinon*), both within kinds and across kinds, while he addresses the traits of species that are one of a kind, like that of man, as "specific" (*kat' eidos*).[92] For instance, individual species of birds differ in terms of the qualities of their parts, by "the more and the less": a crane has bigger, and likely tougher, wings than a sparrow, but both species still have wings. As for the common attributes across kinds, a man's bone corresponds to the spine of a fish. On the other hand, the fact of walking erect is an attribute peculiar and unique to the human species (*kat' eidos*).

It is in the context of the delineation of animal kinds[93] that Aristotle refers explicitly to a common nature (*physis koinē*) making it the condition for a species' membership to the same kind and for a consideration of animals' parts according to "the more and the less." In this respect, through an analysis of relevant passages in *History of Animals*, Charles has shown that "the idea of a *common nature* rests on the thought of one organized collection of methods of moving, reproducing, feeding and breathing";[94] consequently, the sharing of similar bodily parts, like the bigger or smaller wings of birds mentioned in *Parts of Animals*, involves a larger spectrum of shared features that include not only the morphology and physiology but also the social habits—such as feeding—of those animals that share a common nature. Yet importantly, for Aristotle, a sort of commonality also exists across animals' kinds, and in this we clearly face an influence of Empedocles' approach to living beings as expressed in frag. 82: "Hairs and leaves and the dense feathers on birds are the same and the scales on stout limbs."[95] This type of commonality is more remote than the one embodied by the creatures that share a common nature (*koinē physis*), but it is still compelling and pervasive and consists in analogies, like the one relating a man's bone to a fish's spine or the roots of a plant to the head of a human being.[96]

In fact, being receptive to this pivotal feature of Aristotle's method—namely the search for, and focus on, what is common among animals—helps us avoid the risk of anachronism, which so often and for so long, as Pellegrin has pointed out, has tinted the exegesis of Aristotle's biological studies and in particular his evaluation on the basis of a classificatory intent.[97] In respect to this feature too—the exploration of the common—it becomes evident that Aristotle did not intend to formulate a classification of animals, but that division among animal kinds and species was applied, at least on one level of his inquiry, in order to bridge it, letting transpire the articulation of animal commonalities via differences. Taken in this way, then, in Aristotle the systematization of animals could not be complete, nor rigidly "monovalent," as if laid out by modern standards.[98] For it was not meant to be so. Rather, in Aristotle classification became a flexible strategy for finding common ground among living beings, while avoiding repetitions. In *PA* 1, however, Aristotle distinguishes between his discourse on method turning around *ta koinēi* and an ontological discourse in which each animal species has its own, irreducible *ousia*, a discourse to which he remains in principle faithful despite the methodological choice. For at the end, when reviewing the aporia between treating animals species by species or according to common traits, Aristotle admits that

> it would be best, if one could do, to study separately the things that are particular and undivided in form—just as one studies mankind, so too bird; for this kind has forms. But the study would be of any one of the indivisible birds, e.g. sparrow or crane or something of this sort.
>
> (*PA* 1 644a30–34)

Aristotle here voices a tension that remains unresolved,[99] For he admits that, in a discourse on animals situated "outside" the constraints required by the method, animals should be considered in their ontological specificity. The animal world is fundamentally plural: each sparrow and crane, and likewise every living being, independently of membership to a larger group, has an ontological reality, just as man does.[100]

This decision to discuss what is common among animals (*ta koinēi*) initiates in de-anthropocentric terms a systematic exploration of that "more kin nature" (*oikeiotera physis*) which, we saw earlier, Aristotle presents in his apology as an incentive to devote his attention to the mortal animals of this earth and value their study. Yet the overall perspective presented in the foundational discourse of *Parts of Animals*, from the emphasis on sensation and the exile of reason to the focus on different modes of "the common" (*to koinon*) that connects animals' lives, displaces the human animal from a central, self-referential position. Overall, as it will be argued in the final part of this chapter and at various intersections of this book, Aristotle's focus is zoocentric, not anthropocentric.

An extreme example of this perspective framing Aristotle's approach is his presentation of the beak of birds within a wider treatment of the lips in blooded animals, worth mentioning at this point.

> For in the birds, as we said, for nourishment and strength their beak is bony. It has been joined together into one, in place of teeth and lips, just as if someone who had removed the lips from a human being were both to fuse the upper teeth together, and separately the lower teeth, and then were to draw them both out to a point; in fact this would be already a bird-like beak.
>
> (Arist. *PA* 2 659b20–27)

In order to reveal the nature of a bird's beak, and the analogy that connects it to human teeth and mouth taken together, in this passage Aristotle asks his readers to envision an experiment of bodily manipulation where a human being loses his soft, flexible lips and find his teeth, both the upper and lower ones, respectively welded forward. In this living being two parts of the human body are exchanged for another that characterizes, instead, a different kind of animal, forcing, at least hypothetically, the new human so manipulated to embrace a physical difference and therefore be different. Thus Aristotle asks his readers to conceive a human being that, in respect to nutrition, "interaction," and language articulation, feeds and feels—for, in the context of animals, feeding is a form of feeling—and acts like a bird, endowed as he is with a beak that allows him to eat in a certain way and defend himself, but not to speak to the degree of articulation humans naturally do. Certainly, the primary force of this example in Aristotle's discourse is explanatory and works at the level of morphology and function, illuminating a common trait of animals' bodies by analogy (*kat' analogian*) and showing, in this case, the relation between human teeth and lips, on the one side, and a bird's beak, on the other. In other words, with this example Aristotle explores the potentialities of a differently composed body without raising the question of subjectivity. There is no allusion, for instance, to what it means for a living being as such, endowed with the rational soul, to be constrained in a physical form that does not allow him or her to fully express him- or herself via articulate speech, a tragic mismatch that will be at the core of many stories of metamorphosis from Homer on.[101] Nor, conversely, is there any allusion to what it means for a bird to feel like a bird beyond the range of activities allowed by a beak. Yet this example is also emblematic of Aristotle's overall unbiased, classless perspective on animals, predicated in *PA* 1 and open to understanding them through the interplay of commonalities and differences. Ultimately, in Aristotle's biological treatises, under the comparative method—which he so strongly advocates and follows—and in light of the first basic common trait connecting all animals, namely sensation, the human animal becomes one of the many, despite its erect posture, superior intelligence, and portable divinity; the last two qualifications are in fact irrelevant in a study of nature. Like some other creatures that feature specific traits,[102] man too has a few idiosyncrasies. In the course of *Parts of Animals* and elsewhere, Aristotle discusses these idiosyncrasies to follow soon again the track of "the common" (*to koinon*), which he articulates via degree (*kata genos*) and via analogy (*kat' analogian*) and on the basis of which the points of contact among animals, in respect to morphology, physiology, and life, become manifest.

1.3 A new beginning

In line with the Presocratic tradition, Aristotle considers animals in the context of the inquiry about nature (*peri physeōs historia*).[103] But for him, as Pellegrin has remarked, nature has lost the grand scope it had in the past; it is not a total science anymore. The list of areas constituting its realm, mentioned at the beginning of the *Meterologics* and culminating with the treatment of animals and plants, indicates this circumscription clearly. Aristotle's inquiry on nature extends from the principles of the natural beings and a theory of natural movement discussed in the *Physics* to the consideration of the ultimate components of matter and their transmutation of the *On Generation and Corruption* and of the celestial movements of *On the Heaven*,[104] but much is now excluded: the immobile being, all that concerns human deliberation, the ethical and political domains, and the technical activities and mathematics.[105] This reduction of the natural realm, significantly, is accompanied in Aristotle by a radical change of perspective. So far for the Presocratics, the inquiry into nature was aimed at retracing the process of formation of the cosmos from the very beginning up to the present. Even Plato in the *Timaeus* adhered to this chronological model, despite the subversive introduction of a managing demiurge. To know the origin (*arkhē*), which was tracked down to the primordial matter and the process of becoming it underwent, meant to know the true nature of something whether the object under inquiry was the world or one of the creatures that inhabited it. In this respect, as Kahn has remarked,

> *physis* can denote the true nature of a thing, while maintaining its etymological sense of "the primary source or process" from which the thing has come to be. "Nature" and "origin" are combined in one and the same idea.[106]

The Presocratic inquiry unfolded as a history of nature that advanced without a preordained design, and therefore under the overall influence of chance (*tykhē*), but not without rules. Each theory included channeling "forces" and devices that granted structure and harmony—and therewith an explanatory basis—to the organization of the primordial matter into physical entities, big and small, animate and inanimate. While embracing chance (*tykhē*), the Presocratics still accounted for an orderly history, balanced changes, and a present harmony. In its unpredictability nature proceeded systematically. Everything in its history happened for a rational, somehow intrinsic reason wherever this was situated, whether in the combination of Empedocles' four elements under the synergy of Love and Strife, or the "law of attraction" among similars affecting the aggregation of Democritus' atoms, or the separation of the elements in Anaxagoras under the rule of mind,[107] or, again, in the physical changes of air by rarefaction and condensation supporting Diogenes' cosmology.[108] And the theoretical principles that guided the conception of the universe were consistently adopted to account for that of its creatures:[109] a homology of treatment that reveals the homology intrinsic in nature and its processes, and the pervasive unity of its embodied manifestations. Ultimately, physiology and, in particular, embryology were used to counterbalance the influence of chance and to explain the continuity of animal species and the preservation of the genders.[110] Yet all along the focus remained on an open-ended process—one dictated by explainable mechanisms, it is true, but one whose direction developed step by step along with nature, not in any preordained way that aimed at a definite goal. This implied for the rise of living beings not a few mishaps along the way, like the diversely shaped zoogonies of Empedocles, aborted attempts to live by creatures unable to survive.[111] And for cosmology at large, it entailed the collapse of this world, literally and also, so to speak, ideologically. Empedocles' cosmology contemplated different stages of the world, from the complete unity of the elements in the *Sphairos* under the tyranny of Love to their ultimate disaggregation under the regime of Strife, while Democritus envisioned the presence of alternative, coexisting worlds and featuring exclusive ecosystems, with the inherent likelihood that ours might not be the best after all.

Aristotle breaks with this approach to natural philosophy, dramatically. The order of the universe is now eternal and ingenerated.[112] As Kahn has remarked,

> The traditional attempt to construct it [the order of the universe] from a hypothetical starting point (*arkhē*) is systematically rejected in favor of a new kind of *arkhai*, the "fundamental principles" into which cosmic movement and change are to be resolved.[113]

Aristotle's *Physics* presents a discourse on the method based on principles (*arkhai*), causes (*aitiai*), and elements (*stoikheia*) that dissolve the chronological trajectory of the earlier students of nature (*physiologoi*) into a set of heuristic tools for understanding its processes. As for the Presocratics, so in Aristotle nature is intrinsically connected with change. Notoriously, in *Physics* he claims that all "beings by nature" are those that contain in themselves the origin of movement (*arkhē kinēseōs*) and rest (*stasis*); he lists among them "animals and their parts, plants and those bodies that are simple, like, for instance, earth, fire, air and water."[114] Displaced from a cosmological discourse where it indicated a "hypothetical starting point," and along with it the original matter,[115] now the *arkhē* is anchored in each being that exists by nature. Each natural being is, by definition, the source of its own "movement" (*kinēsis*) and it is on the exploration of this exclusive relation that, breaking new ground, Aristotle focuses as an aspect of his inquiry on nature, thereby giving rise to his study of animals.[116] No longer considered in relation to the cosmos and its origin, now animals deserve attention in themselves, as autonomous, self-sufficient beings that carry in themselves the origin of their movement. It is significant that in the very passage of *Physics* where he defines "beings that are by nature" Aristotle introduces the same range of movements (*kinēsis*) that he mentions in *Parts of Animals*:[117] locomotion, growth (*auxēsis*), and, conversely, decay (*phthisis*) and alteration (*alloiōsis*).[118] But while in *Physics* he introduces these movements to illuminate the membership and filiation of specific beings—elements, animals, and their parts, plants— from nature, in *Parts of Animals* he refers to them in the context of discussing what parts of the soul the natural philosopher has to take into consideration, whether all parts or only those responsible for such movements. In treating the movement as the common denominator of related subjects—nature, the animals, the soul—this cross reference traces Aristotle's inquisitive itinerary and underscores, in particular, his stance on subsuming the study of (parts of) the soul into that of animals,[119] a feature that was addressed earlier in this chapter, while, at the same time, he forcefully places animals along with their psychological aspects as elements of inquiry in the realm of nature.[120]

1.4 Animals, tykhē, and the logos of nature

In fact, animals enjoy a privileged relation with nature. "The animal is the natural (physical) being par excellence," writes Pellegrin,[121] but not only because, as we saw before, "in them nature as the internal principle of change manifests itself to the highest degree." It is also, and fundamentally, because, in being goal oriented, the change they undergo emblematically represents nature's own "original" proceeding, which is always directed toward an end and never happens at random.[122] Aristotle radically displaces chance (*tykhē*) from animals as works of nature and in so doing severs a fundamental tie that defined it for the Presocratics. In *Physics* where he discusses the possibility of whether *tykhē* could be a cause of natural phenomena and realities, he denies it on the basis that what happens regularly and normally cannot be due to chance, which is responsible, instead, for extemporary and unpredictable results. If in a given animal teeth always grow in the same way and are arranged so that the front teeth are sharp and the molars broad, the first ones fitted to bite the food, the others to grind it, this cannot be the product of chance but reflects the finality inherent in nature.[123] In fact, interestingly, in *Physics* Aristotle even redefines the field of *tykhē* and makes it pertain to the sphere of moral choice (*proairēsis*) and therefore applicable only to humans, not to animals, whose actions with fortuitous outcomes fall, instead, under spontaneity (*to automaton*).[124] At any rate, his disavowal of chance in the works of nature unfolds as a critique of Empedocles, on two fronts. On the one side, Aristotle criticizes the random combination of parts that characterized Empedocles' zoo-gonies and, by extension, his flora—Aristotle mentions, in particular, "the men-headed ox-progeny" and "the olive-headed vines' progeny"—and, on the other, he rejects the idea that entire parts or whole-natured (*oulophyeis*) living beings were born first, as abrupt, already-formed creatures, completely disconnected from a process of formation.[125] For Aristotle, instead, exercising philological authority, "seeds must have come into being first and not straightaway animals: the words 'whole-natured first' (*oulophyeis*) must have meant seeds."[126] And as he breaks down animals and their parts to the original seeds from which each living being ultimately derives via a process of growth,[127] Aristotle also traces a natural development that is orderly and unidirectional.

> For those things are natural which, by a continuous movement originated from an internal principle arrive at some completion: the same completion is not reached from every principle; nor any chance completion, but always the tendency in each is towards the same end, if there is no impediment.[128]

Thus organic development follows a precise, uninterrupted itinerary, step after step.

The completion toward which, as subjects of natural generation (*genesis physikē*), animals develop, coincides with their essence (*ousia*), a notion that was regularly absent in the reflections of the Presocratics, as Aristotle recognizes in the brief historical outline presented in *Parts of Animals*.[129] And while this type of movement from *arkhē* to *telos* partially overlaps and intersects with those types that are identified at work in the domain of nature at large and are functions of the soul—*auxēsis* and *phthisis*, *alloiōsis*, locomotion—[130] it has profound ontological and gnoseological import. If each living being follows a biological trajectory from birth to completion (*telos*) in an orderly, systematic way,[131] it is only when reaching the stage of completion, its *telos*, that, endowed of a full-blown body made of its requisite parts, any animal lives to the fullest of its capacities in accordance with the functions of the soul that are proper to it. Only at this stage, then, does an animal fulfill its nature and perfectly fit its definition (*logos*). In this perspective, which is at the foundation of Aristotle's inquiry on animals as beings by nature and an interface between ontology and biology, form (*morphē*) and end (*telos*) coincide, and represent the very nature of a living individual, as Aristotle[132] clari-fies in the *Protrepticus*:

> The end conformed to nature, *physis*, is what is reached as last in the process of becoming when this develops with continuity until completion.[133]

Aristotle bases this innovative perspective identifying form and end, and both of them with nature (*physis*) on a major critique of the Presocratics, which leads in *Parts of Animals* to the establishment of two more interconnected methodological boundaries.[134] One addresses which natural cause (*aitia*) is to be considered prior, among the others, in his inquiry on animals, and the other discusses whether to follow the process of animals' development or, instead, its final result.[135] Causes (*aitiai*) are for Aristotle pillars of knowledge, and to understand fully an organism's biological trajectory from beginning to completion requires an aetiological discussion. In nature there are four causes—material, efficient, formal, and final—and these constitute complementary modalities in which nature "produces" living beings and which are, in turn, facets of their identity.[136] In trying to discover "the material principle (*hylikē arkhē*) and such a cause (*aitia*)" for Aristotle the Presocratics reduced all their study of nature to an account of matter and its mechanisms.[137] For them animals and plants were seen in relation to their formation under the working of matter: air, for instance, opens up the respiratory channels, while water carves the body's interior receptacles for food and liquids. By contrast, in Aristotle's study of nature, matter (*hylē*) is still relevant inasmuch as physical reality is coextensive with bodies[138] and the constitution of bodies requires diversified material,[139] but it becomes secondary. Indeed, form (*morphē*) and essence (*eidos*) are more congruent with nature than matter is. For, as Aristotle explains in *Physics*, "nature is rather form than matter because every reality is said nature when it is in actuality more than when it is in potentiality."[140] In his view, then, for each individual organism change moves reassuringly toward a definite end, which, we saw, constitutes at once its form and nature. Applying these theoretical principles to the discourse on method in *Parts of Animals*, since as for nature at large so for animals the final cause is more fundamental than the other causes, animals must be considered when they have reached their completion (*telos*) and not merely according to the processes of formation:

> We should not forget to ask whether it is appropriate to state, as those who studied nature before us did, how each thing has naturally come to be (*pephykenai*) rather than how it is (*einai*). For the one differs not a little from the other. It seems we should begin, even with generation (*genesis*), precisely as we said before: first one should get hold of the phenomena concerning each kind, then state their causes. For even with house-building, it is rather that these things happen because the form (*eidos*) of the house is such as it is, than the house is such as it is because it comes to be in this way.[141]

In other words, for Aristotle it is fundamental to discuss animals in their actuality, namely when they have fulfilled their end, and not in relation to the processes that lead to their formation. Thus, Aristotle leaves in the background the process of becoming, which his predecessors have followed in its unfolding, to focus on the *physis* of animals in their finally reached perfection. The solution to this aporia too represents a significant departure from the Presocratics. Both the inclusion of the soul and the priority of essence (*ousia*) at the expense of process determine Aristotle's general outlook in the zoological treatises, where he investigates animals with respect to their physiology, activities, characters, and lives as fully formed individuals rather than giving a step-by-step account of their growth, and related activities, from the moment of their formation. This topic will receive separate treatment in *On the Generation of Animals*,[142] which features their development as a *diakosmēsis*, a coming into order,[143] but in the other treatises animals are dealt with in their crystallized *physis*, each as perfect embodiments of their natural ends, with no exception.[144] When considered in this way, namely from the perspective of their completion, animals are creatures full of *logos*, where *logos* represents the rationality immanent in their bodies' composite nature (and likely differentiated material)[145] and is visible and operative in all aspects of their existence that, rooted in such bodies, express finality. Aristotle lets us understand it explicitly in *Parts of Animals* when, in advocating the priority of the final cause versus the efficient one, he asserts:

> Now it is apparent that first is the one we call for the sake of which (*heneka tinos*); for this is *logos* and the *logos* is an origin (*arkhē*) alike in things composed (*ta synestēkota*) according to art and in things composed by nature.[146]

Significantly, animals are here referred to with a periphrasis combining the dative of agent referring to nature (*physei*) with the substantive perfect participle *ta synestēkota*, "the things which have been composed." This participle stresses the complex constitution of living beings' bodies, which (in respect to "having been composed" by nature and what this entails) are (we may add for the sake of clarification) inherently different from the elements' natural simple bodies (and even the bodies of plants),[147] allowing a legitimate parallel between nature and art. In Aristotle's words, both the works composed by nature and those composed according to art "come into being for something," and this something, their *telos*, is *logos*. Considered in terms of finality, *logos* is also an origin (*arkhē*)[148] in that it preexists the process of "composition" for the products of nature and art alike.[149] Hence, in the realm of nature, *logos* pertains to animals (as opposed to elemental bodies) inasmuch as they result from a process of formation that follows "an original plan," step by step, and that brings about a form, conceived of in terms of structure. For in being "things that have been composed," animals present a composition of parts, which amounts to structure.[150] Aristotle illustrates animals' orderly body by analogy with the attainment of health and the construction of a house, for which a doctor and a builder can respectively account by giving the causes (*aitiai*) and rational grounds (*logoi*) for everything each one of them does and why it must be done in that way.[151] And for living beings too, that the order of composition and final (composite) result can be explained in terms of causes and rational grounds calls for the priority of the final cause and a study that focuses on them in their completed form, which correspond to the two interrelated methodological boundaries discussed above.[152]

True, plants too are composed by nature and as such they also embody a structure and *logos* in terms of finality.[153] But it is noteworthy that, as far as I know, Aristotle speaks of *logos* in these terms and resorts to a parallel with the proceedings of art only when laying out the foundations of his study of animals,[154] a fact which incidentally explains why he devotes to them a separate treatment. Indeed, as we still learn from *Parts of Animals*, plants do not exhibit a great variety of nonuniform parts. Their *telos* is, so to speak, "simple," and their *logos* less complex than that of animals. With hardly any functions to perform, plants' bodies possess a definite proportion and size[155] but a few organs[155] and engage in only one mode of living, namely nutrition, growth, and decay.[157] By contrast, in being sentient creatures, animals have a "more multiform shape" (*polymorphotera idea*) and engage in a wider range of activities, both internal to the organism itself and in the external world.[158] Their body architecture and life functions present an artful complexity that is alien to plants.[159]

1.4.1 Animals' logos from speech to body and life

Used widely in the biological treatises and elsewhere in Aristotle's work, *logos* is a polyvalent word whose meanings range from "definition," "reason," and "proportion" to "speech" and "word."[160] In the passage quoted above, *logos* has been translated in English as "definition" and "account," in German as "plan."[161] Earlier translations feature in English "reason," in French "raison," and in Italian "essenza."[162] On the other hand, Peck (and after him Torraca), whose rendition I follow, has merely transliterated it to avoid adopting "an inadequate or misleading word." In this respect, Peck observes:

> Here is a term of very varied meanings, a term which brings into mind a number of correlated conceptions, of which one or another may be uppermost in a particular case. It is an assistance if we bear in mind that underlying the verb *legein*, as it is most frequently used, is the conception of rational utterance or expression, and the same is found with *logos*, the noun derived from the same root. *Logos* can signify simply, *something spoken or uttered*; or with more prominence given to the rationality of the utterance, it can signify *a rational explanation expressive of the thing's nature*, of the *plan* of it; and from this comes the further meanings of *principle*, or *law*, and also of *definition*, or of *formula*, as expressing the structure or character of the object defined.[163]

As speech, in Aristotle's zoology, *logos* is a faculty that in its extreme versatility is exclusive to humans, whose movable tongue, front teeth, and mouth are so conformed as to enable them to speak to the degree of articulation they do.[164] Birds too, however, speak, and in *History of Animals* Aristotle even goes so far as to recognize dialects' regional variations for the same species.[165] In the zoological treatises Aristotle does not stress in *logos* as speech the correlated conception of reason, which by contrast is dominant in the ethical treatises, and particularly *Politics*, where he notoriously links speech and reason, making man the most political of animals. It is by means of *logos* that man reveals the advantageous and harmful, and based on these, the just and unjust, which in turn enable him to live together with other humans in households and cities.[166] Speech that conveys reason is the fundament of political communities. The neglect of reason in the discussion of humans' speech in the zoological context, however, should not surprise us given that, as we saw earlier, Aristotle clearly limits the parts of the soul that are pertinent to the inquiry on nature to the nutritive, sensitive, and locomotive ones while he exiles the rational part. The student of nature (*ho physikos*) must approach animals in relation to their capacity to grow and feel and move, not their capacity to think. *Logos*, however, we have seen, is still pertinent and operative in a discourse about animals, but at a more pervasive and fundamental level that involves them, and equally each one of them, as (composite) products of nature. Not an intrinsic faculty of the soul, which only the human animal possesses, *logos* is coextensive with animals' very lives, in that it coincides with their immanent finality and—unsurprisingly given the coincidence of the two—with the form they attain when reaching completion.

In *Parts of Animals* 2, while reverting to a comparison with art as he did in the discussion of the priority of the final cause in book 1, Aristotle deals once more with the form (*morphē*) attained by living beings; identifies it with the end (*telos*); and refers to *logos* but in new complementary applications that, while compatible with *logos* as *telos/arkhē* discussed earlier, further clarify his approach to the study on animals.

> In generation things are opposed to the way they are in substantial being; for things posterior in generation are prior in nature, and the final stage in generation is primary in nature. For instance, a house is not for the sake of bricks and stones, but rather these are for the sake of the house—and so it is with other matter. Not only is it apparent from a consideration of cases that this is the way things are, but it also accords with *logos*; for every generated thing develops from something and into something, *i.e.* from an origin into an origin, from a primary mover which already has a certain nature to a certain shape, *morphē*, or other such end. For a human being generates a human being, and a plant a plant, from the underlying matter of each. So the matter and the generation are necessarily prior in time, *khronōi*, but, *logōi*, in *logos* the substantial being, *ousia*, and the shape, *morphē*, of each thing. This would be clear if someone were to state the *logos* of the generation of something; the *logos* of house-building includes that of the house, while that of the house does not include that of house-building. And so it is in the other cases as well. Thus the matter of the elements is necessary for the sake of the uniform parts, since these are later in generation than the elements, and later than the uniform are the nonuniform parts; for these have already attained their ends, *telos*, and limit, *peras*, having achieved a constitution of the third sort, as often happens when generations are completed.[167]

In the field of nature that pertains to living beings—as well as in that of art *tout court*—the process of formation follows an order that is opposite to the essence (*ousia*): the last things (*ta hystera*) are in fact the first (*ta prōta*) and that which comes at the end of the process (*teleutaion*) is at the beginning (*prōton*). As a house precedes its construction with bricks and stones, so any fully formed animal[168] is prior to the entire process that leads to its formation.[169] Hence Aristotle presents two perspectives from which to look at the "causes" of animals as products of nature: one perspective in respect to time (*khronōi*) and another in respect to *logos* (*logōi*). Of the two it is the second that reflects the proceedings of nature (and that Aristotle wants to follow). In considering living beings in respect to time (*khronōi*), matter and generation come first; in respect to *logos* (*logōi*), however, the essential being (*ousia*) and shape (*morphē*) are prior. Indeed, this last perspective constitutes, as we saw in the last section, one of Aristotle's methodological boundaries in *PA* 1, according to which animals must be considered not in their developments from formation, but as perfect embodiments of their natural end (*telos*), corresponding in turn to their (composite) form as fully developed individuals.[170] Thus, in this application, *logos* is contrasted with time (*khronos*) and relates to the formal nature of the living being, which bridges, so to speak, efficient and final cause. On this view, an orderly and systematic relation frames the generation of living beings and involves the transmission of shape (*morphē*).[171] In this respect, Aristotle transcends the particularism of Empedocles, who followed the process of animal formation in materialistic terms, and *in fieri*,[172] and pinpoints, instead, in the relation between the adult and the new organism—the chronological and logical priority of the first over the latter, and its transmission of movement and form—the key to understanding the regularity informing the birth of living beings. "A man begets a man and a plant plant," he remarks, thereby reframing in a new theoretical context an observation that had already been made, in other words, by Democritus.[173] As Aristotle privileges the *telos* in the process of animals' formation, the linear chronology so dear to the Presocratics becomes secondary. Still important in the consideration of nature and its products and itself a manifestation of *logos*,[174] it yields to an alternative chronology that starts from the end and that is fundamentally ontological.

But in the passage from book 2 quoted above, *logos* is not only used in contrast to time (*khronos*) to qualify a perspective that looks at animals in terms of *telos* (and formal nature), it also subsequently features in another compatible sense. As Peck notes, when Aristotle claims that the *logos* of house-building includes the *logos* of the house, but that the *logos* of the house does not include that of house-building, the meaning of *logos* is close to "definition" and, we may add, serves Aristotle to reinforce by means of an example the methodological boundaries so far presented. For transposed in the context of his zoological project, this claim means that as objects of study animals have a *logos* that does not include their formation (*genesis*), but is restricted to them *qua* composed, living (ensouled) beings that (we may now "safely" add) embody *logos* in the sense of *telos*.[175] Hence, Aristotle's zoology turns around animals' bodies, structures, and functions, culminating, as we will soon see, in the body's complex action (*polymerēs praxis*) that is life itself, the ultimate *telos* and *logos* of animals.

Indeed, in Aristotle's study, shape (*morphē*) is not just intended outwardly as it seems to have been for Democritus,[176] who, as discussed earlier,[177] claimed that a human being, and by extension any other animals, can be identified with *morphē*, or *skhēma*, indicating only the contour and surface of the body, and with color (*khrōma*).[178] The analogy between the living being and the house, which Aristotle refers to in the passage quoted above, illuminates his position in this respect. For him, the shape of an animal is teleologically intended, namely the end result of a process, and, importantly, as with a house made of timber, bricks, and stones, a composition of parts that have functions.[179] More specifically, the shape (*morphē*) of animals derives from the progressive intersection of three sorts of compositions (*synthesis*) involving in turn three different types of constituents, from the elements to the parts of the body itself, uniforms and nonuniforms. In Aristotle's view, the composition of elements, so important for Empedocles,[180] represents only the "first," most basic level of *synthesis* among a progression of three discrete levels leading, in a hierarchically arranged teleological sequence, to the formation of animals. Each body constituent from the elements to the uniform to the nonuniform parts,[181] from the more basic to the less, has a function and is instrumental to the formation and/or working of other parts and ultimately to the complex action (*polymerēs praxis*) of the animal's body as a whole, which is life itself.[182] In other words, "in some way the body exists for the sake of the soul,"[183] and the specific functions of the bodily parts exist for the body's *ad hoc* function of life. Thus, Aristotle "deconstructs" the complex architecture of animals' bodies, identifies different, progressive orders of composition ("elements," uniform, and nonuniform parts), and explains each of them in terms of finality, the ultimate *telos* being life. He therefore gives us in this (albeit here general) way for the animals as well that rich fabric of causes (*aitiai*) and reasons (*logoi*), which leads to their ultimate *telos* and which, to return to the analogy he has himself used,[184] a doctor and builder alike would be able to provide when accounting respectively for the attainment of health and the construction of a house.[185]

Significantly, for Aristotle the finality of body composition (*synthesis*) extends into animals' mode and space of existence.[186] That is, combining their different *dynameis*, the uniform parts enable the body's nonuniform parts, such as the eye, the nose, and the face as a whole, to do their work for the sake of the multiform range of activities and movements in which the entire animal engages in order to live (*polymorphai praxeis kai kinēseis*). In this way, Aristotle sees a continuum connecting the functions of the uniform parts with those of the nonuniform ones and, in turn, with animals' practices. A passage of *Physics* is in this respect illuminating. In it Aristotle complements and corroborates the arguments presented in his discourse on method in *Parts of Animals* 1, and in particular the finality and consequently the *logos* of animals' lives, but focuses this time on their activities in the external world, as agents that live, work, and make things. This testimony is of special interest to our discussion especially because it addresses the *telos* of the other animals (*ta alla*) in comparison with humans where the major gap between the two groups is reducible to the presence or lack of *logos* as rational speech. In other words, the other animals in *Physics*, although they are not explicitly labeled as such, are the nonhuman animals lacking speech, the *aloga zōa*.

> It [the end, *telos*] is especially visible for the animals other than men (*alla zōa*), who do not act (*poiein*), neither by art (*tekhnē*), nor by research (*zetein*), nor by deliberation (*bouleuesthai*). Henceforth, some are at a loss whether the spiders, the ants and animals of this sort work (*ergazesthai*) with intelligence (*nous*) or something similar. Going forward in this direction, it seems

that also in the plants themselves useful things are produced for an end, as, for instance, the leaves in order to protect the fruit. If therefore it is by nature and on account of something that the swallow builds its nest and the spider its web, and if the plants produce their leaves on account of fruit, and direct their roots not upwards, but downwards on account of the nourishment, it is clear that this sort of causality exists in those that come into being and exist by nature.[187]

In this passage Aristotle temporarily breaks the overall unifying outlook characteristic of his own method and, for the most part, that of the Presocratics to focus on *ta alla zōa*, the other animals. As Ross has remarked, in referring to the bewilderment of "some" when accounting for the work of animals like spiders and ants, Aristotle may be thinking of Democritus, who paid attention to the works of the *aloga zōa* and attributed to them more than five senses and a status and capacity equal to diviners, seers, and the gods.[183] Aristotle sidesteps the question of what specific faculty enables animals to attend to their particular tasks; instead he explains it in terms of the finality of nature, which is, we saw, at the core of the arguments presented in his apology for the study of animals in *Parts of Animals* and the ensuing treatment. In *Physics*, however, while he suggests a set of differences between human and the other animals he still acknowledges some common ground. Nonhuman animals make things (*poiein*); they work (*ergazesthai*), but not by the same means that humans adopt. The other animals, the *aloga zōa*, do not have art, they do not research or deliberate; they are not involved in activities that are proper to humans and that derive to them from the possession of the rational soul, which is excluded, as we have seen, from Aristotle's treatment of nature and its products. Yet importantly, the other animals still act and work like humans. And it is in this synchronicity of the poietic dimension of their lives and the parallel absence of human-like activities—such as art, deliberation, and research—that the finality and therefore *logos* of nature is supremely visible.

It appears, then, that it is this practical, poietic dimension of animals' lives which is at the core of Aristotle's interest and of which the webs of spiders and the organized activities of ants are only the most symptomatic and blatant manifestations. For in actuality every animal, its body, and in turn its parts all working in masterful coordination, accomplishes a function and "works," thereby fulfilling its end. And while *Parts of Animals* is ultimately dedicated to exploring in a comparative framework the working of animals' bodily parts as activities essential to the life of a living being, from the majestic elephant to the modest mole, in *History of Animals* our philosopher will ultimately follow animals' differences tracing them in the particular dispositions (*ēthos*), activities (*praxeis*), and modes of life (*bios*)[189] that characterize animal species, kinds, or groups vis-à-vis other living beings. In sum, therefore, Aristotle's inquiry on animals' lives in the external world is tied to the description of their morphological differences. This external dimension of animals' lives, from their environments and feeding habits to their interpersonal relations to the practices that are peculiar to the political animals (*politika zōa*) and conducive to their definition as political, will be the topic of Chapter 6.

1.4.2   On the birth of zoology and animals' equality (and not)

Animal shape, for each living being with no distinction, is a structure functional to life. Under this light, as Aristotle has announced in the apology to the study of animals discussed earlier, there is no worthless animal; every one of them, when considered in terms of physical form and the ensuing aptness to live, may become—and in fact is, from a philosophical perspective—an object of contemplation (*theōria*) and an expression of beauty (*to kalon*).[190] Indeed, the purpose that drives the formation of animals has "its place among what is beautiful," he explains in the apology.[191] Forcefully, Aristotle states that he will not leave out any of them, "neither more worthless (*atimōteron*) nor worthier (*timiōteron*), because, when contemplated (*kata tēn theōrian*) each animal shows in its body the actual causes of its existence."[192] Thus under the philosopher's gaze all animals equally embody the causes of their life. Significantly, in the claim for animals' beauty and the erasure of traditional categories of worthless and worthy, Aristotle seems to echo Anaxagoras' language in fragment 12,[193] where the Presocratic philosopher advocated the presence of *nous*, mind, in every animate being, either bigger (*meizō*) or smaller (*elattō*). From Anaxagoras to Aristotle the two sets of adjectives switch emphasis, from size to worth,[194] but both philosophers use them to dispel discriminatory parameters when considering animals as creatures of nature that depend, respectively, on *nous* and *logos*. In *Metaphysics* itself Aristotle bursts into a eulogy of Anaxagoras as being the only sane man, in contrast with his predecessors who "said that there is mind (*nous*) in nature (*physis*), just as in animals (*zōa*), and that this is the cause of all order (*kosmos*) and arrangement (*taxis*)."[195] Here Anaxagoras' mind can be taken as a notion parallel to Aristotle's *logos* in that, despite their different nomenclature, each of them is considered the cause of the visible order and arrangement that inform animals' bodies and, ultimately, their lives.[196] Indeed, on the basis of Aristotle's praise of Anaxagoras, who attributed *nous* not only to nature but also to animals, it is possible to argue that for the Stagirite too the rationality at work in the designing nature (*dēmoiurgousa physis*)[197] is transferred to, and embodied by, the animals themselves in their act of being alive.

Conceived of in this way, zoology emerges as the science that, within the wider field of nature, studies all animals *qua* living, that is, in terms of their body structure and related activities that constitute (and contribute to) their life. Indeed, considered from the perspective of the "general" study of nature (*ta physika*), animals require an *ad hoc* treatment inasmuch as, alone among the other natural bodies, they are the source (*arkhē*) of a wide variety of linked movements. Besides growth, nutrition, and decline, which Aristotle includes under *auxēsis* and further defines in "biological" terms in *On Generation and Corruption*,[198] animals are also the origin of a peculiar form of alteration (*alloiōsis*), namely sensation (*aisthēsis*).[199] The basic minimum is a sense of touch but some animals are also able to move across space (*kinēsis kata topon*). The key to understanding animals' unique range of movements is their body structure, which Aristotle considers, in teleological terms and in analogy with art, as *logos*, making it (along with the movements it entails and the consequent activities in the external world) the proper object of his study of animals. And as he takes animals' structure as a manifestation of *logos*, Aristotle banishes *logos* as speech and reason (which only the human animal possesses) from being the proper object of zoology. Thus, based on movement and life, the perspective animating his study of animals is fundamentally zoocentric, not anthropocentric.

Aiming at living forever, but constrained by the limits of their nature, animals can continue to live only through their offspring. In "physical" terms this implies the capacity to perpetuate (projected into another natural complex body) the constellation of movements of which each of them is the source, a fact that may explain why in discussing the *telos* for both products of nature and art Aristotle claims that there is more finality (and "goodness") in entities composed by nature (the animals) than in those composed according to art.[200] For, we may well argue, while the finality of works of art is confined to the static accomplishment of the works themselves, that of works of nature is realized in the dynamic embodiment of their own movements and life into another living being for eternity.[201] And it is, further, in sharing this feature that animals for Aristotle are quintessentially equal: whether conscious or not, they all aspire to the immortality that characterizes divinity, and they all act according to their own individual nature in order to attain it. It is true that reproduction manifests such desire for all living entities, even in passive plants, the living things (*zōnta*) devoid of any cognition.[202] But animals are endowed with sensation, and this implies a significant discontinuity from plants in psychological and bodily terms:[203] not merely and tacitly assumed on the basis of reproduction,[204] animals' desire is a positive, object-oriented affection, and related to pleasure the search for which fulfills for all, in the different practices that materialize it, their recondite desire to live.[205]

Certainly, Aristotle's zoology is not the place for radical egalitarianism. All animals are equal as objects of study, namely insofar as they are composed living bodies, embodying knowable causes, and, as such, partaking of a common *logos*. And they are equal in their desire to live and the ensuing pleasure. So while he considers each animal equal in possessing a structural capacity to sense and desire, to move and live, Aristotle occasionally still maintains a hierarchy of Platonic legacy: he calls the human being divine and considers him, alone among the other animals, aligned with the orientation of the cosmos and conforming to nature.[206] Besides, Aristotle also sees a hierarchy at a more extensive level that involves animals' material natures. In *On the Generation of Animals*, for instance, he distinguishes animals in terms of elemental composition and inherent heat, claiming that animals have different levels of perfection, which in turn is reflected (explaining it) in their different embryogenesis.[207] In Leunissen's words, "The more perfect the animal, the more perfect its product of generation is, and more perfect according to Aristotle means hotter, less earthy (i.e. purer) and moister."[208] And further, if some animals possess all the five senses others are imperfect because they only possess touch.[209]

Nonetheless, it should be stressed, these hierarchical statements emerge from context-based discussions (animals' posture and number of legs in *PA*, their reproduction in *GA*, and *phantasia* in *DA*), and do not inform Aristotle's foundational discourse for the study of animals in *PA* 1, in which, as seen earlier, and perhaps influenced by Anaxagoras, the distinction between "worthier" and "unworthier" animals (which must involve the hierarchical gaps just outlined) is overcome by the recognition of the causality equally informing their different bodies. So in Aristotle's zoological study hierarchical perspectives coexist and are consistently subdued by an overall approach,[210] which rather than focusing on the radical break between orders of living beings, strives to pursue their common nature as opposed to all insentient natural beings (not only the elements but also plants and the celestial animals). One should underscore at this point Aristotle's method of considering differences (*diaphorai*) as articulations by degree and analogy of sameness (*to auto*), a method which he applies not only to animals' body parts (based on their functions) in *Parts of Animals* but also in *History of Animals* making it possible to cover the animals' activities (*praxeis*) and lives (*bioi*) discussed in that treatise.[211] Ultimately, no matter how different their degree of perfection and the sporadic eruptions of hierarchical perspectives, in Aristotle's discourse animals emerge as equal in that, in acting and living for the sake of living, each holds in itself the principle of life motivation, i.e. the heart or its analogous (which is foreign to plants). It may be important here too to underscore Aristotle's departure from Plato and his model of tripartite soul, on which more will be said in the next chapter. Suffice to mention that in *Timaeus* Plato locates the rational soul in the head, the spirited soul in the chest, and the appetitive one in the belly, thus establishing a compartmentalized, divided, and hierarchical view of the human body that will define the gap between man as the sublime rational creature and the degenerated, irrational (women and) animals.[212] In *Republic* this tripartition mobilizes a conflict of motivations in a discussion aiming to assess the superiority of reason over appetites and desires (and spirit, *thymos*).[213] By contrast, in his study of living beings' complex phenomenon of life, Aristotle elects the heart (and its analogous areas) in bloodless animals as the primary seat of sensation, nutrition, and movement,[214] and hence as the center of life motivation and, we should add, of the different forms animal life is manifested. Both a uniform and a nonuniform part, the heart is also the starting point of the blood vessels and holds the power (*dynamis*) by which the blood is first produced.[215] Not an emissary of reason as in Plato, thus, for Aristotle the heart with its related capacities becomes the unifying, defining, and superior feature of all animals' bodies and lives.

Notes

1 This section of Aristotle's discourse on method in book 1 of *Parts of Animals* does not merely present in a positive way the reasons one should study the living beings of this world. Indeed, it embodies, buried within, an anticipation of the critiques and resistance to a study of animals, thereby unfolding also as a defense of it. The main points to which Aristotle replies are 1) the unworthiness of the sublunary animals

(*ta atimōtera zōa*) when compared to the "ungenerated and imperishable substantial beings constituted by nature," namely planets and stars, and 2) their unattractiveness—both of bodies and their organic constituents—to human perception (see the expression *ta ou kekharismena* and, in the part subsequent to the one cited above, Aristotle's reference to the disgust *dyskhereia*, that the consideration of animals' organic parts bodies elicits in those looking at them, *PA* 1 645a26–31). In respect to this second point, Aristotle insists on the "more akin nature" (*oikeiotera physis*) of nonhuman animals with humans, which constitutes the main counterpoint to point 1 (see below) and promotes a new legibility of animals' bodies, one that transcends the materiality of their parts, subsuming it into a consideration of the bodies' whole conformation (*holē morphē*).

2 See Lennox (2011), who reviews the position advocated earlier in his commentary (2001) and addresses the integral role of the apology in the treatise's narrative. It is true that, as many critics have observed, *PA* 1 is composed of different essays, but they still construct a systematic discourse. For instance, this section of *PA* may well have a more rhetorical style and tone than what precedes and follows it, but it is not a mere interruption of Aristotle's layout of his method. Rather, it seems a rhetorically charged, climactic closure of the discussion of criteria (*horoi*), which up to now have been presented aporetically. After this "apology" Aristotle will resume his discussion of the method in a positive way, giving guidelines of interpretation, and not alternatives to be assessed as before.

3 Aristot. *PA* 1 644b23-645a25, with a slight modification. All translations of this treatise are by J.G. Lennox.

4 Cf. Plat. *Leg.* 12 966d–67 where stars are presented as gods; cf. Moreau, 1959.

5 Lennox, 2001, 172.

6 Plat. *Tim.* 30B, cf. Zatta, 2019, 142–3.

7 See respectively Aristot. *Cael.* 2, 1, 284 a, 17 and 2, 12 292b, 2; in the first passage Aristotle appeals to the authority of the tradition (*arkhaioi kai patrioi logoi*), which identified something immortal and divine in those bodies that possess limitless motion; in the second he claims that "the action, *praxis*, of the planets are analogous to that of animals and plants." Now Aristotle frames the "commonality" between celestial animals and their terrestrial counterpart in terms of analogy showing that they belong respectively to different kinds, but however elusive that analogy may be, it is undeniable that for him the celestial beings are animals. Just a little earlier Aristotle has rectified the conception of stars by attributing them action and life: "We think of stars as mere bodies (*sōmata*) and as units with a serial order indeed, but entirely inanimate; but we should rather consider them as enjoying action (*praxis*) and life (*zōē*)" (292a19–22); cf. also fragment 18 (Rose) of *On Philosophy*, where to be atheist in front of the visible divinity of the sun is considered an extraordinary fact. On the heavenly bodies as living beings with a soul, see Johansen, 2009, 22–3.

8 An interesting parallel can be found also in the spurious treatise *On the Cosmos*, where the author discusses the cosmos as divided in the upper and lower regions along with their respective inhabitants. To the first region he assigns the immortal gods, to the second the "creatures of a day" (*ta ephēmera zōa*), inclusive of all living beings subject to mortality (Pseud-Aristot. *Mu.* 393a3–6). Ultimately this distinction is inscribed in Aristotle's general statement about different orders of inquiry pertaining to "things that are not subject to change, things that are changed but imperishable" and "things that are perishable" (*Phys.* 198a29–31). On the constitution of the celestial bodies, see Falcon, 2001, and for a general discussion of natural bodies in terms of tridimensionality and movement, Falcon, 2005, 31–42.

9 Moreau, 1959, 60.

10 It is true that occasionally Aristotle does compare human beings with the divine cosmos, and in turn with the other animals (see, i.e. *PA* 4 686a24–87a2 and 2 656a3–13; cf. *de Resp.* 477a22), but these references speak of the particularity of the human

kind vis-à-vis the other animals rather than offering a standard criterion of analysis and evaluation for all. Indeed, because, as he states in this discourse on method, his study of animals lies fundamentally on their definition as sentient beings rather than rational beings, the comparison with the celestial animals, which are entirely rational, can only intersect his study of living beings when he discusses the human being, who is partly rational, but it is peripheral to his treatment of the forms of life as a whole. When considered in this light, Aristotle's overall approach ultimately tends to grant autonomy to sublunary animals as subjects of inquiry in themselves, from the unworthier to the worthier (cf. *PA* 1, 5), and apart from both a consideration of the celestial beings and the uniqueness of human kind (see discussion below).

11 Lennox identifies two scales of value at work in Aristotle's consideration of the "objects of natural science": intrinsic value and divinity and knowability (2001, 172).

12 On the limitations on the knowledge of planets and stars, see, for instance, *Cael.* 2 291b 24 and *Met.* 12 1074 a 14.

13 *PA* 1 645a10–12. The philosophically oriented student will discover in animals (and each one of them, even the more worthless) something natural and beautiful (*ti tou physikou kai kalou*). Significantly Aristotle exemplifies this discovery with the so-called Heraclitus anecdote. One day the philosopher was warming himself up at the oven in a kitchen when some visitors showed up, but they were hesitant to enter. Heraclitus invited them to come in stating that even there, in a modest kitchen environment, there were gods. The immediate reference for "gods" is the "more worthless" terrestrial animals, mentioned by Aristotle earlier, in opposition to "worthier" animals (for a thorough discussion of the Heraclitus anecdote, see Gregoric, 2001, 1–13). Yet, given that Aristotle's apology develops balancing out the study of worthy, celestial, divine living beings (planets and stars) with worthless, terrestrial, mortal living beings (the animals) on account of their different knowability, it is compelling to extend the Heraclitus anecdote to the wider context of the apology and see in the reference to the gods in the kitchen an allusion to the worthy nature of terrestrial animals when looked at from a philosophical approach investigating the causes of their existence (*PA* 1 645a19–23). In this respect, it should be remarked that for Aristotle every living being and thing reproduces another like itself because it has an intrinsic desire for immortality and divinity (*DA* 2; see Chapter 3); cf. *GA* 2 731b33–732a2 where we learn that "what is generated" (the animals) becomes eternal in the way it is open to it, namely "specifically," but not "numerically." Thus while maintaining the hierarchy between higher (celestial)/lower (terrestrial) animals and that between higher and lower animals of this earth, Aristotle also softens them by recognizing in every lower animal the attainment of immortality in the way its nature allows it to.

14 Given the priority of the nutritive soul in Aristotle's discussion of animals' life (see Chapters 2 and 3), it is tempting to read in *to syntrophon* the notion of "eating together," as still evoking physical proximity, but with an accent on the function of nutrition (as accompanied by sensation, and pleasure and pain).

15 Aristot. *HA* 8 629b8–12. Cf. Plu. *Aem.* 10; S. *Aj.* 861; Xen. *Mem.* 2, 3, 4. *History of Animals* offers an interesting parallel involving the sacred crocodiles of Egypt, wild animals that become tame on account of being fed (*HA* 8 688b30–3); see Chapter 6.

16 In this respect, in the passage quoted above Aristotle calls his object of study the "living nature" (*zōikē physis*), by which we need to understand the nature of the terrestrial living beings (*zōa*), humans included.

17 Lennox, 2001, 172.

18 Porph. *Abst.* 3, 25; cf. Sorabij, 1993, and Cole, 1992. It seems, however, that a partial or absent account of Aristotle's biological treatises, and in particular the apology in *PA*, have led to a displacement of Aristotle's position in respect to the philosophical tradition. For instance, Sorabji ignores the acknowledgment of kinship between humans and animals and its development in the biological treatises, and overem-

phasizes the intellectual distinction between men and animals, thereby associating Aristotle with the Stoics, who would deny reason to animals, and exile them from the extended community of humans and gods (1993). On the other hand, also Cole, while recognizing Theophrastus' kinship between humans and animals situates him in strong opposition with Aristotle, who, she claims, considered humans and animals decisively separated (1992, 52–5). For a discussion of the "Aristotelian premises" in this passage and the actual closeness of Theophrastus to Aristotle, see by contrast Brink, 1955, 129–31.

19 On the discontinuity between animals and plants, see Thphr. *HP* 1, 1–4. Cf. *PA* 2 646a14–24 and 655b4–656a7, where Aristotle announces a distinct treatment for plants and their formation inasmuch as they do not have the faculty of sensation, which is essential in the definition of animals (*ta zōa*); see also *PA* 4 680b10–29.

20 Cole, 1992, 55.

21 I intend by "to feel" the experience of a range of sensations (*aisthēseis*) and affections (*pathēmata*) that the possession of the sensitive soul enables animals to have—each according to its nature—i.e. from the feeling of hot and cold and the other sense perceptions (see *DA* 2, 2) to that of fear, courage, and similar emotions (see *HA* 7 588a16–588b4 and 8 608l1–608b20), along with the pain or pleasure by which each sensation and affection are accompanied. See the discussion in Chapter 5.

22 Not all animals possess all senses, but all possess at least touch. On senses—their "allocation" and "architectures"—see Chapter 4.

23 Le Blond, 1945, 41. On the difficulty to assess the purpose of Aristotle's biological treatises, see Balme, 1987b. A fundamental feature of Aristotle's method, to be discussed later in the chapter, unequivocally reveals the common nature of animals. Indeed, the first, and therefore prior, methodological dilemma Aristotle confronts is whether animals should be discussed species by species or by common traits (*PA* 1 639a–639b6). He opts for the second path and returns on it multiple times, a sign that it was a difficult, and perhaps, controversial choice, yet methodologically sound.

24 *DA* 2 412a6. More specifically, Aristotle is concerned with finding a definition that may suit every (part of the) soul and hence every being possessed of soul.

25 In book 1 of the *On the Soul* Aristotle calls the nutritive soul the first principle (*prōtē arkhē*) and claims that animals and plants have in common (*koinōnein*) this alone (*DA* 1 411b29–30).

26 See Aristot. *DA* 2.2; on the nutritive soul, see below Chapters 2 and 3.

27 Manuli, 1977.

28 *PA* 3 666a34–b1; cf., *PA* 2, 647a, 653b21–35; for the crucial role of the heart as principle of animals' "organization" (*diakosmēsis*) in the process of embryological formation, see *GA* 2 740a6–24.

29 In fact, Aristotle locates the soul in the heart only in *Parts of Animals* (2 655b36, 656a38), while in *On the Soul* he considers it the functional organization of bodies, without anchoring it to a specific organic part. See Chapter 3.

30 Aristot. *GA* 2 740a2–9.

31 Aristot. *PA* 2 647a4–8; cf. *HA* 1 486a5–8.

32 Aristot. *PA* 2 647a14–24.

33 On the coextensiveness of animal life and sensation, see also the discussion of Aristotle's general treatment of sensation in *DA* 2.5 in Chapter 4.

34 Aristot. *PA* 1 640b30–2; see Zatta, 2019, 50–2. The same critique of Democritus as privileging animals' body structure and "surface" rather than "movement" (and life) appears also in *On the Generation of Animals*, but in reference to the development of the embryo. According to Aristotle's testimony, Democritus believed that "the external parts" (*ta exō*) of the animal would be the first to become visible and were then followed by the internal ones (*ta entos*), but for our philosopher this model pertains to the animals of art, carved out of wood and stone, and not to the animals of nature as

having in themselves the principle of movement and rest (*GA* 2 740a13–17; see also 740a37–38).

35 Aristot. *PA* 1 640b36–641a2. Cf. also *DA* 2 412b11–413a3, where, in the context of the examination of the soul, Aristotle discusses the eye whose action is that of seeing, and in the failure of sight considers the eye as existing only in an equivocal sense. The same argument is then applied to the whole body, whose activity—life—depends on the possession of the soul.

36 See, for instance, Johnson, 2008; cf. Tipton, 2014.

37 Leunissen, 2010.

38 On the order of exposition, see Lennox, 1999b.

39 Aristot. *PA* 2 650a3–4.

40 On the flexibility of nature, see Leunissen, 2011. For a discussion of conditional necessity, see Chapter 3.

41 Preservation (*sōteria*) figures widely in the treatise to qualify bodily parts and their role. See, for instance, *PA* 2 652b8.

42 Johnson, 2005.

43 Cf. *PA* 1 645b15–20, where Aristotle forcefully places the finality of bodily parts in the soul, which in its basic meaning, inherited from the Presocratics, corresponds to the activity of life. "Now, as each of the parts of the body, like every other instrument, is for the sake of some purpose, viz. some action (*praxis*), it is evident that the body as a whole must exist for the sake of some complex action. Just as the saw is there for the sake of sawing and not sawing for the sake of the soul, because sawing is the using of the instrument, so in some way the body exists for the sake of the soul (*psykhē*) and the parts of the body for the sake of those functions to which they are naturally adapted." Cf. also *PA* 1 641a19–21, where Aristotle establishes the coincidence of soul and life. On the soul, its basic meaning and functions, see Johansen, 2012, and below Chapters 2, 3, and 4.

44 Aristot. *PA* 641a33–641b10.

45 On the homogeneity of the soul for the Presocratics, see Chapter 2.

46 See, for instance, Plat. *Tim.* 69c–70d and *Resp.* 4 435e–441c. For an extensive discussion of Plato's partition of the soul and Aristotle's departure from it in his study of living beings, see Chapter 2. Here let it suffice to remark that despite some analogies between Plato's and Aristotle's conception of the soul, namely its composite nature and the identification of the rational part, there are also fundamental differences that pertain, for instance, to its immortality—completely irrelevant for Aristotle, who speaks, by contrast, of its "departure" from the body—the "psychological affections and activities" each part of the soul represents, and, finally, the very relation it entertains with them and the body. Cf. Aristotle's passage quoted in this page and Plat. *Tim.* 69c–71.

47 Cf. Aristot. *DA* 2 413a21–414.

48 Cf. *Phys.* 2 192b16 and next section.

49 Within this range of changes, it should be remarked, it is growth (*auxēsis*) and conversely decay (*phthisis*) that properly overlap with the Presocratics' very notion of *physis*, as encompassing all stages in a given process of formation, inclusive of beginning, growth, and final result (Naddaf, 2005). On the definition of growth as a strictly biological phenomenon, see Chapter 3.

50 This centrality, in the consideration of the soul, of a movement that is intrinsic to the very fabric of the body—whether movement is intended as growth, sensation, or locomotion—is in line with the definition of animals as beings that exist by nature because, unlike those produced by art, they have in themselves the origin of "movement (*kinēsis*) and rest (*stasis*)" (*Phys.* 2 192b9–23). See below. By contrast, reason is excluded from this array of movements because as, Aristotle claims in *On the Soul* within a critique to Plato's *Timaeus*, thinking (*noēsis*) seems to correspond more to a state of rest (*ēremēsis*) or a halting (*epistasis*) than a movement (*kinēsis*) (*DA* 1 407a33–4).

51 In *On the Soul* Aristotle attributes the cause for "traveling movement" (*poreuomenē kinēsis*) to appetite (*orexis*) and mind (*nous*), intending for this last one the practical mind (*praktikos nous*), which in contrast to the theoretical one (*theoretikos nous*) pursues an end outside itself (Aristot. *DA* 3 432b7–433a10).

52 Aristotle gives two reasons for the exclusion of the rational soul. On the one hand, treating of mind (*nous*) along with its objects (*noeta*) would undermine the systematization of Aristotle's corpus, on the other, mind and related activities are uninvolved with nature (*physis*). For a discussion of these two reasons in relation to one another and to Aristotle's pursuit of the "animal as such," and the crucial importance of this "ultimate object" of study, see Chapter 3. Aristotle is completely faithful to this proscription (the exclusion of the rational soul) in his discussion of animals in *Parts of Animals* (Lennox, 1999b). For a discussion of Aristotle's position on the rational soul in the context of his inquiry on the soul as the principle and cause of life in *On the Soul*, see Chapter 2.

53 In *History of Animals*, it should be noticed, when describing animals' behavior Aristotle acknowledges the presence of psychological abilities that have to do with reason (see Labarrière, 1990) and explicitly attributes *dianoia* to the swallow inasmuch as a skillful architect (*HA* 8 612b18–27). These abilities have to be considered as natural capacities (*physikai dynameis*), disconnected from moral ones, and are explainable via animal physiology—namely the blood quality—rather than as endowments from the rational soul. See below note 63 and Chapter 6.

54 See n. 10 and below.

55 I discuss animals' physical degeneration as emblematic of their progressive removal from reason (and man) in *Timaeus*' account of the origin of animal kinds in Zatta, 2019, Chapter 6.

56 Aristot. *PA* 4 686a24–686b22; cf. Lennox, 2001, 317.

57 Aristotle explains humans' upright nature by the intensity of the heat around the heart pushing growth upward (2 653a9–32), but leaves unexplained why a heavy upper part of the body would hinder the faculty of reason. As Lennox notes (2001, 218), this question has to do with the material basis of reason and *de Somn.* 3 may shed some light on it. In this treatise the digestion-induced upward movement of the blood to the region of the head is the cause of sleep (and the ceasing of the process of cognition that accompanies it) (456b32–457a6).

58 In Aristotle's view, the weight (bending an animal down) makes thought (*dianoia*) and common sense (*koinē aisthēsis*) (*PA* 686a21–32) sluggish and it is on account of their body structure that nonhuman animals are less intelligent than human beings.

59 *PA* 4 686b21–3, transl. by J.G. Lennox.

60 On this point, see again Lennox, 2001, 319.

61 The upper part extends from the head to the anus while the lower parts encompass the members of the body that supports it and affects locomotion. For the evaluative use of the compound *nanophyēs* (with the stature of a dwarf), see Aristoph. *Pax* 790.

62 More specifically, in *PA* 4.10 Aristotle discusses the earthbound structure of nonhuman animals in two "movements," first addressing the quadrupeds (*PA* 4 686a32–35) and, subsequently, the blooded many-footed and footless animals (*PA* 4 686b29–31). In the case of quadrupeds, he combines the mechanistic explanation invoking the elemental composition of their body with a teleological explanation that stresses ambulatory functionality and ignores reason. Indeed, it is for the sake of "security against falling" (*asphaleia*) that nature has provided animals' earthbound body with two forelegs instead of arms and hands (*PA* 4 686a34–35).

63 In sum, animal posture is indicative of a specific physiology, more corporeal and earthier for those that are more inclined to the ground, a fact that has consequences on the intellectual soul, but not, as it seems from this passage, on the sensitive one. Further, in his excursus Aristotle discusses blooded animals from different types of quadrupeds to snakes, ending up surprisingly with plants, an inclusion that stands in

contrast with his separation of plants from animals on account of their lack of sensation. But he includes them here because he is looking at living beings' differences from the perspective of body structure, limb number, and locomotion. Further, in this passage, the association of animals and plants may also be due to the overall influence of *Timaeus* where plants are considered animals, endowed with sensations but of a different type than those of man (*allai aistheseis*) (Plat. *Tim.* 77a).

64 The decentralization of reason in Aristotle's discussion is also apparent from the comment he makes about dwarf adults as compared with other adults: they are inferior in intelligence (*nous*), but they may have other characteristics in which they are superior (*PA* 4 686b25–26).

65 *PA* 4 686a27–32.

66 Indeed, in moving from the human to the nonhuman animals, Aristotle also departs from a Platonic-based interpretation, turning around reason, to a consideration of animals' structure and nature which is purely "mechanic."

67 Indeed, as we have seen, Aristotle focuses on animals' growth (*auxēsis*) and locomotion, which pertain respectively to the nutritive and locomotive parts of the soul.

68 See n. 52 and Chapter 3.

69 Likewise, in *On the Generation of Animals* human embryogenesis guides the discussion of animals' generation (see, for instance, books 2 and 5).

70 On the discussion of these points, see Lennox, 1999b, 6–8. To the important discussion of the spatial coordinates of "up" and "down" on the basis of the natural order of things, which ultimately corresponds to the order of the universe, it should be added that this cosmological reference in *Parts of Animals*, the only one of this kind in the treatise, is perhaps a residue of the cosmological framework within which animals had been ordinarily treated and from which Aristotle moves away (see below section 1.4. *A New Beginning*).

71 For a difference between *Parts of Animals and History of Animals* in the order of treatment of animals' parts, see again Lennox, 1999b, 16 n. 10.

72 Aristot. *Phys.* 1 184a17–19.

73 For a discussion of the unique features of the human animal, see Lloyd, 1983, 30.

74 Lennox, 1999b, 9; see also Lennox, 1985 and 2001, 141, where he adds that "Aristotle treats mankind as he treats every other animal—whatever we share in common with a wider group is treated as a feature of the wider group. Humans are singled out for separate discussion, just as elephant or camels are, only when there is some part that distinguishes them from the other animals." In other words, humans' attributes when unique to human species are treated as "specific" (*kat'eidos*). See below.

75 For instance, the human being has the larger brain for his size than other blooded animals—and man has larger than the woman—because of the great heat in the region of the heart and the lungs. Indeed, the function of the brain, which is cold, is that of balancing the heat of the body, thereby granting the animals' good health and preservation (*PA* 2 652a24–653b9).

76 *PA* 4 686a29.

77 It is noteworthy that Aristotle appeals to the teleological principle also later on in *Parts of Animals* when he replies to Anaxagoras that the human being has hands because it is the most intelligent animal and not the other way around as the Presocratic had instead claimed (Aristot. *PA* 4 687a8–10; DK 19 A 12).

78 When considered in this respect, all animals, in spite of the specific degree of sophistication they may embody, are perfect (*teleia*). See Aristot. *DA* 3 432b22–5: "Then seeing that nature does nothing in vain, and omits nothing essential, except in maimed or imperfect animals (and the sort of animal under consideration is perfect and not maimed; this is proved by the fact that they propagate their species and have a prime, *akmē*, and a decline, *phthisis*), they would have also parts instrumental to progression." This passage, which belongs to a longer section devoted to inquiry on what part of the soul enables the living being to move across space, contains in fact an

important general assertion. Aristotle lets us understand that there is a standard of perfection inherent in each animal, regardless of species or genus, inasmuch as it is able to reproduce and follow the curve of life.

79 Significantly, Aristotle breaks the assignment of the parts of the soul to distinct parts of the body that sustained Plato's hierarchical and anthropocentric view. He adopts, instead, an abstract model and conceives the soul like a figure that progressively encompasses its different parts, from the nutritive to the sensitive to the locomotive to the rational (*DA* 2, 2–3). See Chapter 2.

80 See p. 17.

81 Aristot. *PA* 2 655b33–656a2; see also *GA* 1 731b4–8.

82 Aristotle, however, remains cautious in assessing consistently secure membership, aware as he is of the difficulty of establishing rigid divisions among beings, from the lifeless ones to plants to animals. Indeed, some cases present ambiguities. Elsewhere, in *Parts of Animals*, when discussing the Ascidians, commonly known as sea-squirts, with which he probably became familiar during his permanence in Lesbos (Lee, 1985, 3–8), Aristotle acknowledges that they are very similar to plants (*phyta*) and yet more akin to animals (*zōtikotera*) than the sponges. For, he goes on to explain, "nature passes in a continuous gradation from lifeless things (*apsykha*) to animals (*zōa*) and on the way there are living things (*zōnta*), which are not actually animals, with the result that one class is close to the next that the difference seems infinitesimal" (*PA* 4 681a10–15).

83 See Chapter 2. On the status of plants as animals in Presocratic doctrines, see Zatta, 2019, Chapter 4.

84 Cf. Solmsen, who connects the parts of the soul and their functions to the scale of nature, but considers the locomotive soul in this context irrelevant (1955, 149).

85 For all these treatises dealing with animals as living beings are based on the notion of commonality articulated via difference. *History of Animals* addresses animals' differences with regard to distinct aspects of their lives (see 6.1.1), *Parts of Animals* accounts for the causes of differences with respect to body parts, *On the Soul* aims at finding the most common definition (*koinonatos logos*) of the soul, while *Parva Naturalia* discusses those attributes that are identical among different groups, i.e. sleep, respiration, growth, decay, death.

86 *PA* 1 639a15–23; cf. Aristot. *PA* 1 639b4–6; 644a24–644b10; 645b1–14.

87 Cf. Lennox, 2001, 122, who detects a tension between being and universality, which Aristotle expresses elsewhere in *Metaphysics* (2 1003a5–17; 6 1038b10–12) and *Posterior Analytics* (1, 24).

88 On division in Aristotle's method in *PA*, and in relation to Plato, see Lloyd, 1961, Balme, 1975, and Pellegrin, who offer a state of the question and emphasize its definitional rather than classificatory value (1986).

89 On the attributes, see Lloyd, 1981, 162 and n. 4, and Granger, 1981.

90 Stenzel, 1929, 1652–3. In the treatise *Likes* Speusippus engages in the study of plants and animals, offering accounts similar to those found in Aristotle's *History of Animals* and directed at exploring, as the title itself suggests, similarities among living beings. Although we do not know exactly the criteria by which these similarities were established, Speusippus' groupings indicate the presence of the division between blooded and bloodless animals that will cut across the animal kingdom for Aristotle (Lang, 1964 repr., 9–15). For instance, in one of the 25 fragments preserved by Athenaeus, we read that "trumpet-shells, purple-fish, snails and clams are similar," whereas as in many other fragments similarity is based on physical constitution (Ath. *Deipn.* 3, 86c, trans. by Gulick). Tarán (1981), however, pinpoints in the adoption of the similarity based on locale a major difference between Speusippus and Aristotle's methods.

91 See Aristot. *Post. An.* 2 96a7–11, where Aristotle declares the impossibility of a method, attributed by the ancient commentators to Speusippus, for which each being is fully determined by the totality of the relations—of both identity and difference—that connect it to other beings.

92   Aristot. *PA* 1 644a24–644b10; cf. Aristot. *GA* 1 715a1–3, where, reviewing the sub-
     ject of *PA*, Aristotle underscores the method based on discussing the parts in common,
     koinēi, to all animals, and then those to distinct groups.
93   In *History of Animals* Aristotle distinguishes ten major groups, "major genera"—man,
     birds, fishes, cetaceans, viviparous quadrupeds, oviparous quadrupeds (the blooded
     kinds), crustaceans, testaceans, mollusks (i.e. cephalopods), and insects (bloodless
     kinds); see Balme, 1987a, 80.
94   Charles, 1990, 158.
95   DK 31 B 82/Aristot. *Mete.* 4 9, 387 b4; for a discussion of this fragment in relation to
     the use of analogies as a tool to identify connections among different kinds of living
     beings, see Zatta, 2019, 53. So the great intuition on the "nexus between anatomi-
     cal structure and physiological functionality" which Lanza and Vegetti attribute to
     Aristotle (1971, 39–40) and consider at the basis of his zoological project can in fact
     be traced back to Empedocles.
96   On analogies of animals' parts see also *PA* 4 692b15–18, where birds' beaks are analo-
     gous to other animals' teeth and lips (see below), the elephant's trunk to humans'
     hands and the tongue of certain insects to the mouth of creatures belonging to other
     kinds.
97   Pellegrin, 1986; cf. Balme, 1975.
98   The two major differences with Linnaean taxonomy are the lack of intermediate
     groups between kinds and species (Balme, 1975) and the shifting levels of generality
     attached to kinds (*genos*) and species (*eidos*) (Pellegrin, 1986).
99   Lennox, 2001, 170; cf. Lanza, Vegetti, 1971, 556, n. 4. On the resolution of this ten-
     sion see also Kullman, 1974, 75, who speaks of a "compromise between entitlements
     of theory and praxis," and also Carbone, 2002, 560, who sees it as an urgency for
     "a point of contact between the ontological point of view, according to which the
     substance is intended first of all and principally as individual, and the practical needs
     connected to the research in the zoological field."
100 This point finds further support if we consider that for Aristotle each living being is
     both the *arkhē* and *telos* of its own movement. See below and Chapter 3.
101 In the *Odyssey* the episode of transformation of Odysseus' companions into pigs
     encapsulates the tension of a human mind trapped into an animal body, and once the
     heroes achieve their original shape they burst into tears (*Od.* 10, 198–399).
102 Among the blooded animals, for instance, also elephants, lobsters, and moles pos-
     sess specific traits. See respectively *PA* 2 658b34–659a37, where Aristotle presents
     a detailed description of the elephant's most extraordinary part (*idiaitaton morion*)
     among the other animals; *PA* 4 684a32–6, which discusses the uniqueness of the lob-
     ster in having a claw larger than the other; and *HA* 1 491b27–36, where the mole
     is introduced as the only life-bearing animal that does not possess eyes; cf. *HA* 4
     533a2–3.
103 Aristot. *PA* 1 639a14.
104 For the different treatises and relative objects of Aristotle's research plan on nature,
     see also Chapter 3.
105 Pellegrin, 1995, 14, although, it should be remarked, that while excluded from the
     study of nature, in Aristotle art becomes an analogue, and imitator, of nature.
106 Kahn, 1960, 202–3.
107 These examples are taken from Aristotle's *Parts of Animals* (1 640b5–13) and dis-
     cussed in Zatta, 2019, 35–7.
108 The prototype of a conception of the world based on the view of nature as cosmos, dis-
     cussed here in relation of the pluralist philosophers, goes back in fact to Anaximander
     (Kahn, 1960).
109 Longrigg, 1993.
110 See Zatta, 2019, 66–71.
111 Cf. Aristot. *Phys.* 2, 199b5–8.

112 In the spurious treatise *On the Cosmos*, for instance, the cosmos and its organization along with its dynamics and resulting harmony are discussed as a reality already established, and not in the process of formation (Pseud.-Aristot. *Mu.* 391b9–13).

113 Kahn, 1960, 202–3; on Aristotle's revision of the Presocratics' position on "change," his endorsement of first principles and the development of "a *theory* of contrariety," see also Falcon, 2005, 23–6; cf. *Metaphysics* (5 1013a17–9), where Aristotle states that "all things can be *arkhai* ... the common character of all principles is to be the first factor from which the thing is or comes into being or is known."

114 Aristot. *Phys.* 2 192b8–14, 192b9–23.

115 For the Presocratics, *arkhē* was used as not only a chronological notion, but also a material one.

116 In this respect, see the fundamental critique that Aristotle poses to the philosophers of nature in the *Metaphysics*. After having reviewed the progressive development of their respective doctrines from the consideration of matter to the explanation—albeit partial—of movement, once he arrives at Leucippus and Democritus, Aristotle acknowledges that "the question of movement—whence, *hothēn*, or how, *pōs*, it is to belong to things that are, *ousi*, these thinkers, like the others, lightly neglected" (*Met.* 1 985b20). Cf. also Aristot. *DA* 2 41215–8.

117 On the rich intertextuality between *PA* 1 and *Physics* 2, and the dating the two to the same period after the composition of the rest of *Parts of Animals*, see Balme, 1987b, 17.

118 Cf. Aristot. *Phys.* 2 192b14–5 and *PA* 1 639b15, where Aristotle discusses the finality of nature by analogy with art, which, also like nature, produces its works according to design. On living being's movement in the sense of growth see Chapter 3, in the sense of alteration, Chapter 4.

119 For a review of the presence of Aristotle's general psychological theory and specific psychological doctrines in his zoological inquiries, see Lloyd, 1992, 147–57. For Aristotle's treatment of the soul in *On the Soul* as an introduction to his study of animals, see also Pellegrin, 1996, and the introduction above. Yet, Aristotle's outline of the fields of inquiry encompassed by the study of nature, presented in *Meteorologics* (1 338a20–339b9, see p. 23 above), ends with animals and plants without mentioning the soul and it remains, I believe, problematic whether the study of the soul conceived of as a nonmaterial and immobile entity strictly belongs for Aristotle to the study of nature. So much is certain: the study of the soul greatly contributes to the study of nature, and in particular of living beings as creatures of nature inasmuch as the soul is the principle of animals (*DA* 1 402a1–7). And conversely, the science of animal nature, i.e. zoology, should treat the parts of the soul that make animals move, but not of the soul in its entirety (*PA* 1 641a18–641b10, see above and Chapter 3).

120 See Chapter 3.

121 Cf. Pellegrin, 1995, 19.

122 Also, art is goal oriented, but in being so it imitates nature. Cf. Aristot. *PA* 1 639b19–21, where Aristotle claims that there is more finality in the works of nature than in those of arts. At any rate, among the products of nature, animals feature as those, which (literally) embody finality to the highest degree. See discussion below.

123 *Phys.* 2 198b18–199a8. In the same passage Aristotle also mentions the case of air becoming water as a counterexample that, happening by necessity (*anankē*), may illustrate the working of nature and deny that the growth of teeth (and by extension of living beings/things and related parts) may happen "for an end" (*Phys.* 2 198b18–21); see discussion in Chapter 3. On the different involvement of causes (material versus final) in natural phenomena, see Charlton, 1970, 115, and 120–121.

124 Aristot. *Phys.* 2 196b10–198a14.

125 This notion of birth of complete, fully formed living beings evokes parallel formations accounted for in Greek myth like that of the Spartoi, who also sprung from the earth as full-blown individuals.

126 Aristot. *Phys.* 2 199b8–10.
127 For Aristotle's definition of growth as a key phenomenon of life, see Chapter 3.
128 Aristot. *Phys.* 2 199b15–19.
129 Aristot. *PA* 1 642a24–31.
130 See above.
131 See *PA* 1 640a33–640b4, where Aristotle states, "Hence it would be best to say that, since this is what it is to be a human being, on account of this it has these things; for it cannot be without these parts. If one cannot say this, one should say the next best thing, i.e. either that in general it cannot be otherwise, or that at lest it is good thus. And these things follow. And since it is such, its generation necessarily happens in this way and is such as it is. (This is why this part comes to be first, then that one)." It should be remarked, however, that an interesting, and "exceptional" case is that of the so-called deformed animal kinds, which do not use their bodily parts to the fullest and are, so to speak, naturally incomplete. Among them, for instance, features the mole, qualified as *ateles*, incomplete (*HA* 1 491b27–36; Witt, 2012).
132 For a recent assessment of Aristotle's authorship, see Johnson and Hutchinson, 2005.
133 Aristot. *Protrept.* Fr. B 18 Walzer; cf. *On the Generation of Animals* (1 715a7–9), where Aristotle also affirms the identity of the final and formal causes.
134 So far we have discussed the *horos* pertaining to which parts of the soul the natural philosopher has to consider and the aporetic alternative between individual species and common attributes; see above.
135 See respectively Aristot. *PA* 1 639b12–21 and 640a10–28.
136 Aristot. *Phys.* 2 192b9–194b34. Cf. *PA* 1 639b12–639b28 and 1 640b27–9. A backbone of Aristotle's study of nature in *Physics*, the theory of the four causes is still relevant in a study of animals inasmuch as they are beings that exist by nature. But as scholars have remarked, in the context of animals the emphasis is different than when discussing causes in nature at large (on this aspect, see also Chapter 3). In *Parts of Animals*, rather than discussing the four causes in specific combinations as he does in *Physics*, Aristotle first discusses the efficient and final cause (*PA* 1 639b12–15) and, later on, the material and the formal cause (see note below and *PA* 1 642a3–18).
137 Aristot. *PA* 1 640b5–29. For Aristotle, overall the Presocratics ignored form (*eidos*) apart from some extemporary exceptions in Empedocles and Democritus. See *PA* 1 642a.19–29.
138 Cf. *Physics*, where Aristotle states that "all natural entities are substance … they have a certain substrate" and discusses the differences between mathematics and physics, which may well deal with the same notions such as "surfaces and dimensions, length and points," but the first in abstraction, the second in bodies (2 193b22–35). Further, matter is relevant also in terms of its qualities and availability, and as such discussed in relation to conditional necessity, which affects both natural beings and technical products (see *Physics* 2 199b33–200a16, *PA* 1 639b21–30 and Chapter 3).
139 Bodily materials must possess different *dynameis* that allow an animal and its parts to accomplish specific actions and movements (Aristot. *PA* 2 646b15–19).
140 Aristot. *Phys.* 2 193b6–8; cf. *PA* 1 640b27–9, where Aristotle considers form (*morphē*) of more fundamental importance (*kyriotera*) than matter (*hylē*).
141 *PA* 1 640a10–7.
142 At the outset of the *On the Generation of Animals* (*GA* 1 715a1–18) Aristotle remarks on this shift of perspective and how in *PA* he has addressed the parts in relation to all the four causes—final, formal, material, and efficient—while the present treatise, in dealing with the parts that serve the animals for the purpose of generation, turns around the efficient cause (*aitia kinousa*), and, therefore unfolds, one could add, a linear account of animals' formation. As Aristotle continues, "Consideration of this cause [the efficient] and the consideration of the generation of each animal come to the same thing." It should be remarked, however, that while Aristotle's discussion follows in this case a chronological order, it is still informed by the same theoretical

apparatus including form and end. For animal generation is ultimately conceived as a transmission of form and, in this respect, *On the Generation of Animals* shows "how teleology actually works" (see Balme, 1987b, 10).

143 *GA* 2 740a6–24.
144 Difficulties in reaching the final end may happen, and Aristotle discusses them in terms of monstrosities and aberrations, see *Phys.* 2 199a35–8; *GA* 4 769b11–770b27.
145 On the possible inclusion of the differentiated material in *logos*, see *Phys.* 2 200b5–10, and n. 180 below.
146 *PA* 1 639b15-7, with slight modifications. Cf. *PA* 2 646b1–10 and see discussion below.
147 On animals' bodies (*qua* ensouled bodies) as "other" (*heteros*) and more divine than the elements, see *GA* 2 736b30–32.
148 It is noteworthy that for Aristotle *logos* is the embodied *telos* in works "that have been composed by nature" (as in the passage we are discussing here) as well as the disembodied *telos* that preexists the work's formation. In this respect, see what looks like a marginal, clarifying note to the proceeding of art at *PA* 1 640a32–33, "as for those things whose producing agent is preexistent, e.g. the art of statuary, no spontaneous formation occurs. Art is the *logos* of the work without matter."
149 In the case of nature, Aristotle is likely thinking at the regularity of form that characterizes the process of generation. See, for instance, *PA* 1 640a19–27 and discussion below.
150 As it will be seen below, this structure is a highly sophisticated one that is in fact the result of three orders of compositions: elements, uniform, and nonuniform parts.
151 *PA* 1 639b17–19.
152 See pp. 26–27.
153 See *Phys.* 2 199b10–11 where Aristotle claims that also plants manifest purpose (*to heneka tou*), "though they are less elaborately articulated (*hetton diarthreisthai*)" (than animals). On plants' soul and living activities, see Chapters 2 and 3.
154 *Logos* is equally absent in the discussions of plants that feature in Aristotle's zoological treatises as well as in the treatment of plants' parts in Nicolaus Damascenus' *On Plants*, which was likely based on Aristotle's homonymous treatise and Theophrastus' work. For an overview of the text history, see Drossaart Lulofs and Poortman, 1989, 1–4. In fact, as Frey points out, there are significant differences between natural bodies and artifacts (2007, 174–5, 198–200), but what interests Aristotle in *Parts of Animals* is to underline the complex structure that characterizes equally animals and artifacts when completed.
155 *DA* 2 416a10–18. In this passage, Aristotle discusses the growth of animals and plants as naturally composed beings remarking that they have a limit and proportion (*logos*) of size, on account of *logos*, intended as form/end and not matter.
156 See *PA* 2 656a1–3.
157 Even the function of nutrition, which plants share with animals, is carried out as a minimal process and with a basic body equipment: plants do not have a stomach with which to process ingested food, but absorb from the earth already concocted nourishment (*PA* 2 655b33–37). The fact that plants do not concoct their food likely implies that they do not possess a standard body temperature against which to discriminate the tangibles (see n. 159 below and Chapter 4). On Aristotle's treatment of plants in the works devoted to animals, see Falcon, 2015; for plants' only mode of life, see Chapter 2. It bears noting that if animals share an *oikeia physis*, plants possess a *physis* of their own (*idia*) (Aristot. *de Long.* 467a13).
158 *PA* 2 656a3–5.
159 Ultimately, plants' bodies lack the sophisticated form of receptivity to the sensibles that characterizes animals and that Aristotle defines as the capacity "to receive forms without matter" (*DA* 2 424a17–21); see Chapter 4.
160 See Bonitz, 1955. Lanza and Vegetti in the "terminological note" dedicated to *logos* in Aristotle present its general meaning as "discourse," "reasoning," or "argument" and

proceed to highlight its different connotations in logic, ontology, and mathematics, and in the presence of the adverb *eulogōs* (1971, 1251), discussed later. On the other hand, Lennox stresses the "discursive core" of *logos* and claims that it "can refer to a variety of linguistic units (words, definitions, reasons, arguments, books), as well as to mathematical relationships, such as ratio; ... to the *content* of a definition, or to the *relationship* denoted by a ratio" (2001, 126). On the polysemy of *logos*, see also Shields, 2016, 381–2, who identifies a "semantic" and a "nonsemantic," metaphysical meaning.

161 See, respectively, Balme, 1992, Lennox, 2001, and Kullmann, 2007, for whom "plan" well captures the identification of *logos* with the form or the definition of the structure, which equates with the final cause rather than with the formal one.

162 For the English translation, Ogle, 1882; for the French, Louis, 1956; Le Blond, 1945, who in the note *ad loc.* specifies it as "contenu intelligible;" and Gain, 2011, who, also in the note to this passage, points out that "the term *logos* can designate the explanation of a thing as well as its definition;" for the Italian, see Lanza and Vegetti, 1971, Torraca, 1961, and, finally, Peck, 2006 (first ed. 1937), who further adds to the transliterated *logos* "rational grounds" as an apposition.

163 Peck, 2006, 26–7 under *logos*.

164 See, respectively, Aristot. *PA* 2 660a14– 28, 3 661b8–14, 662a17–23.

165 See Aristot. *PA* 2 660a26–660b4; cf. *HA* 536b11–13; cf. Zirin, 1980; Long, 2011, 79–81.

166 Aristot. *Pol.* 1 253a8.

167 Aristot. *PA* 2 646a29–646b10, with a slight modification.

168 For "fully-formed animal" here I do not only refer to the end (*telos*) of the animal's process of formation, but, in light of Aristotle's inclusion of the soul in his study of animals, I also refer to the movements, namely the living activities, that such an end enables a given animal to possess.

169 Aristot. *PA* 2 646a25–30. On these two chronologies, see also *Metaphysics* (1 989a15–20) and discussion below.

170 See above.

171 As Carbone notices, in the process of formation movement is always associated with the transmission of form (*morphē*) (Carbone, 2011, 98).

172 However, also Empedocles investigated the process of generation and discussed the contribution of the male and female parents to the formation of the embryo (see DK 31 B 63/*GA* 1 722b10–12; cf. Zatta, 2019, 68 n.41), but for Aristotle in his focus on the "sculpting effect" of matter Empedocles neglected to consider the transmission of form.

173 Aristot. *PA* 2 646a34–5; cf. *PA* 1 640a25–6.

174 Indeed, that a doctor or a builder are able to account for the causes and rational grounds of everything they do in order to achieve their respective *telos* means that also the stages encompassed by the process of formation of given animals are teleologically intended and, as such, included in the *logos* as *telos/arkhē* (see *PA* 1 639b17–19 and discussion above).

175 By contrast, the *logos* of animals' formation includes both process of formation and *telos*, an object of study, which, as mentioned earlier, Aristotle deals with in *On the Generation of Animals* (see above).

176 On the other hand, Empedocles identified analogies among living beings' parts based on function rather than constitution (DK 31 B 82/Aristot. *Mete.* 4 387b4–6), cf. n. 180 below.

177 See p. 15.

178 Aristot. *PA* 1 640b30–36.

179 That form involves composition rather than simply matter, is explicitly stated toward the end of *PA* 1: "Just one who discusses the parts (*moriai*) or equipment (*skeuai*) of anything should not be thought of doing so in order to draw attention to the matter,

nor for the sake of the matter, but rather to draw attention to the overall shape (*holē morphē*) (e.g. to a house rather than bricks, mortar, and timbers); likewise, one should consider the discussion of nature to be referring to the composite (*synthesis*) and the overall substantial being rather than to those things which do not exist when separated from their substantial being" (*PA* 1 645b30–6).

180 Empedocles—under the guidance of truth (*alētheia*) as Aristotle ironically puts it—felt compelled to introduce the notion of *logos* in the understanding of the nature of the organic parts of the body. Bones, for instance, derived from an exact formula: two measures of water, two of earth, and four of fire. Empedocles calls it the proportion of the mixture (*logos mixeōs*) (Aristot. *PA* 1 642a21–24; DK 31 A78), where *logos*, in Furth's words, "evinces *no* appreciation of the difference between *mixture* and *structure* (1988, 91).

181 For a discussion of the three levels of composition and, in particular, the novel range of relations that, in Aristotle's view, the elements are liable to entertain on account of their double qualities, see Mingucci, 2015, 40–54.

182 This follows a specific order of composition, first of the elements into the uniform parts—the blood, tissues, bones, and other organic "ingredients" of animals' bodies—and then of the uniform parts into the nonuniform ones—including noses, eyes, members, and similarly well-defined parts—and, finally, of these last parts into complete bodies, apt to fulfill their functions as wholes and in their parts. On the purpose of the whole body and each of its parts, see also *PA* 1 642a10–4.

183 *PA* 1 645b19.

184 See above.

185 In the rest of the treatise Aristotle will then proceed to discuss animals' bodies, part after part. For a list of the parts and the order in which they are arranged, see Peck, 2006, 15–18.

186 *PA* 2 646b11–19. Cf. the above discussion on how the common nature as the condition for a species' membership to the same kind involves, for instance, social habits (p. 21).

187 Aristot. *Phys.* 2 199a20–9; all translation of this treatise are by P.H. Wicksteed and F. Cornford.

188 Ross, 1936, 529; cf. DK 69 A 116/Aët. 4, 10, 4; see Zatta, 2019, 65–66.

189 Cf. Aristot. *HA* 1 487a11–2.

190 Aristot. *PA* 1 645a4–29.

191 *PA* 1 645a24–27.

192 It should be remarked that Aristotle stresses on the visual approach to animals. Not only does he use *theōria*, and introduce a contrasted analogy with the images produced by artists; he also adopts the verb *kathoran* as the action that allows the student of animals to see their causes (*PA* 1 645a15).

193 DK 59 B 12.

194 Cf. *On the Soul* where Aristotle paraphrases Anaxagoras' fragment 12 juxtaposing the categories of size and nobility: "for mind he [Anaxagoras] regards as existing in all living things, *apanta zōa*, great and small, *megala kai mikra*, noble and base, *timia kai atimotera*" (*DA* 1 404b4–7, transl. by W.S. Hett). The direct quotation in *On the Soul* strongly supports the implicit reference to Anaxagoras in *Parts of Animals*.

195 *Met.* 1 3 984b15–8.

196 While Aristotle praises Anaxagoras for recognizing the presence of mind (*nous*) in nature and animals alike, in *On the Soul* he criticizes him for confusing at times *nous* with soul, and thereby, we can infer, for attributing what in Aristotelian terms is the rational part of the soul to all living beings, big and small (*DA* 1 404b1–3; cf. Zatta, 2019, 89).

197 For the use of *dēmiourgein* as qualifying nature's "creating activity," see, for instance, *PA* 1 645a9, 2 654b32, *GA* 1 731a24, 2 743b23.

198 See Chapter 3.

199 See Chapter 4.
200 *PA* 1 639b19–21.
201 *DA* 2 415a27–415b3. As we will see in Chapter 3, under the scientist's contemplative glance animals' *telos* appears to include for each of them not only the capacity but also the desire to live: all the natural functions animals perform are for the sake of (their own) life.
202 Plants unite in a single individual the male and female sex, while in animals male and female are separate (*GA* 1 731a2–10).
203 *DA* 2 414b3–6; cf. 2 413b21–23 and *GA* 1 731a30–731b8; see above, and Chapters 2 and 4.
204 While in his treatment of the nutritive soul Aristotle is ready to admit in general terms that all living entities desire (*oregesthai*) to live forever (*DA* 2 415a27–415b3), by denying that plants possess sensation he also denies them desire.
205 *NE* 10 1175a11–22 and discussion at 5.2.2.
206 Aristot. *HA* 1 491a24–26; cf. *PA* 2 656a9–13 and 4 686a24–29 and discussion above on pp. 17–18; on man's cosmic and natural orientation, see Preus, 1990, 473–477; cf. also Aristot. *Resp.* 477a15–25, where animals with more heat are called "worthier" (*timōtera*) and are attributed a "worthier" soul, and a "worthier" nature.
207 See *GA* 2 732b8–733b17, and especially 732b28–33 and 733a34–733b17.
208 Leunissen, 2018, 60.
209 *DA* 3 433b32–434a3.
210 In this respect, see *GA* 1 731a34–731b3, discussed in the introduction, in which Aristotle acknowledges two perspectives from which to look at animals *qua* sentient, knowledgeable beings: the intelligence (*phronēsis*) of human kind, on the one hand, and the kind of lifeless entities (*genos tōn apsykhōn*), on the other. When compared to human intelligence, the possession of touch and taste is nothing, but if we compare it to the entire lack of sensibility, "it seems a very fine thing." On the other hand, in *On the Caelo* Aristotle calls all animals endowed with the capacity of locomotion "perfect" (*teleia*) (*Cael.* 284b21–4).
211 See 6.1.1.
212 See Plat. *Tim.* 69C–71D, 89E–90A, 90E–92C, and Zatta, 2019, Chapter 6.
213 See Plat. *Resp.* 4 439b–442a and discussion below at 2.2.1.
214 Aristot. *PA* 2 647a25–30, 3 665b10–5, *Somn.* 456a1–7.
215 *PA* 2 647b4–8. On the heart and the blood vessels as receptacles of the blood and its body-nourishing function, see *PA* 2 650b2–13. For Plato too the heart is the junction of blood-vessels and the source of blood, but it collaborates with reason by conveying to the organs of sense "threats and injunctions" forcing them to obey (*Tim.* 71B).

2 From reason to life

Aristotle on soul division

A lofty and challenging undertaking, the study of the soul greatly contributes to the study of animals as creatures of nature. In *On the Soul* Aristotle calls the soul the principle of animals (*arkhē zōōn*).[1] As we understand later on in the treatise, the soul plays this role on account of being "the form of a natural body, which potentially has life."[2] It is what makes a body capable of living and actually live.[3] And it does so by enabling it to engage in the array of activities that constitute life, first and foremost the capacity for self-sustenance, growth, and decay.[4] Conceived of in terms of his general study of nature (*ta physika*), such activities correspond to the movements originated by animals *qua* natural complex bodies. Hence the investigation into the soul is an exploration of the principle of life intrinsic to each living being that has the capacity to move and rest and its definition is dependent on the kinds of movements that characterize animals as creatures of nature: growth and decay (*auxēsis*), alteration (*alloiōsis*), and locomotion (*kinēsis kata topon*).[5]

Central to the understanding of the soul as the principle of life is the problem of partition. If in *Parts of Animals* nutritive, sensitive, locomotive, and rational parts are presented as a well-established fact,[6] in *On the Soul* Aristotle raises the questions of whether the soul is partite and of what parts constitute it offering a version of partition that supports the claim in *PA* and stands as an original response to Presocratics and Plato's psychological doctrines. In this new psychological model he breaks down life into a set of fundamental and complementary functions, internally related by order of importance in respect to a "physical" (i.e. nature based) conception of life that matches the kinds of natural movements animals possess *qua* creatures of nature and in a way that (clarifies and) articulates that *koinē physis* we have discussed in Chapter 1 in relation to *PA*.

2.1 Understanding ensouled bodies: Soul partition and homogeneity

In *On the Soul* psychological partition features as a possibility to be assessed,[7] one that has found some supporters[8] and different articulations,[9] and whose validation is fundamental to the comprehension of the nature of the soul. At the beginning of

DOI: 10.4324/9780367816001-3

book 1 after asking to which highest genus the soul belongs and whether it exists
in potentiality or actuality, Aristotle writes,

> We must inquire whether it [the soul] has parts (*meristē*) or is without parts
> (*amerēs*) and whether every soul is homogeneous (*homoeidēs*) or not; and
> if not, whether the difference is one of species or of genus. For speakers and
> inquirers about the soul seem today to confine their inquiries to the human
> soul. But one must be careful to evade the question whether one definition
> of "soul" is enough, as we can give a definition of "living creature", or
> whether there must be a different one in each case; that is, one of the horse,
> one of the dog, one of man, and one of God, and whether the words "liv-
> ing creature" as a common term (*to koinon*) have no meaning, or logically
> come later.[10]

In this passage Aristotle pairs soul partition with a related problem, that of soul
homogeneity.[11] One has to investigate whether the soul is partite or not and
whether "every soul" is homogeneous (*homoeidēs*), namely whether it is same or
different, and if different, whether the difference pertains to species or kind. This
is a pressing set of problems for Aristotle. For if his contemporaries (i.e. Plato and
the members of the Academy) admit partition they still focus on the human soul.
Partition is related to homogeneity because if the soul is partite different living
beings may possess different psychological parts and therefore a nonhomogene-
ous soul, with repercussions on the very conception of living beings. For lack of
psychological homogeneity would compromise living beings' membership to an
all-encompassing genus of living (*zōion*).[12] Intent to understand the soul as the
principle of life and not merely human life, Aristotle rejects a narrow focus to
undertake a psychological inquiry that unfolds on a universal basis.

Besides, the use of *homoeidēs* in association with the problem of soul partition
suggests another issue at stake in Aristotle's discussion, still connected to the elu-
siveness of the category of living being, mentioned earlier. *Homoeidēs* appears in
On the Soul and elsewhere to qualify air, water, and fire[13] and its attribution to the
soul likely points to a materialistic view as that held in different versions by the
Presocratics. A nonpartite soul is also homogeneous and conceived in this way the
same soul is liable to be attributed to a disparate number of bodies with the consequent failure in understanding which ones are actually ensouled and living as well
as what types of psychological capacities they may possess. In this case too (as
with the problem relating to soul species and kind addressed above), the category
of living being (*zōion*) as one that embraces all creatures that live would be com-
promised. Thus much of Aristotle's critique of earlier psychological doctrines in
book 1 targets a series of aspects that stemming from the belief in a monolithic
notion of the soul led to a confusion about living beings and their capacities: the
soul's materiality and movement and undifferentiated capacity to feel and think.
For our philosopher the early Greek philosophers were right in making the soul
responsible for movement (*kinēsis*) and sensation (*aisthēsis*),[14] but they failed to
understand the nature of the soul and its relation to the bodies that possess it, and

likewise they failed to understand fully the soul's range of differentiated capacities.[15] In being tied to the material origin (*arkhē*) of the cosmos (whether air, fire, or atoms),[16] the soul could well dwell in—or even coincide with—different types of natural bodies nor did such conception make it clear what such bodies could actually do. For instance, if for Empedocles the elements were soul, elemental bodies such as mountains or seas could be living and endowed with sensation and thought,[17] while in considering the soul identical with the particles of air or with what moves air,[18] some of the Pythagoreans mistakenly pictured it entering by chance any kind of body.[19] Besides, as long as a "body" manifested one of the attributes imputed to the soul—movement and/or sensation—it could be ensouled. In this respect, for Thales even a simple magnet qualified as having a soul on account of its capacity to move the iron.[20]

Avoiding such shortcomings in *On the Soul*, Aristotle is concerned with assigning the soul to the appropriate kind of body and proceeds to discriminate at first between life and nonlife. Only, then, after this preliminary discrimination, will Aristotle articulate living beings' capacities. To put his way of proceeding in a question, if the realm of nature encompasses bodies that are the origin of their movement and rest, which are the ones that "move" (and therefore live) by virtue of the soul? At the beginning of his positive account of the soul in book 2 Aristotle clarifies "where" to look for it. Of the natural bodies (*physika sōmata*) some have life (*zōē*), others do not,[21] and those that do have life are composed of matter (*hylē*) and form (*morphē/eidos*). Matter is the material substratum, form is soul,[22] while life is conceived of in its most basic but essential aspect: "the capacity of self-sustenance, growth, and decay," which all fall under the umbrella term of *auxēsis* (growth).[23] Thus armed with technical notions (i.e. matter and form but also *auxēsis*) and dismissing the materiality of the soul, which often led his predecessors to attribute carelessly psychological faculties to disparate bodies, Aristotle restricts the number of natural bodies that host a soul to those that have a form enabling them to nourish themselves, grow, and decay. In this way, he preliminarily and fundamentally qualifies the movement (*kinēsis*), which his predecessors had credited the soul for being responsible for, pinning it down to a basic set of life faculties and operations that require a suitable body[24]—one provided with organs (*organikon*)[25]—and that is the condition for other psychological faculties and movements to exist. Hence it becomes clear in what sense at the beginning of *On the Soul* the soul is predicated as "the principle (*arkhē*) of living beings (*zōa*)."[26] Such sense has to be traced back to the body and its functions as enabled by its organic articulation due to the soul, and the entire treatise may be seen as an effort to uncover the soul's obscure relationship to embodied life in all its different (and complementary) forms and manifestations. Key to such an enterprise is the definition of the soul as a unity of parts, arranged in a serial order, of which the part responsible for living beings' nutrition, growth, and decay is considered to be prior to the others inasmuch as it provides the most basic condition for life without which other capacities cannot subsist. Aristotle will establish the soul's parts in *On the Soul* 2, 2–3, and will subsequently discuss part after part with relevant capacities individually. To appreciate his original solution to the problem of soul partition and its relevance to the comprehension of life, the next section will turn to the current soul divisions and discuss the reason of their inadequacy for the type of universal inquiry Aristotle has set out to undertake.

2.2 Problematic divisions and attributions: The bipartition and tripartition of the soul

Aristotle confronts the problem of soul partition in *On the Soul* 3.9 in the context of the discussion of animals' ability to move about. The fact that this problem resurfaces in a fully developed form after Aristotle has assessed that the soul has parts and which ones they are not only indicates its key importance in Aristotle's thought, but also suggests an intrinsic problematic nature never to be completely resolved.[27] He writes,

> A problem at once arises: in what sense should we speak of parts of the soul, and how many are there? For in one sense they seem to be infinite (*apeira*) and not confined to those which some thinkers describe, when they attempt analysis, as calculative (*logistikon*), spirited (*thymikon*), and appetitive (*epithymētikon*), or, as others have it, rational (*logon ekhon*) and irrational (*alogon*). When we consider the distinction according to which they classify, we shall find other parts exhibiting *greater difference* (*meizōn diastasis*) than those of which we have already spoken; for instance, the nutritive part (*threptikon*), which belongs both to plants and to all living creatures, and the sensitive part (*aisthētikon*), which one could not easily assign either to the rational or irrational part; and also the imaginative part (*phantastikon*), which appears to be different in essence from them all, but which is extremely difficult to identify with, or to distinguish from any one of them, if we are to suppose that the parts of the soul are separate.
>
> (*DA* 3 432a22–432b7, with some modifications)

Soul division is rife with problems: it is difficult to come up with an appropriate notion of part as well as to reckon how many there are.[28] In a sense the soul could encompass infinite parts (*apeira*) so as to reflect the bottomless complexity of the living experience whose innumerable aspects transcend linguistic definitions.[29] At the beginning of the treatise Aristotle has remarked that his contemporaries focused on the human soul and this passage presents the two models of division such focus led to. The one encompasses a calculative (*logistikon*), spirited (*thymikon*), and appetitive part (*epithymētikon*), while the other distinguishes between a rational part (*to logon ekhon*) and an irrational one (*alogon*). Scholarship has conventionally called them tripartition and bipartition assuming the presence of an ongoing debate on soul division at the time of Aristotle,[30] without however reaching a conclusive paternity for these models. Since Antiquity there has been some consensus in considering Plato the advocate of soul tripartition (already Simplicius and Philoponus in their commentaries mention *Republic* 4),[31] while bipartition has had a more controversial attribution, with candidates ranging from the Academic thinkers to Aristotle himself.[32] And indeed, in both *Nicomachean Ethics* and *Politics* Aristotle does use the model of soul bipartition, which appears already at an early stage of his philosophical production in *Protrepticus*.[33]

The situation is however more complex because in *Magna Moralia* Aristotle attributes soul bipartition explicitly to Plato,[34] a reference that indicates, on the one hand, a likely filiation of Aristotle's bipartition from Plato and that makes us question the degree of compatibility between the tripartite and bipartite models. In other words, if Plato uses both tripartition and bipartition, and Aristotle himself seemingly adopts a Platonic-style bipartition how do the two models of soul division overlap? In the passage above Aristotle himself offers a short answer to this question by pointing out that current soul divisions have been unable to represent a "more significant difference" (*meizōn diastasis*) among parts. What would this be? The upcoming discussion of soul tripartition and bipartition in Plato and Aristotle's relevant texts will tackle their flaws in addressing the difference among psychological parts which Aristotle is determined to pursue.

2.2.1 Under the rule of logos: From Plato's Republic to Aristotle's Ethics

Plato discusses soul tripartition in a number of treatises. In *Phaedrus* he refers to it with the composite imagery of horses and charioteer,[35] while in *Republic* and *Timaeus*[36] he uses almost the same nomenclature cited by Aristotle in *On the Soul* 3.9—the rational, spirited, and appetitive parts.[37] *Republic* is of particular interest to our discussion, not only because it was identified by the ancient commentators as the source alluded to by Aristotle when mentioning tripartition, but also because Socrates' articulate discourse on justice appeals to different models of soul division: in book 4 we find the tripartite model, while book 10, confirming Aristotle's reference in *Magna Moralia*, mentioned earlier,[38] presents a bipartite model. A consideration of whether and how these models intersect seems therefore crucial. It will clarify whether (and how) tripartition and bipartition are compatible[39] and show Plato's influence on Aristotle (in respect to his model of bipartition) leading us to understand the basis of Aristotle's critique of these concurrent soul divisions in *On the Soul* 3.9.

Initially conceived of as an investigation into the nature of justice at the scale of the individual, Plato's *Republic* takes a detour: it first analyzes justice in the composition of a just city, and then traces the same components back in the human soul. As the just polis is comprised of three classes—the philosophers-rulers, the guardians, and the producers—so the soul is tripartite, and in both cases justice is achieved and maintained when each class/part does its own work (*ta heautou prattein*).[40] Whether in the case of individuals or communities, justice relies on—in turn defining it—a set of virtues that in the ideal condition inform respectively each a class of citizens along with the corresponding part of the soul. Wisdom informs the class of philosophers-rulers and the rational part, courage the class of warriors and the spirited part, and temperance the class of producers and the appetitive part. This tripartition however is not given at once and Plato reaches it via a multistep argument, attention to which will now contribute to clarify the "smaller difference" among psychological parts along with the anthropocentric focus and ideological framework current soul divisions for Aristotle entail. In book 4 Plato first establishes the division between rational and appetitive part by using, to adopt Shield's terminology, the "principle of opposite relations," according to which "the same thing will never do or suffer opposites (*tānantia*) in the same respect in relation to the same thing and at the same time."[41] In other words, the same psychological faculty cannot desire and reject contemporaneously the same thing, and contradictory states like this must depend on a clash—and therefore distinction—of psychological faculties, namely of reason and appetite. Significantly, this clash is staged against the backdrop of an irreducible human/ animal charged alterity. Plato writes,

> The soul of the thirsty then, in so far as it thirsts, wishes (*boulesthai*) nothing else as than to drink, and yearns (*oregesthai*) for this and its impulse is toward this (*horman*) ... Then if anything draws it back when thirsty it must be something different in it from that which thirsts and drives it like a beast (*hōsper*) to drink.[42]

Thirst is here refrained by reason and in giving in to it the human being is like a beast. Note that the issue is not with excessive drinking but with a desire that has to do with the body and what we could call a physiological need. Hunger, love, and the feeling of flutter and titillation due to other desires are soon added to thirst and attributed to the appetitive part of the soul (*to epithymetikon*), called "the companion of various repletions and pleasures," implying that whoever follows them becomes assimilated to the beasts but is a human being if s/he resists through the calculation of reason on account of the rational part (*to logistikon*). With the appetitive and rational part established,[43] Plato proceeds to identify the spirited part (*to thymoeides*) by distinguishing it from the rational one. Each psychological part embodies the character of the corresponding class in the just city, and identifies autonomous motivations for action—appetites, reasoning, and "spirit"[44]—a plurality of concurrent motivations that manifest the soul's intrinsic disposition to conflict. Delineating a "politics of the soul"[45] and interpreting intrapsychological conflict in terms of stasis,[46] Plato considers the human soul in constant tension between the rational part and the appetitive one, and ready to fall into a beast-like condition. While appetites bid, reason forbids as it holds power (*kratein*) over them. To succeed reason needs to form an alliance with *thymos*, spirit, which becomes its subject (*hypēkoos*) and ally (*symmakhos*).[47] In the ideal condition, which is what grants a harmonious constitution of the soul (and asserts human over animal nature and is achieved through an education based on music and gymnastic), reason and spirit form a team that will control the impelling desires of the appetitive part, preventing it from taking over. Strengthened by the right education, the alliance between the two may become so solid that to express it Plato recurs to a persistent use of duals.[48] But without a good education *thymos* is intrinsically volatile, and when it is corrupted by evil nurture (*kakē trophē*), it can support the appetites against reason.[49] Not only descriptive, soul tripartition in book 4 is normative and ideological, and prescribes a political mode of interaction. For the individual to attain justice the rational part of the soul must rule (*kratein*) and the other parts obey.[50] Man has to suppress his animal-like desires and embrace reason. Without question, nurture is evil.

The principle of opposite relations returns in book 10 of *Republic* in the context of a long discussion on the mimetic nature of poetry and its effects on the audience.[51] Here too the principle of opposite relations is used to identify different parts of the soul, but instead of appealing to desires and their censorship as in book 4, it involves the senses and simultaneous contradictory opinions. Plato gives the example of things in water whose submerged parts appear to sight crooked[52] while, in the same perceiving subject, reason judges them straight, and this compresence of contradictory judgments testifies to a bipartition of the soul,[53] with one part encompassing reason (*logos*) and another part that is irrational (*alogon*). Poetry appeals to the second,[54] triggering emotions that, in the wise man, are kept in check and restrained by reason.[55] As Burnyeat remarks, in book 10 Plato is less interested in identifying the opponents of reason than in describing their effects on the soul, and it is probably a mistake to try to fit them into the earlier scheme of tripartition. "Reason—this author continues—is the only constant "[56] But if we take into account Aëtius' insight into Plato's psychology there is even more to this compresence of soul divisions in *Republic* than the constancy of reason. Aëtius identifies two parallel discourses on the soul: an "immediate (or proximate) and precise one" (*prosekhēs kai akribēs logos*) that presents the soul as tripartite (*trimerē*) and one "at the topmost" (*anōtatō logos*) that considers it bipartite (*dimerēs*).[57] As to why he labels them so ("immediate and precise" versus "at the topmost") it arguably has to do with their relative pertinence to the soul embodiment in the physical world. *Republic* may be illuminating in this respect if Aëtius was indeed referring to this dialogue.[58] For throughout the books we find a progressive narrative on psychological parts ultimately designed to "land"—with the myth of Er later in book 10—on a metaphysics of the soul.[59] Before reaching this climax dealing with the disembodied, pure soul and its experiences, the earlier references to the soul in books 4 and 10 analyzed above present alternative soul divisions that suit specific contexts in the realm of the sublunar world—the isomorphism of justice in city and individual, on the one hand, and the effects of mimetic poetry on the audience, on the other. It appears then that in *Republic* soul divisions are context-related and definitory inner partitions may collapse—as it happens for the spirited part in book 10 when Plato addresses the effects of mimetic poetry,[60] or as it will happen later on in the same book for the features of the irrational part, simply demoted into mortal accretions that supervene with the rational soul's embodiment.[61] Yet despite different contexts and suitable articulations of psychological parts, these soul divisions—tripartition and bipartition—are still compatible with one another. Referring to the human soul, they revolve around *logos* intended as reason (and its expressions),[62] claiming rational supremacy over other psychological affections and addressing its challenges related to the body in a normative discourse that aims to overcome them and pursue virtue.[63]

Now, also in Aristotle's ethical treatises we find a soul bipartition gravitating around *logos* in a discourse directed to promote moral action and ultimately human happiness.[64] Already present in *Protrepticus*,[65] this model of soul division receives its most elaborate account in *Nicomachean Ethics* 1.1:

Now on the subject of psychology some of the teaching current in extraneous discourses (*exoterikoi logoi*) is satisfactory and may be adopted here: namely that the soul consists of two parts, one irrational (*alogon*) and the other possessing reason (*to logon ekhon*) ... Of the irrational part of the soul (*to alogon*) again one division appears to be common to all living things, and of vegetative nature (*phytikon*): I refer to the part that causes nutrition (*trephesthai*) and growth (*auxesthai*); for we must assume that a vital faculty of this nature exists in all things that assimilate nourishment, including embryos— the same faculty being present also in the fully-developed organisms (this is more reasonable than to assume a different nutritive part in the latter). The excellence of this faculty therefore appears to be common to all animate things and not peculiar to man; for it is believed that this faculty or part of the soul is most active during sleep, but when they are asleep you cannot tell a good man from a bad one ... This is a natural result of the fact that sleep is a cessation of the soul from the functions on which its goodness (*spoudaia*) or badness (*phaulē*) depends ... We need not however pursue this subject further, but may omit from consideration the nutritive part (*threptikon*) of the soul, since it exhibits no specific human excellence (*anthrōpinē aretē*). But there also appears to be another element in the soul, which, though irrational (*alogos*), yet in a manner participates in rational principle (*metekhein logou*). In self-restrained (*enkrateis*) and unrestrained (*akrateis*) people we approve their principle, or the rational part of their soul (*to logon ekhon*) because it urges them in the right way and exhorts them for their good; but their nature seems also to contain another element beside that of rational principle, which combats (*makhesthai*) and resists (*antitinein*) that principle ... nevertheless it cannot be doubted that in the soul also there is an element beside that of principle, which opposes (*enantiousthai*) and runs counter (*antibainein*) to principle (though in what sense the two are distinct does not concern us here). But this second element also seems, as we said, to participate in rational principle: at least in the self-restrained (*enkratēs*) man it obeys (*peitharkhein*) the behest of reason (*logos*)—and no doubt in the temperate (*sōphrōn*) and brave (*andreios*) man it is still more amenable (*euēkoōteron*) for all parts of his nature are in harmony with reason (*homophōnein tō logō*). Thus we see that the irrational part, as well as the soul as a whole, is double (*ditton*). One division of it, the vegetative (*threptikon*) does not share in rational principle at all; the other, the set of appetites (*epithymētikon*) and desires (*orektikon*), in general, does in a sense participate in principle, as being amenable (*katēkoon*) and obedient (*peitharkhikon*) to it.

(*NE* 1 1102a26–1102b34, transl. by H. Rackham,
with slight modifications)

The soul encompasses a rational and an irrational part.[66] The irrational part is divided in turn into two: a nutritive "element," which is responsible for all living beings' nutrition and growth, embryos included, and an "element" that encompasses appetites and desires.[67] Concerned with a soul division that is instrumental to define and reach human excellence (*anthrōpinē aretē*), Aristotle considers the nutritive element irrelevant (it works especially at night when the rest of the soul is dormant!) but still stresses that it is the same for all human beings and rescues it from Plato's contempt. He then proceeds to discuss the other elements of the irrational soul—appetites and desires— that are relevant to moral action and compatible with reason. Of all human types, Aristotle brings up the self-restrained (*enkrates*) and the unrestrained man (*akrates*),[68] whose souls experience a conflict between appetite and desires, on the one hand, and reason, on the other,[69] with the irrational element opposing (*enantiousthai*) and running counter (*antibainein*) to the rational one. In the self-restrained man, reason is mastering its antagonists, while in the unrestrained man it is succumbing to them.[70] The model Aristotle mobilizes in this passage is, as Solmsen has noted,[71] rich in Platonic resonances and unmistakably political, although some shifts are also noteworthy. By identifying the nutritive element of the soul (*threptikon*) and attributing to it autonomy from conscious appetites and desires Aristotle is also, and significantly, breaking off from Plato.[72] As for the dynamics involving rational and irrational parts, they present some changes too. That participation in reason, which in Plato characterized *thymos* making it a subject and ally of reason,[73] is now transferred upon the conscious desires and appetites (*orektikon kai epityhymētikon*), which are presented as amenable (*katēkoon*) to the rational principle.[74] In fact, Aristotle's bipartition even features a scale of degree in participation of the irrational in the rational. If in the self-restrained man, the irrational part complies (*katēkoon*) with reason, it is even more "obedient" (*eukooteron*) in the brave (*andreios*) and temperate (*sophrōn*) man.[75]

Introduced as belonging to extraneous discourses (*exoterikoi logoi*),[76] soul bipartition is at the basis of Aristotle's view of moral action. Choice (*proairesis*), which is the key process leading to moral action, involves both the rational and irrational parts: it results from the interaction of thought (*dianoia*), which belongs to the rational part, and disposition of character (*ethos*),[77] which is a feature of the irrational one. To act morally the human being needs to deliberate about the means to achieve the objects s/he desires,[78] while character needs to be trained through habituation according to reason.[79] Inaccessible to the other animals (*tālla zōa*)—or children—who do not possess rational faculties,[80] choice is restricted to adult humans and so is Aristotle's moral discourse. Indeed, nonhuman animals only have sensation (*aisthēsis*), which, in another passage of *Nicomachean Ethics*, Aristotle claims to be uninvolved in moral action.[81] Thus in his moral discourse on the soul, like Plato so Aristotle too focuses on the human soul and grants value and supremacy to reason over the other psychological activities which are defined, and assessed, in relation to it. The irrational elements of the soul, such as character and desires, must "listen" to reason and follow it, meeting its standards. Significantly, to express the subjection to reason of *thymos*, on the one hand, and of appetites and desires, on the other, Plato and Aristotle use cognate adjectives: *hypēkoos*, *katēkoos*, and *euēkoos*. All related to the "action" of hearing (*akouein*), these adjectives present the dynamics of the irrational part of the soul within a "discursive" framework. For both philosophers moral action depends on *logos* in the sense of rational deliberation,[82] although, as noted earlier, Aristotle allows those desires and appetites, which in Plato had to be dominated or repressed, a more active participation in the rational element.[83] In this respect, significantly, at the end of his extensive presentation of soul division in *Nicomachean Ethics* 1, quoted above, Aristotle even raises the doubt that the appetites and desires may in fact belong to the rational soul. In this case, there would be a different type of bipartition, namely one that does not divide between a part that has reason and one that does not have it. Rather the rational part would be divided into two: a part that has reason in the proper sense (*kyriōs*) and another "in the sense in which a child listens to his father."[84] While Aristotle presents this reason-enmeshed bipartition of the soul as a possibility, he also shows that the participation of desires and appetites in reason makes him hesitant both about considering them as radically detached from reason and, likewise, about labeling the part of the soul that opposes reason as absolutely irrational. Hence we can understand why in *Nicomachean Ethics* Aristotle presents the doctrine of soul partition as belonging to "exoteric discourses." He finds bipartition effective as a framework to discuss moral philosophy, but he does not completely endorse it because it is somehow superficial and incomplete. For it leaves open the problem of explaining how "the purely irrational" may participate in the rational, an issue that according to modern commentators already afflicted Plato's psychological dynamics in *Republic* 4.

A full account of the relation between Plato and Aristotle's moral psychologies, their developments, and educational applications is out of the scope of this section, which aims more generally at pointing out the common ground between soul divisions of different types in the field of ethics concerning the human being. But one more passage from *Politics* deserves some reflection inasmuch as it continues the political interpretation of soul dynamics of *Nicomachean Ethics* 1 revealing, however, Aristotle's progressive detachment from Plato's doctrine (and also from his own position in *Protrepticus*), seemingly in the direction of his treatment of the soul in *On the Soul*.[85] Aristotle uses a bipartite model consisting of a rational and irrational part[86] to support at first the naturalness of slavery.

> But in the first place an animal (*zōon*) consists of soul and body, of which the former is by nature the ruling (*to arkhon*) and the latter the subject factor (*to arkhomenon*). And to discover what is natural we must study it preferably in things that are in a natural state, and not in specimens that are degenerate. Hence in studying man we must consider a man that is in the best possible condition in regard to both body and soul, and in him the principle stated will clearly appear—since in those that are bad or in a bad condition it might be thought that the body often rule the soul because of its vicious and unnatural condition. But to resume—it is in a living creature (*zōon*), as we say, that it is first possible to discern the rule both of master and statesman: the soul rules

the body with the sway of a master and of a statesman (*despotikēn arkhēn kai politikēn*), the intelligence (*nous*), the appetite (*orexis*), with constitutional or royal rule (*politikēn kai basilikēn*); and in these examples it is manifest that it is natural and expedient for the body to be governed by the soul and for the emotional part (*pathētikōn morion*) to be governed by intelligence and the part possessing reason (*nous kai to morion to logon ekhon*), whereas for the two parties to be on an equal footing or in contrary positions is harmful in all cases.

(*Pol.* 1 1254a22–1254b10, transl. by H. Rackham
with slight modifications)

Power structure is a universal truth of nature[87] and features in every ensouled being as well as in the soul itself. To become a "single common entity" every composite being necessitates a complementarity of roles between the parts that form it, namely a ruling and a subject factor. While there are different forms of power and subjection, Aristotle deems methodologically necessary to focus on the form of power which is natural and not degenerated, and which occurs when a given being is in the best condition. In the case of the human soul, such a condition is obtained when intelligence (*nous*) rules desire (*orexis*) and emotional part (*pathētikon morion*)[88] with constitutional and royal rule (*politikē kai basilikē arkhē*). This is an important qualification. On the one hand, with regard to this particular context, it differentiates the rule of the soul's rational part over the irrational one from the rule of the soul over the body, which in the same passage Aristotle qualifies as "despotic" (*despotikē arkhē*).[89] On the other hand, however, this qualification of reason's power is significant also in terms of soul division and dynamics in that, in the ideal condition, the power of the rational part over appetites and desires is not conceived in hegemonic terms as it was in earlier treatments, by Plato or in Aristotle's *Protrepticus* itself, for instance.[90] Reason has lost its hegemony, and holds a power that is conceived as "over free and willing subjects,"[91] with important repercussions on the understanding of the nature of "these subjects." As Saunders emphasizes, this political interpretation of soul dynamics shows that for Aristotle "rulers and subjects," that is reason, on the one hand, and appetites and desires, on the other, possess a rational nature[92]—a reappraisal of psychological capacities that well fits Aristotle's hesitancy to label the irrational part of the soul as strictly irrational.

So to return to *On the Soul* 3.9,[93] why does Aristotle consider current tripartition and bipartition inadequate to understand the soul? In a compelling article Corcilius and Gregoric have pinpointed in the criterion that determines the soul's division into parts the reason for the inadequacy of current psychological partitions for Aristotle. The bipartition and tripartition Aristotle criticizes, without however dismissing them, merely reflects different (and, ultimately, one could add, always partially selective) psychological capacities, while Aristotle in *On the Soul* supports a soul division on the basis of how parts separate (either *aplōs* or in account, *logōi*),[94] a strategy that holds repercussions on the conception of the relationship between the soul's capacities and parts and, in turn, of the very notion of the soul and its parts.[95] But if we approach the inadequacy of current soul divisions from the inquisitive path undertaken in this chapter, namely from a discussion of what they consisted of in relation to whether, and how, they overlapped, we find an additional reason that intersects the technical rationale for soul division argued for by Corcilius and Gregoric,[96] and has to do with Aristotle's goal of understanding the soul as the principle of life conceived of in its continuity and phenomenological diversity. At the outset of *On the Soul* Aristotle remarks that his contemporaries were restricting their inquiries to the soul of man (*anthrōpinē psykhē*),[97] and from the above comparison between tripartition and bipartition[98] we can now understand what this focus entailed. Whether conceiving of the soul as divided into three or two, both these versions of partition offered a biased (let alone limited) vision that turned around *logos*, reason, and its ideal domination.[99] In both cases, the field of reference was that of the human being—the only creature possessing reason—and his moral life[100] to the exclusion of the other living beings. Current soul divisions disregarded psychological functions such as those of the "nutritive part" (*threptikon*), which humans do share with all the other forms of life, or those of the sensitive part (*aisthētikon*), which equips living beings with the capacity of sensation and equally defines them (in Aristotle's view) whether fish, horse, or man.[101] Founded on such exclusions, they involved psychological parts that, to paraphrase Aristotle's language, were "less distant" from one another and did not present the characteristic of *meizōn diastasis* (more significant difference) we mentioned earlier in relation to Aristotle's critique.[102] Emblematic of the "less significant difference" was also the relational mode between psychological parts. In Aristotle's bipartition by communicating with the rational part, desires, and appetites—which allegedly belong to the soul that does not have *logos*—participate in reason (*logos*).[103] So it was, as we have seen, also for *thymos* in Plato's tripartition. In each case, such participation of the irrational in the rational meant for Aristotle a psychological proximity which was imposed by the human frame of reference and which attracted into the sphere of discursive reason (or simply dismissed) capacities and functions that in forms of life other than human were autonomous from reason and, in his view, even more fundamental to life than reason. We will discuss Aristotle's positive account of the soul consisting in a new partition based on a "more significant distance" in the next section. Here let it suffice to mention that of the two parts of the soul that in *On the Soul* 3.9 Aristotle cites as "more distant"—the nutritive (*threptikon*) and sensitive (*aisthētikon*)—the latter, as he admits in the same passage, is not easily reducible to neither the rational nor to the irrational part of the current bipartite division.[104] Thus, in *On the Soul* too, he presents the same doubt we have discussed earlier in relation to soul dynamics in *Nicomachean Ethics*[105] and particularly in relation to the participation of human desires and appetites in the rational element.[106] In *On the Soul* 3.9, however, Aristotle is confronting the problem of how to divide the soul to support a biological perspective that holds philosophical implications for understanding life, its processes, and the continuity among its forms. Aristotle contests the validity of *logos*, reason, as the ultimate referent in the identification of the parts of the soul in the effort to fathom the complex relation that the soul entertains with life and to understand life in its fullness, as rooted in the body and independent from human *logos*.

2.3 A new model: The geometry of the soul

Being fundamentally anthropocentric, the political, reason-based model of soul division—and body and soul interaction—does not find a legitimate place in Aristotle's study of the soul. In it Aristotle claims that he wants to find the soul's most common definition (*koinotatos logos*),[107] thereby informing his psychological inquiry with the search for the common (*to koinon*), which, let us recall, frames also his study of animals' comparative anatomy in *Parts of Animals* and in that treatise he declines in terms of analogy and degree. Yet in this refusal of the "political model," we should not see an effort to revise the psychological partition voiced in the ethical treatises: the criterion of *logos* as reason remains well suited to address the human being and its life, and the moral dimension that teleologically characterizes it. But in *On the Soul* reason loses its centrality. Here the soul is conceived not in the context of a pathway to virtue, but as the entity that allows to distinguish living beings from those not living and to understand the multiformity of embodied life, in terms of the greater difference discussed earlier.[108] Aristotle writes,

> The word living (*to zēn*), however, is used in many senses, and we say that a thing lives if any of the following is present in it—mind (*nous*), sensation (*aisthēsis*), movement (*kinēsis*), or rest (*stasis*) in space, besides the movement implied in nutrition (*kata trophēn*) and decay (*phthisis*) or growth (*auxēsis*).[109]

Intent to dissect the soul's relation to life, Aristotle divides it into parts that are responsible for, and at the same time reflect, living beings' activities, from nutrition, growth, and decay, to sensation, locomotion, and reason. Hence he identifies the nutritive (*threptikon*), sensitive (*aisthētikon*), locomotive (*kata topon*), and the rational (*noētikon*) parts of the soul.[110] In this division, something is certainly old, but something else is strikingly new. Soul specialization along with the distinction of the faculty of thought was claimed before Aristotle, and by Aristotle himself, we have seen, but locomotive, sensitive, and nutritive parts are Aristotle's own creations. Plato's *Timaeus* with its reference to food (*trophē*), body's growth (*auxanein*) and decline (*phthinein*),[111] influxes (*prosoi*) and effluxes (*apioi*)[112] hinted at the faculty of nutrition as belonging to the part of the soul that desires food and drink (*to sitōn kai potōn epithymētikon*),[113] but this faculty did not acquire a canonical placement among the parts of the soul claimed in that dialogue. On the other hand, the Presocratics expressed a kin interest in the process of nutrition with regard to the type of food and in relation to growth and vitality (among other effects), particularly, as the extant evidence revealed, in the case of plants.[114] It stood to Aristotle to develop the capacity of nutrition into a full-fledged, independent part of the soul and to make it the foundation of his psychological establishment.[115]

But besides the identification of "more distanced" psychological parts representing different functions, and manifestations, of life, what is also strikingly new in Aristotle's soul division—and suits the biological perspective of his inquiry—is the order he applies to them. Aristotle identifies the nutritive, sensitive, locomotive, and rational parts in a division that is classificatory of forms of life and fundamentally "continuous," and where what was once central is now peripheral.[116] In other words, Aristotle's new soul division explains and reflects a *scala naturae* in which the nutritive soul constitutes the foundation of living beings' existence and the progressive possession of the other parts of the soul identifies all forms of life, from plants to animals.[117] Plants (*phyta*) only have the nutritive soul, which they share with the rest of living beings.[118] They are able to nourish themselves, grow, and reproduce, but they do not have sensation. Aristotle labels them *zōnta*, living things.[119] As for animals (*zōa*), besides the nutritive soul and the capacity to nourish and reproduce themselves as plants do, they all possess the sensitive soul and hence the capacity to sense, at least with touch if not with the other senses. On this view, sensation emerges as the common denominator of all animals, humans included.[120] In addition to the sensitive part of the soul and the capacity to feel, some animals have also the locomotive soul,[121] and are therefore able to move across space. The human animals further possess the rational part and are consequently endowed with the faculty of reason.

Significantly, to exemplify the way the parts of the soul relate to one another and determine different forms of life in *On the Soul* Aristotle adopts a new model of the soul, which is alternative to the political one used in the ethical treatises and which we may well call "geometrical." The passage featuring this model is worthy to be quoted in full since it sheds a new light on *logos* as reason by relating it to the other psychological parts differently than one familiar with its discussion in the moral discourse of *Nicomachean Ethics* and *Politics* would expect.

> The facts regarding the soul are much the same as those relating to figures; for both in figures and in things which possess soul, the earlier type always exist potentially in that which follows; e.g., the triangle is implied in the quadrilateral, and the nutritive faculty by the sensitive. We must then inquire in each several case, what is the soul of each individual, for instance of the plant, the man, and the beast. But we must also consider why they are thus arranged in a series (*ephexēs*). For without the nutritive faculty the sensitive does not exist, but in plants the nutritive is divorced from the sensitive faculty. Again without the sense of touch none of the other senses exist, but touch may exist without any others; for many animals have neither vision nor hearing nor any sense of smell at all. Again, of those which have sensation, some have the locomotive faculty, and some have not. And lastly, and most rarely, living creatures have the power of reasoning and thought. For those perishable creatures which have reasoning power have the other powers as well, but not all those which have any one of them have reasoning power; some have not even imagination, while others live in virtue of this alone.[122]

The parts of the soul relate to one another as more complex rectilinear figures relate to basic ones. As a quadrilateral always contains a triangle, and—one may continue—the pentagon, which is more complex than a quadrilateral and a triangle, contain both a quadrilateral and a triangle in succession, so the sensitive soul always implies a nutritive soul, and the locomotive, the sensitive and nutritive.[123] Clearly, in this model the soul parts are not seen in antagonism or in an optimal, but fragile, balance, always liable to be broken as it happens in the political model presented in *Nicomachean Ethics* and *Politics*. Rather, now the parts of the soul are complementary to, and, importantly, build on one another. They form a nestled sequence and entertain a serial relationship in which the nutritive part detains ontological priority over the others, and where in turn the sensitive part does the same over the faculties of the soul "above it."[124] In Leunissen's words,

> The unity between the capacities of the soul is one of order (*taxis*), which is here rendered as being ontological in nature: whereas the lower capacity can exist separately and does not need the higher capacities, the higher cannot exist without the lower ones.[125]

In (mortal) living beings each part of the soul, from the more basic up to the least, is necessary for the next one in the series to exist. And while it is possible for the nutritive soul to exist by itself, separated from the sensitive part, the sensitive one—and the successive ones—cannot exist without the nutritive.[126] As for the faculties of the sensitive soul, while touch can exist without the other senses, the other senses cannot exist without touch.[127] At any rate, no matter what parts of the soul and faculties a living being partakes of, they result into a set of harmonious and integrated activities conducive to life. Thus in Aristotle's vision, from the simplest to the most complex, all living beings share some basic psychological functions (and suitable organic bodies to carry them out).

Significantly, in this model the relationship among psychological parts is expressed with the notion of *ephexēs*, "next in succession," whose ontological implications and temporal ramifications shed light not only on Aristotle's conception of the continuity among forms of life but also of life development. In *Physics* Aristotle defines *ephexēs* as the qualification of something that comes after the "point of departure" "in an order determined by position, or 'form,' or whatsoever it may be" and soon adds that what is next in succession "must succeed something and be a thing that comes later."[128] Applied to the geometrical model of the soul, this expression indicates that the parts "next in succession" to the nutritive soul (sensitive, locomotive, and rational) encompass the earlier ones. But transferred (with some liberty) upon the conception of living beings, as natural and hence moving bodies, it also indicates that those activities that pertain to the "formally later" parts of the soul emerge and are "lived out" later in time. Looked at from this angle the notion of succession adds a temporal dimension to Aristotle's conception of continuity among the different forms of life, which he has broadly identified by appealing to soul partition. That is, all living beings (*zōa*) share the same psychological parts, i.e. the nutritive and sensitive souls, and hence live by the same functions (at least, nutrition, reproduction, decay, and sensation, with the last accompanying the first ones). Because all the parts of the soul, including the locomotive and rational ones (for those living beings who possess them) are supervenient upon the nutritive soul, all animals have lived at one point the life of "plants."[129] This means that they have developed into animals with the acquisition of sensation as their body was progressively becoming more articulated, i.e. growing while fulfilling its *logos*.[130]

In the geometrical model of the soul, *logos* as reason is neither ruling nor under threat. Now, the rational part—at least for sublunar creatures[131]—depends, first, on the nutritive part and, then, on the sensitive one in order to exist. From a teleological perspective the supervenience of the rational soul may well indicate the ultimate fulfillment of a human being's nature, hence differentiating him/her from the other animals. But life can be full and fulfilled without reason—i.e. when the rational soul is no part of a living being's psychological makeup. Besides, in the context of the psychological faculties and their relations as described in *On the Soul*, reason is not merely a "disembodied" activity,[132] fighting against—among others—bodily desires and appetites, but requires the sensitive soul, which is central to understand animal life. That *logos* as reason, which constitutes a human privilege in the ethical treatises, is dismissed as irrelevant in *Parts of Animals*,[133] and finds little elaboration in *On the Soul*,[134] could not exist without the functions of the senses and *phantasia* (imagination),[135] which human beings share with the other animals. Likewise, locomotion too involves sensation.[136] As for the nutritive part, it is indeed the most common (*koinotatē*) part of the soul and its work (*ergon*) the most natural (*physikotaton*) as it is shared by all living beings, from plants to humans.[137] None of the other parts, we have seen, could exist without it.[138]

In the end, in this novel partition of the soul, exemplified by the geometrical model, when coexisting in a given creature each part of the soul corroborates the next one in sustaining animal living (*zēn*), and eventually leads to a good living (*eu zēn*).[139] For if the nutritive soul and the possession of touch enable a given animal to live,[140] the possession of the other senses—sight, smell, and hearing—enables it to live well. The range of sensorial experiences together with the self-preservation they provide, the capacity to communicate of pain and pleasure[141] and orderly motion,[142] contribute all for Aristotle to a "good life" (*eu zēn*). For in possessing the locomotive soul and the range of faculties of the sensitive soul, nonrational animals live to the fullness of their potentiality as sentient beings. They lead a life whose goodness is made of relational connections, richer sensations, and deeper awareness than animals endowed with only contact senses have—features that will be discussed more in depth in Chapters 5 and 6. Suffice here to point out that, as Polansky remarks, the notion of animals' good life shows that for Aristotle ultimately "goodness belongs in natural things even apart from human evaluations"[143] and standards. And it also shows that goodness is inherent in the action and experience of living intended as a natural end. Thus, as in the moral discourse on the soul, so in the biological one too there emerges a notion of well-being. In the first, well-being pertains to humans, involves *logos* as reason, consists in action according to virtue or contemplation (*theōria*),[144] and is attained in the polis.[145] By contrast, in the latter it involves living beings *tout court* and rests on the actualization of the potentialities of a complex body whose functions go beyond the mere vital activities of nutrition and touch, and is attained in animals' space of existence.[146] In this case too, as in the treatment of the human being in the ethical treatises, well-being, or a "good life" (*eu zēn*), is species-specific, each animal pursuing its own.[147] And while Aristotle's moral discourse on the soul is (like Plato's) anthropocentric and hierarchical, pinpointing in the possession, and conversely lack of reason, the human/animal difference, the biological one is fundamentally zoocentric and in tendency egalitarian and reflects his program to study all animals.[148] In it Aristotle dissolves human self-referentiality into the search of the most common definition of the soul and redeems the body, privileging sensation over reason and striving to capture life in its multiple, related manifestations.

Notes

1 *DA* 1 402a5–8.
2 *DA* 2 412a20–1; see below. As Polansky remarks, while these technical notions are fully explicated in *Metaphysics*, they also appear in *Physics*, 2010, 149, and are therefore pertinent to understand the physical world (and living beings as belonging to it).
3 See Hamlyn, 1968, 83–4.
4 On this capacity and its fundamental role in life, see Chapter 3.
5 See *Phys.* 2, 1 and discussion in Chapter 1. It should be stressed, however, that not all animals for Aristotle are able to move across space, the crucial definitory capacity being that of sensation.
6 Aristotle mentions the nutritive, sensitive, locomotive, and rational soul, which are responsible, respectively, for living beings' nutrition, growth, and decay, their sensations and emotions, their locomotion, and reason. See Chapters 1 and 3, and discussion below.
7 In the series of questions regarding the soul that Aristotle considers important to address, there is whether the soul has parts or not (*meristē/amerēs*) (*DA* 1 402b1, see below). Aristotle returns to this question again in *DA* 1 411b5–30, 2 413b11–414a1, 432a22–b7. For the metaphysical implications of this question and Aristotle's overall coherence in dealing with it, see Hicks, 1965, 183.
8 *DA* 1 411b2–7 where Aristotle mentions the presence of some who "say that the soul has parts, and thinks (*noein*) with one part and desires (*epithymein*) with another." As Polanski notes, in this passage the reference is especially to Plato, 2007, 136.
9 On the tripartition and bipartition of the soul, see *DA* 3 432a22–432b7 and below.
10 *DA* 1 402b2–5, with slight modifications. All translations of *On the Soul* are by W.S. Hett.
11 The relation between these two orders of problem appears from the fact that they are raised in the same sentence and are related by a conjunction (*kai*) without any intervening particle [i.e. *de* as in the previous questions relating to the soul's categorical membership and potentiality/actuality (*DA* 1 402a23–6)].
12 In this respect, see Themistius' commentary on this passage (3, 16): "The fourth inquiry described is: is every soul in relation to every soul of the same kind, or not, and if they are under distinct kinds, are they also under a single genus? Take the soul of a human being and that of a horse: if they do not have the same kind of soul, is their genus even the same? Are a human being and a horse under [the genus] animal, although their souls are not under the single genus of soul? These problems cannot

be decided without considering every soul, something earlier thinkers overlooked," transl. by R. Todd; cf. Hicks, 1965, 183.

13 For *homoeidēs* in reference to air, see *DA* 1 411a21, water *Metaph.* 5 1024a30ff, and fire *Cael.* 1 276b5ff; Hicks, 1965, 183.

14 *DA* 1 403b25–31, 404a21–5, 407b34–408a1, 409b19–25, 410b16–8.

15 Aristotle's review of his predecessors' theories of the soul occupies sections 2–5 of book 1. While his critique does not follow a linear, systematic path, it can be pinned down to the denial of the soul's materiality and movement, with the aim of both reconfiguring the relationship between body and soul and reassessing animals' bodies as organic wholes and loci of sensation (on this aspect, see below and Chapter 4). Indeed, according to Aristotle, the Presocratics identified the soul with the material origin (*arkhē*) of the cosmos and, at any rate, in association with movement. For instance, for Democritus the soul was constituted by very tiny, fiery atoms, for Empedocles, and Plato, by the elements; Anaxagoras considered the soul as mind (*nous*)—although Aristotle admits that his view is not entirely consistent; Heraclitus as fire, Diogenes as air. For Critias, instead, it was blood (*DA* 1 405a–11). For a recent discussion of the identification of the soul with the *arkhē* in the doctrines of the Presocratics, see Zatta, 2019, 107–132. On the other hand, however, Aristotle also refuses the consequences for an understanding of animals' cognitive processes that a monolithic notion of the soul leads to, namely the coincidence and/or close proximity of thought and sensation (Empedocles and Democritus) and of mind and life (Anaxagoras). For each of these positions fails, in its own way, to take into account the presence, or absence, of truth.

16 In fact, Anaxagoras stands out in respect to the other Presocratics because he identified (or at least associated) soul with mind, the entity setting the primordial chaotic mass into an orderly movement, and not what he thought to be the material origin of the world, the *homoemeries* (DK 59 A 100, DK 59 A 99).

17 *DA* 1 404b10–14, DK 31 B 109.

18 *DA* 1 404a17–21. Aristotle introduces this information after discussing the atomist doctrine of the soul, and it is unclear whether the Pythagoreans who conceived the soul as air did it also in cosmological terms. Yet, the subsequent remark that some other Pythagoreans considered the soul as responsible for the movement of the particles of air leaves this possibility open.

19 See Aristotle's remarks at *DA* 1 407b14–27 where he criticizes previous interpretations for associating the soul with a body and placing the first inside the latter, without specifying anything about that particular body; for a thorough analysis of Aristotle's argument targeting the connection between the soul and body in terms of some commonality (*koinōnia*), see Shields, 2016, 131–4.

20 See, respectively, *DA* 1 404b11 and 405a19.

21 *DA* 2 412a13–4.

22 Form, matter, and their compound are here importantly discussed in terms of substance (*ousia*). On this aspect and its implications in Aristotle's study of animals, see Chapter 1. On these notions, see, for instance, *Metaphysics* (7 1029a1–8) and in *Physics* (2 193a10–194b15). On Aristotle's application of the diaeretic method to reach a definition of the soul, see Movia, 2001, 13–14.

23 *DA* 2 412a14–5.

24 Polansky remarks that the linking of soul to life is not an obvious point and that Aristotle's task will be that of securing it (2010, 37–8; cf. also p. 151 where the author acknowledges the connection soul/life in Plato, but points out the overall obscurity as to what things have life and the role of the body in it). In making the soul responsible for nourishment, growth, and decay, Aristotle pins down the most fundamental set of functions of living beings—without which no living being (or thing) can exist. As we will see in Chapter 3, such set of functions is subsumed under a refined conception of *auxēsis* (growth), one of the movements that affects the creatures that are by nature—the others

being alteration (*alloiōsis*) and movement across space (*kinēsis kata topon*) (*Phys.* 2 192b9–23; *PA* 1 641b4–10; cf. Chapter 1). Importantly, however, Aristotle's linking of the soul to life determines, and is conversely dependent upon, a definition of the bodies, which are ensouled versus those that are not.

25 The qualification of the body as "organic" appears in Aristotle's more refined definition of the soul in *DA* 2 412b, which explains the preceding assertion that the soul is the first actuality of "a body potentially possessing life." Such an instrumental body is one composed of parts that are in turn instruments. Aristotle gives the example of plants the different parts of which (leaf, pericarp, fruit, and roots) offer specific services thereby allowing the soul to perform its functions (Polansky, 2010, 160–1). For a view that stresses instead the notion of the whole body as an instrument of the soul, see Everson, 1997, 64, and Shield, 2016, 172, who recognizes, however, that these two interpretations (body made of parts that are instruments and body as instrument) need not to be taken as competitors. For a discussion of living beings' organic body in terms of *logos* and in comparison with nonorganic natural bodies, see Chapter 3.

26 *DA* 1 402a7–8.

27 See below on the possibility of the soul's infinite partitions.

28 On psychological parts and how they separate, see also *DA* 2 413b13–33, and for an extensive discussion of these issues Whiting, 2002, 141–200, and Corcilius and Gregoric, 2010, 81–120. The latter stress the difference between soul capacities and parts and show that Aristotle effects his division on the basis of two ways, namely by separability "*tout court*" and by account. See below.

29 This possibility was already voiced in Plato's *Theaetetus* where Socrates mentions aside from linguistically identifiable sensations (*aisthēsis*) such as "sight, hearing, smell, cold, heat, pleasure and pain, etc.," the presence of innumerable others which do not have a name (*Tht.* 156b). As Polansky remarks, with his statement Aristotle seems to deny fixity of the soul partition (2010, 503–5), and also manifests (we could add) the empiricism and the variety of angles from which he looks at the living world.

30 For the presence of this debate, which engaged members of Plato's Academy, and relative bibliography, see Vander Waerdt, 1987, 628, 642–3, and Corcilius and Perler, 2014, 4–9, who restrict it to the consideration of the positions of Plato and Aristotle.

31 See respectively *CIAG* 11.287.26–9; 289.7–17, and 15.571.18–574.2; for Plato's *Republic*, see 4 435a–436b).

32 For a discussion of possible referents, see Rees, 1957, 112–8, Fortenbaugh, 1970, 233–50, and Waerdt, 1987, as in n. 34, with relative bibliographies. As for specific attributions, Fortenbaugh attaches bipartition to the Academic thinkers, denying that with bipartition Aristotle may in fact be referring to a position he advocated elsewhere in his corpus. By contrast, Waerdt considers bipartition an allusion to Aristotle's own doctrine in the ethical treatises. Rees attributes tripartition to Plato in *Republic*, but indicates a less clear commitment to it in other dialogues foreshadowing the soul bipartition presented in *Laws* (*Leg.* 9 863e–864a) and circulating in the Academics' circle (1957, 112–8). In fact, the general label "*tines*" under which Aristotle situates these theories has probably to do not only with their diffusion but also with their overall convergence, as I argue below, in assigning a privileged, dominating role to *logos*.

33 Examples of bipartition are found at *NE* 1 1102a26–1102b34 (discussed below), 5 1138b6–13, 6 1138b35–1139a; *Pol.* 1 1254a38–b10, 1260a5–17, 7 1333a17–30, 7 1334b7–28; Iamblicus *Protr.* 7 41, 20–24= Düring B60/fr. 6 Ross.

34 *MM* 1 1182a23–5.

35 *Phaedr.* 246a–249.

36 See, respectively, *Republic* 4 435e–441c and *Timaeus* 65a, 69c–d, 70a, 72d.

37 The only difference pertains to the label of the "spirited" soul, which Plato names *thymoides* (*Resp.* 4 440e) and Aristotle *thymikon* (*DA* 3 432a25).

38 See above.

39 The compresence in *Republic* of these two versions of the soul has puzzled scholars who have tried to reconcile them with various strategies, for a discussion of the literature, see Moss, 2008, 35 n. 1; on the problem of soul's partition in Plato with a focus on the unity of the soul in conflictual motivation for action; see also Shields, 2014, 15–38.

40 Plat. *Resp.* 4 433B.

41 Plat. *Resp.* 4, 436E–437A; 438B; cf. 10, 602E–603. For an analysis of Plato's argument for soul's division in *Republic*, see Stalley, 1975, 110–28 (on the division between appetitive and rational souls), and Shields, 2014, 19–34.

42 *Resp.* 4 439b, all translations of *Republic* are by P. Shorey.

43 *Resp.* 4 439d. The fact, as we will soon hear, that the appetitive (and spirited) part of the soul should (and does) listen to the rational part has made Cross and Woozley criticize Plato for self-contradiction, for in obeying to reason the appetitive part shows some degree of reason (1964, 124; for a defense of Plato, see Stalley, 1975, 110–28; on the rational capacities of the appetitive and spirited souls, see also Moss, 2008, 37). The attribution of capacities to the different parts of the soul is indeed not as clear-cut as one would like it to be and there is some overlapping. Also reason, for instance, has desires (see Frede, 1996a, 7). For a solution of this difficulty through the definition of a substantive concept of rationality as based on calculation (*logismos*) and able to transcend appearances both in the ethical and sensory realms (and discrepancies between soul's tripartition and bipartition), see Moss, 2008.

44 Admittedly, *thymos* is of difficult translation. It is a complex psychological feature, holding executive nature, and directed toward a person *qua* agent (see Moss, 2005, 156).

45 See Vegetti, 1998, vol. III, p. 91, n. 97.

46 Vegetti, 1998, vol. III, pp. 92–3, n. 100.

47 *Resp.* 4 441E.

48 *Resp.* 4 442A.

49 Resp. 4 441A; cf. 8 553Cff.

50 While book 4 of *Republic* focuses on the harmonious relation of the soul's parts and the attainment of psychic justice, books 8–9 adopt the same model of soul tripartition to address the various psychological dynamics that inform the souls of deviant men—timocratic, oligarchic, democratic, tyrannical—within a comparison with equally deviant political communities (see Irwin, 1977, 226–33).

51 *Resp.* 10 603A–607A; on contradictory opinions (*enantia doxazein*), see *Resp.* 10 602E–603A; on soul's self-contradictions (*enantiōmata*) in general, 603D; on contradictory impulses (*enantia agōgē*), 604B.

52 *Resp.* 10 602C. For the illusions of sense, and measurement as a means to correct them, cf. Plat. *Phileb.* 41C–42A, 55E, *Protag.* 356C–D, *Euthyphro* 7C.

53 Plato's reference to the emergence of contraries that oppose "that which has measured and indicated" is obscure, and scholarship has attempted to illuminate it in various ways, see Adam, 1902, vol. II, *ad loc.*; Penner, 1971, 96–118; Halliwell, 1988, *ad loc.*; Naddaf, 2002, 107, n. 2. Here I agree with Adam who identifies the periphrasis under discussion with the rational part of the soul, the judgments of which are challenged by the irrational part.

54 As Gastaldi remarks, the goal of Plato is that of reaching a "homology" on the basis of which all mimetic arts appeal to the irrational part and are removed from truth (1998, 127).

55 As in the tripartite model presented in book 4, also in bipartition, reason has a censoring and inhibiting role: it exhorts the man suffering misfortune to resist (*antitenein*) while passion urges him (*helkein*) to give way to his grief (*Resp.* 10 604B).

56 Burneyat, 1976, 34. Besides the commonality of reason, soul bipartition overlaps with the earlier scheme of tripartition in that the irrational element conflates bodily

desires (606D) and a degenerate form of the *thymoides* (604E, 606A; cf. 606D) (see Adam, 1902, vol. 2, 406, *ad* 602Cff).

57 Aët. *Plac.* V 4.1.

58 In *Republic* Socrates discusses soul bipartition (as consisting of a mortal and immortal part) through the myth of Glaucus and asserts that the soul must be looked at as what it might be "if it were raised out [by its love of wisdom] of the depth of the sea in which it is now sunk" (611E). With such imagery and words this passage seems to correspond to Plato's "highest discourse" on the soul, identified by Aëtius. For it alludes to the spatial coordinates of high and low letting us understand that we can grasp the immortal soul by raising it up from its (lower) natural environment (i.e. its embodiment in the sublunar world) and purifying it of its mortal accretions.

59 More specifically, this narrative develops via the myths of Glaucus (10 611D) and Er (10 614–21), both centering on the soul as detached from the body. The myth of Glaucus is particularly interesting in that it shows the progressive nature of Plato's discourse on the soul shifting focus from the embodied soul to the "true soul" and addressing its immortality. In the end Plato concentrates on the rational part, which alone is immortal, and presents the "other psychological parts"—consisting in "sufferings and forms" (*pathē kai eidē*)—as later accretions (*prosphyein*) that derive from the soul's embodiment; cf. *Tim.* 69C. In fact, only a consideration of the soul in its true nature enables us to understand whether it is multiform (*polyeidēs*) or single (*monoeidēs*) (612A).

60 In this case, spirit (*thymos*) becomes irrelevant inasmuch as Plato is interested in describing the abandonment of the individual to pain and sorrow, a state which, we may well assume, encourages passiveness and is therefore incompatible with the executive nature of *thymos*.

61 Cf. Plat. *Tim.* 69C. That bipartition of the soul can never exclude tripartition has been remarked by Saunders in his discussion of psychological partition in *Laws* (1962, 37).

62 For *logismos* as the driving force of the rational soul, which implies both a narrow, arithmetical sense and a wider one, indicating "reckoning and accounting," see Moss 2008, 36–7.

63 Here I present an interpretation compatible with that of Moss, who identifies in *Republic* 10 a psychological distinction between rationality and nonrationality that is paramount to Plato's understanding of the soul and as such intersecting his different versions of psychological partition (2008, 35–68). Unlike Moss, however, I stress the ideologically reason-bent nature of the Platonic soul to show Aristotle's dramatic departure in his treatment of the soul as the principle of life in *On the Soul*.

64 On the two versions of human happiness proposed by Aristotle in *Nicomachean Ethics*, see Kraut, 1991.

65 A fragment of *Protrepticus* reads, "In the soul there is, on the one hand, reason (*logos*) (which by nature rules (*arkhein*) and judges (*krinein*) in matters concerning ourselves), on the other hand, that which follows (*hepesthai*) and whose nature is to be ruled (*arkhesthai*); everything is in perfect order when each part brings its proper excellence to bear" (Düring B60=Iamblicus *Protr.* 41, 20–24/fr. 6 Ross). For the rule of that which is more fit for rule (*arkhikōteron*) and more authoritative (*hēgemonikon*), see Düring B 61= Iamblicus *Protr.* 41, 24–42.1/fr. 6 Ross.

66 In the course of *Nicomachean Ethics* Aristotle introduces some further distinctions and differentiates, for instance, the rational part of the soul into a contemplative and practical element (6 1139a25–30).

67 For a discussion of this model of soul division as presupposing the soul taxonomy of *On the Soul*, see Shield, 2016, 245–6.

68 For the role of the senses and ensuing pleasure in the configurations of these characters, see Chapter 5.

69 See Gauthier and Jolif, 1959, vol. II, 95.

70 Later on in the same passage from *Nicomachean Ethics* appetites and desire are said to combat (*makhestai*) and resist (*antinein*) reason (*logos*) (cf. *Resp.* 10 604a), but also participate in it, by listening (*katēkoon*) and obeying (*peitharkhikon*).

71 Solmsen, 1955, 150.

72 See below.

73 See above.

74 In Aristotle's moral discourse *thymos* and appetites (*epithymiai*) are in fact causes of noble actions, they supply the motivation in action (*NE* 3 1111a30–2). Indeed, thought (*dianoia*) by itself—claims Aristotle later—does not move anything (*NE* 6 1139a35–6).

75 The mention of these human types—the self-restrained and unrestrained, the brave, and the wise— suggests that Aristotle is discussing soul bipartition incorporating the individual types Plato brings up in his discussion of soul partition (see, for instance, *Resp.* 4 430e–431b where Plato discusses the man "master of himself" or "the unself-controlled, *hēttōn heautou*, and the licentious, *akolastos* one"). This sharing of exemplary types ultimately shows the compatibility of the two models, let alone Plato's influence on Aristotle.

76 By extraneous discourses Aristotle means the works he composed early on in his youth such as dialogues and lost treatises. They were devoted to the divulgation of his doctrines as opposed to the more technical, "esoteric discourses" directed to the members of his school (see Gauthier, 1970, 63–5).

77 *NE* 6 1139a32–6.

78 On *proairesis'* connection to reason, see *NE* 3 1113a10–1, 6 1139a32, b4–5.

79 On habituation, see, for instance, Kraut, 2012, 529–57; on the involvement of pleasure in actions done by habit as a second nature, see Chapter 5.

80 *NE* 3 1111b4–14.

81 *NE* 6 1139a20–1.

82 See respectively *Resp.* 4 441e, *NE* 1 1102b32, and 1102b28.

83 In this way, Aristotle makes explicit the tension inherent in Plato's discussion of soul's dynamics in *Republic* 4, which commentators have interpreted as self-contradictory (see Cross and Woozley, 1964, 124 and n. 44 above).

84 *NE* 1 1103a1–5. In this case, the rational part has reason not in its own right, but by derivation and participation.

85 Certainly, the political model to interpret the soul's dynamics and the body/soul relation is not compatible with the hylomorphic doctrine of *On the Soul*. The continuity I point out here regards the progressive claim for a positive autonomy of psychological functions like desires and appetites from reason.

86 In *Politics* Aristotle continues the study of "human philosophy" (*anthrōpinē philosophia*) undertaken in *Nicomachean Ethics* (*NE* 10 1181b10–6).

87 Simpson, 1998, 32.

88 Newman observes that by *pathētikon morion* here, Aristotle intends what he usually calls *to orektikon* (cf. *Pol.* 3 1286a17; 1887, vol. 2, 145). In this version of bipartition too, emotions and desire are seen as constituting the part of the soul antagonist to the rational part.

89 *Pol.* 1 1254a22–1254b10.

90 See above.

91 Newman, 1887, vol. 2, 144. According to Newman, the qualification of reason's power as royal (*basilikē*) has probably the role of denying such an alternation of rule with desire (*orexis*). This may be certainly the case, although the immediate denotation seems to lie in the contrast between the despotic rule of the soul over the body, and the royal one of the rational soul over the sensitive one, with a consequent reevaluation of the role and nature of desire.

92 Saunders, 1995, 77–8.

93 This passage is quoted above on p. 54.

94 See above.
95 Corcilius, Gregoric, 2010, 108–9; on the capacity-based Platonic division, see also Johansen, 2012, 247–8.
96 To anticipate Aristotle's positive division of the soul (see next section), our philosopher considers the nutritive and rational souls separate *aplōs* (we find the first in plants, the second in stars), but in living beings that possess them equally (the human being) they can be separate only in account. For in human beings (when fully grown) these parts do not exist independently from one another but they coexist (together with the sensitive and locomotive souls). Now, also in the case of animals, who possess nutritive and sensitive souls (and for some of them the locomotive), these parts are only separate in account because they do not exist independently from one another. Still, their soul makeup is separate *aplōs* from the rational soul which they do not possess, hence making their life independent from human reason.
97 *DA* 1 402b4–5.
98 See above pp. 6–20.
99 On the inadequacy of the Platonic soul, based on reason (*logos*), in explaining life in its full range of manifestations, see also Johansen, 2012, 4. In addition, for Aristotle the Platonic model is ineffective also because the centrality of reason leads to the conception of the intrapsychological relation in terms of domination and conflict rather than integration.
100 Aristotle goes as far as to specify that it is not appropriate to call nonhuman animals unrestrained (*akratē*), which suits by contrast human beings in which desire disobeys reason. And the inappropriateness of this label has to do with the nonhuman inability to form universal concepts (*tōn katholou hypolēpsis*) (*NE* 7 1147b4–5).
101 In fact, in the scheme of soul bipartition Aristotle advocates for in ethics, sensations are not involved in the production of action and hence eclipsed by more relevant psychological functions such as desires and reason (which, by contrast, do contribute to action) (see, for instance, *NE* 6 1139a15–1139b5).
102 See above, p. 55.
103 *NE* 1 1102b26–1103a2.
104 So Aristotle writes: "the sensitive part (*aisthētikon*), which one could not easily assign either to the rational or irrational part." Although he does not elaborate on this claim, the reason for it likely lies in the capacity of sensation of grasping the *logoi* and becoming itself *logos*, see Chapter 4.
105 See above.
106 *DA* 3 432a27–432b1.
107 *DA* 2 412a3–6; cf. 2 412b4–5. The most common definition (i.e. the soul as first actuality of an organic body) which Aristotle brings up in his progressive discussion in *On the Soul* 2, 1 pertains also to plants, but he will eventually develop a mereological discussion of the soul that, while embracing all forms of life via the nutritive soul, joins all living beings, humans included, via the sensitive soul. See below.
108 On the soul as the cause "of the different kind of living," see Johansen, 2012, 51–2.
109 *DA* 2 413a21–5.
110 See *DA* 2 413b12–4 where we find a preliminary definition of the soul in relation to living faculties: "But for the moment let us be satisfied with saying that the soul is the origin of the characteristics we have mentioned, and is defined by them, that is by the faculties of nutrition (*threptikon*), sensation (*aisthētikon*), thought (*dianoeitikon*), and movement (*kinesis*)."
111 Plat. *Tim.* 41D.
112 Plat. *Tim.* 42A.
113 Plat. *Tim.* 70D; later this part of the soul, located between the navel and midriff, is said to "feed at the manger" expressing once more a connection with the faculty of nutrition (70E).

114   In his often-neglected article Solmsen stresses the Presocratic influence in Aristotle's psychology, also with regard to nutrition (1955, 154–7); further, he considers the treatment of nutrition in *Timaeus* as physiological and developed under medical influences, but still acknowledges in it "an Aristotelian soul-function" (154–7). For the Presocratics, see, for instance, DK 31 *ad* B 77, DK 59 A 46, DK 68 A 162.

115   For Johansen the introduction of the nutritive soul is Aristotle's "trump card against Plato" because it is irreducible to reason while the other faculties of the soul—for instance, perception and belief— are for Plato degraded forms of reason (2012, 6). Yet, to understand the development of Aristotle's study of the soul, it is also crucial to consider how he replies to the hegemonic role of reason (*logos*) that informs the current versions of soul division criticized in *On the Soul* 3, 9 with his consequent, fundamental reconceptualization of soul dynamics founded on nutrition and centering on sensation (*aisthēsis*) as separate from human reason.

116   Continuity in Aristotle's biology has been a controversial issue, with its disallowance often tied to the attempt to savage Aristotle's metaphysical and logical coherence, on the one hand, and animal kinds' discontinuities, on the other. For a discussion of forms of continuity, supporters, and critics, and the relevant passages in Aristotle's corpus, see Granger, 1985. Surprisingly, however, this discussion omits to consider the "geometrical" model of the soul, which is key to understanding the form of continuity that Aristotle has in mind and that runs through his biology, connecting all forms of life. That the figures with more facets include those that have less indicates that more complex living beings possess the same basic capacities of less complex ones. See below and n. 135.

117   On Aristotle's *scala naturae*, Lovejoy, 1936, 55–9; Solmsen, 1955; Granger, 1985 (with additional bibliography); Lennox, 2001, 87. Recently, Osborne has argued for a continuity between animal capacities and the human capacity to think in conceptual terms their experiences and behaviors, and has consequently denied the presence in Aristotle's biology of a *scala naturae*, if by it we conceive a ladder framed in terms of human value and superiority (Chapters 4 and 5 of her 2007 book). Here, in line with Osborne, I use the expression *scala naturae* to convey a sequence of forms of life from the most basic (plants) to the most complex (humans) transcending any attribution of value. I disagree, however, with Osborne's attribution of Aristotle's approach to "a humane and perceptive attitude to the non-linguistic members of our own species and other creatures" (Osborne, 2007, 63). For such a claim ultimately perpetuates a vision of Aristotle's project as anthropocentric. Rather, the continuity that informs Aristotle's *scala naturae* has to be considered in light of a zoocentric perspective, namely a program directed to find the common (*to koinon*) among the different forms of life of this planet, and the consequent marginalization in such a project of *logos* intended as reason. In this respect, the search for "commonality" coincides with the search for what enables living beings to exist. See Chapter 3.

118   In this attribution Aristotle firmly departs from those Presocratics who attributed sensation to plants. See Zatta, 2019, Chapter 4 and n. 1, and her forthcoming article on plants' physiology and life. Further, Solmsen remarks that Aristotle's concern in *On the Soul* with claiming that there are only five senses has to do with the Presocratics and Plato's attribution to plants of different sensations (*allai aisthēseis*) than those which humans possess (for Democritus see DK 68 A 116 and Plato *Tim.* 77A 5, cf. Aristot. *DA* 3 424b22–425b12).

119   See *GA* 1 731b4–5; cf. *DA* 2 413b1–4.

120   On the centrality of sensation in defining animals and understanding their lives see *DA* 2 413b1–4. "This [the soul] is the principle through which all living things have life, but the first characteristic of an animal is sensation (*aisthēsis*); for even those who do not move or change their place, but have sensation, we call living creatures

(*zōa*) and do not merely say that they live (*zēn*);" cf. *DA* 3 433b30–1. See also discussion in Chapter 1.

121 Recently, elaborating on Aristotle's notions of separability Whiting has argued for the presence of the "locomotive soul" as a part of the soul that integrates the functions of desire, sensation, imagination, and locomotion (2002). While I agree that in Aristotle these functions are integrated in a similar way as the senses are integrated into the common sense, or the parts of the body into a living body, I disagree with the hypostasis of a "locomotive soul," encompassing the sensitive soul. Indeed, for Aristotle the distinction of the soul into parts is functional to the comprehension of different forms of life, from plants to stationary animals to animals that move in space. Further, each part of the soul is responsible for a specific type of movement (*kinēsis*) that living beings possess as products of nature—growth and decay, alteration, and movement in space (cf. *PA* 1 641b4–9). Subsuming alteration under locomotion ultimately implies to overshadow Aristotle's articulation of animals' differences *qua* animals in terms of their psychological activities that involve different types of movement.

122 *DA* 2 414b28–415a14.

123 On the analogies and disanalogies between soul and figure, see Polansky, 2010, 195–6.

124 If we follow Aristotle's model it is not possible to agree with Menn that the serial succession of souls is framed by a hierarchical relationship and, likewise, that the nutritive soul is inferior to the other parts inasmuch as it serves them (2002, 113).

125 Leunissen, 2010, 59.

126 Needless to say, the dependency of "higher parts" of the soul from the nutritive part is a phenomenon that characterizes life in the sublunar world. For divine living beings only possess the rational soul.

127 On touch as the common sensation of all animals, see *DA* 2 413b5. Aristotle calls it "the most indispensable" (*anakaiotatē*) of the senses (*DA* 414a3–4). Touch is prior to the other senses not only because the other senses are forms of touch via a medium (*DA* 2 423b13–27, 3 435a18–9), but also because it is crucial to the animal's survival (*DA* 2 434b12–18). Cf. *DA* 3 435b3–5.

128 Arist. *Phys.* 5 226b35–227a4; see also 227a18; 6 231a23, b8.

129 Aristotle remarks on this feature in *On the Generation of Animals* in relation to the development of animals' embryos (see *GA* 5 779b11, see discussion in Section 3.2.3. cf. *HA* 7, 3; *Pol.* 7 1335b30, and *HA* 8 588b1). On this aspect, see Waerdt, 1987; Menn, 2002; Leunissen, 2010. For a discussion of the human fetus, see Preus (1990, 75–6, with additional bibliography); for the appropriation of movement by the new organism, its growth and the involvement of the sensitive soul in *On the Generation of Animals*, see Section 3.2.3; for the articulation of psychological parts with "life-stages" in the context of Aristotle's analysis of sensation via the analogy with knowledge in *DA* 2.5, see Section 4.2.

130 For this notion of *logos*, see Chapter 1 and on growth, Chapter 3.

131 Aristotle calls them *ta phtharta* (*DA* 2 415a8–10).

132 Indeed, while in the framework of *On the Soul* reason is disembodied in that the intellect (*nous*) is separate from the body and uninvolved in bodily affections as the faculty of nutrition and the senses are (*DA* 3 429a22–429b10), it still presupposes the works of the other parts of the soul to operate and actually exists. Indeed, an animal's dead body does not think.

133 On the consistency of Aristotle's indifference to *logos* as reason in his treatment of animals at large, humans included, see Lennox, 1999b.

134 The discussion of mind and human intellectual faculties occupies only *DA* 3, 4–7, while the core of the treatise is devoted to sensation and related issues (*DA* 2, 5–3, 3).

135 *DA* 3 431b13–17; 431b2, 431b20–432a14; for a discussion of *phantasia*, see
    Chapter 5. Sensation contributes to thought (*noein*) through *phantasia* (imagination)
    which—Aristotle states—does not happen without sensation (*DA* 3 427b16–7) and
    belongs, along with judgment (*hypolēpsis*), to thought (3 427b29–30). The question
    whether rationality is continuous with capacities present in other animals or detached
    has concerned Preus who contextualizes it in terms of biological continuities (embry-
    ological development, anatomical structure of various species, human and animal
    behavior) (1990, 74–84); on this question, see also Mingucci, 2015, especially 191–
    231 and 233–59.
136 *DA* 3 434b25–8.
137 *DA* 2 415a24. On the qualification of the nutritive soul as "the most natural," see
    Chapter 3.
138 On the fundamental nature of the nutritive soul, see also Menn, 2002, 106–7, Johnson,
    2005, 173.
139 *DA* 3 434b9–30, 435b19–26; cf. *Sens.* 436b10–437a3 where Aristotle further clarifies
    that those animals who live well on account of the distant senses are the intelligent
    ones (*phronimoi*) leaving some ambiguity as to whether he might refer to only human
    beings. But given that in *History of Animals* he does attribute *phronesis* to nonhuman
    animals in relation to their actions, this claim should be taken along with his general
    discussion of the finality of the senses in *DA* 3 12.
140 Unlike distant senses—sight, hearing, and smell—touch is indispensable to animals'
    survival inasmuch as it enables them to secure food and avoid tangibles that threaten
    the body (Polansky, 2010, 540).
141 Some animals have speech by which they communicate pain and pleasure and which
    enable them to live well (*DA* 2 420b23; 3 435b25–6; *Pol.* 1,1).
142 The locomotive soul is dependent on the possession of the full range of senses of the
    sensitive soul. Only in this way can an animal be aware of its surrounding from a
    distance, avoid what is harmful, and seek what is beneficial (*DA* 3 434a4–434b1); cf.
    the discussion of the finality of sensation in Chapter 4.
143 Polansky, 2010, 542.
144 Aristotle's doctrine of the human good is not straightforward. In *Nicomachean Ethics*
    he identifies it differently, with action according to virtue (book 1) or the active exer-
    cise of theoretical wisdom (book 10). In both cases, however, reason plays a crucial
    role. See Kraut who integrates the two versions of human good relating them to excel-
    lent rational activity (1991).
145 Even the theoretician, we may assume, must inhabit the polis since, as Kraut argues,
    "the philosophical life is the life of a good person" and requires some primary goods
    such as the virtues of justice, courage, and likes (1991, 6).
146 In order to experience sensation a living being has to have access to the sensible
    objects, see *DA* 2 417b25–6 where Aristotle addresses this condition for sensation in
    relation to man; cf. Chapter 4.
147 On "success" being relative to species, see Osborne (2007). Consistently with this
    view, in *History of Animals* Aristotle qualifies some species of birds as *eubiota*, "with
    a good life"; for their characteristics, see Chapter 6.
148 An illuminating passage to consider in this respect is *NE* 1 1097b29–1098a20. In
    it Aristotle contextualizes the anthropocentric perspective with the zoocentric one,
    which is at the core of his biological treatises, and presents the human being's natural
    end in the context of the natural ends of other living beings (which are common to
    man as well). "Must we not rather assume that, just as the eye, the hand, the foot and
    each of the various members of the body manifestly has a certain function of its own,
    so a human being also has a certain function over and above all the functions of his

particular members? What then precisely can this function be? The mere act of living (*to zēn*) appear to be shared (*koinon*) even by plants, whereas we are looking for the function peculiar (*idion*) to man; we must therefore set aside the vital activity of nutrition and growth (*threptikē kai auxētikē zōē*). Next will come some form of sentient life (*aisthētikē* [*zōē*]); but this too appear to be shared (*koinē*) by horses, oxen, and animals generally (*pan zōon*). There remains therefore what may be called the practical life of the rational part of man (*praktikē tis tou logoa ekhontos*)," transl. by H. Rackham.

3   Animals and nature

At the core of Aristotle's zoocentrism

3.1 Animality and the living body

In the discourse on method in *Parts of Animals* Aristotle denies that the shape of an animal's body and its structure could be enough to understand what an animal is. He claims that a dead human being (*ho tethneōs*) is not a human being (*anthrōpos*) and continues by saying that a hand made of bronze or wood is not a hand, and again that the eye or the hand of a corpse is not an eye or a hand. These remarks expound at once an ambiguity inherent in language and the inadequacy of an approach to animals based on mere morphology. Words referring to living beings and their bodily parts may signify either lifeless sculptures and still representations or realities that exist in nature and actually live. It is on the meaning of the second that Aristotle builds his study of animals and advocates the importance of understanding the living being as such (*toiouto to zōon*).[1] In other words, there is a *suchness* of animals, an animality, so to speak, that transcends bodily shapes and that only the inclusion of the soul can account for. The soul *may indeed be*— he claims in *Parts of Animals*—what makes the living being as such, namely its form (*eidos*). It is its form, however, not only by shaping it, as the form of a bed does with a bed, but also importantly by making it "operate" and live.[2] The formal role of the soul is here presented with a conditional statement that presupposes the more authoritative discussion of *On the Soul*,[3] which we will address in the course of this chapter. In *Parts of Animals* Aristotle merely supports this role with the empirical consideration that when the soul is gone, an animal is not an animal anymore, but a shapely, lifeless body. So were those living beings of myth like Niobe who had been transformed into stone or, we could add, the animals trapped under the morphological investigations of Democritus.[4]

There follows in *Parts of Animals*, a selection of the parts of the soul that are appropriate to the study of the living being as such—nutritive, sensitive, and locomotive—to the exclusion of the rational soul.[5] Balme finds it strange that Aristotle does not include the intellect in man's nature and that he does not consider it a part of the soul that makes "the animal as such."[6] But, as strange as it may appear, it is a programmatic decision whose terms need to be discussed not merely to justify it, but in order to elucidate and define the notion of "animality" Aristotle is pursuing in his project on animals.[7] In Chapter 1 we saw that because

DOI: 10.4324/9780367816001-4

thought (*dianoia*) and the other capacities such as reason (*logos*) and calculation (*logismos*) bestowed to humans by the rational soul were unrelated to physical, bodily movement, there could not be a place for these powers/activities in a study that aims at discussing animals in terms of their filiation from nature and their intrinsic power to move in whatever combinations they possess it.[8] It is now time to explore more in depth Aristotle's basis for excluding the rational power and related capacities and objects from his zoological project.

In *Parts of Animals* Aristotle gives us two interconnected reasons as to why he excludes the rational soul. One is rather general and has to do with issues of method; the other is theoretical and pertains to the relation of animals to nature (*physis*). Method-wise, were the study of nature to include the consideration of the human mind (*nous*) and its objects (*ta noēta*)—besides the other powers and operations of the soul it actually includes—the science of nature would exhaust everything that needs to be studied.[9] This inclusion would destroy, we may add, the systematization of knowledge that supports Aristotle's *corpus* where the study of nature is circumscribed to a set of related subjects approached with a suitable and consistent epistemological apparatus.[10] On the other hand—and this is the second reason for the exclusion of the rational soul—mind (*nous*) along with its objects (*noēta*) may well constitute the quintessence of human nature *vis-à-vis* that of other living beings, but in fact it does not properly reflect nature (*physis*) and its working. Indeed, considered in respect to nature and its phenomena, mind is qualified by a series of negations. While, in Balme's words, "nature" has to do with "the natures of things that come to be, that have matters and movers," mind is not involved with matter nor is it necessary to move the living being.[11] It is not a function of the body, with which it does not mix,[12] nor is there in the living body an organ (*organon*) for the mind.[13] Indeed mental activity (*noētikē energeia*)—we learn from *On the Generation of Animals*—is not a physical activity (*sōmatikē energeia*),[14] and instead of associating it to a form of body movement as other psychological activities,[15] Aristotle considers it a state.[16] In this respect, it is significant that in *On Sense and Sensible Objects* Aristotle lists a series of activities that belong to both soul and body (*koina tēs psykhēs onta kai tou sōmatos*). He includes sensation, memory, passion, desire, appetite, pain, and pleasure but omits thought, which according to the line of argument presented here cannot be common to both in that it does not involve the body. Indeed, even if the objects of mind (*noēta*) presuppose sensation and *phantasia* (imagination),[17] they are abstractions, namely forms and universals.[18] This status of the *noēta* appears evident if we consider *On the Soul* 2.4 where Aristotle programmatically separates thought and sensation by attributing them distinct objects (*noēta* and *aisthēta*).[19] Inasmuch as they are abstractions, forms and universals (*noēta*) are ultimately "products" of the rational soul itself,[20] without tangible external referents affecting the body. Distillations of reality, they are good for humans to think and to speak with, but they do not make up the physical environment that surrounds the living body (while in the womb and after birth), and that stimulates it via its objects (*aisthēta*) to perform the activities—nutrition, sensation, and locomotion—that for Aristotle enable it to live.[21] Thus treating the intelligibles along with mind in his study of animals would undercut the special, and ultimately definitory, connection that animals entertain with movement (*kinēsis*) and matter (namely their body *qua* exposed to the physicality of the world).[22] Hence, we see how the two reasons (methodological and "physical") for excluding the rational soul from the inquiry into animals do in fact intersect. Animals are studied in terms of body movement, which defines them all, and not mental state and activity, which only define the human animal qua human (not qua animal) and leads to an inquiry of its own.[23]

Ultimately it is on this basis that Aristotle marginalizes the rational soul in his study of animals. Animality does not relate to the human-specific, "disembodied," and supervenient abstractions of the mind, abstractions that are after all always self-referential.[24] Rather it relates to what causes the animal to live in relation to its environment and the things it perceives, feels, and performs from conception to birth to any moment of its embodied experience in this world. Conceived of in this way, animality is viscerally connected to the animal's living body as immersed in its space of existence, from the uterus of the mother or the shell of the egg to the habitat in the outside world—an approach that overall aligns Aristotle (with due qualifications) with the Presocratics' inquiry.[25] And it is by pursuing the psychological and at once physical (and further social and even political) relations that constitute such a mode of existence[26] that Aristotle carves out in his study of nature an area devoted to living beings and life pursuing what ultimately makes up, despite their interspecific differences, animals' common nature (*koinē physis*).[27]

3.2 Nature, bodies, movement, and life

If, as we saw in Chapter 2, in *On the Soul* Aristotle considers the soul what differentiates complex natural bodies from simple natural bodies (the first living, the second nonliving) and, as we have just seen in the section above, in *Parts of Animals* he invokes it as what makes the animals move in the different ways movement is accessible to them (*auxēsis, alloiōsis, kinēsis kata topon*), then to understand the animal as such (*toiouto to zōon*) requires understanding the role of the soul as an agent of movement and the specific ways in which living beings "move" *vis-à-vis* other natural beings, namely the elements and plants (which Aristotle considers "living things"). And it implies in turn to follow how Aristotle applies the conceptual tools he uses for the study of nature to explain animal life. Let me further clarify this point. When in *Physics* 2.1 Aristotle defines natural beings as having in themselves the origin of movement and rest, he gives us a list of such beings. Besides animals, plants, and their parts, he also includes simple bodies (*apla sōmata*), namely the elements—earth, fire, water, and air. In this discussion, he breaks down the movement of which natural bodies are originated into different types: growth and decay (*auxēxis/phthisis*), alteration (*alloiōsis*), and locomotion (*kinēsis kata topon*).[28] Yet here, as elsewhere in *Physics*, Aristotle is not concerned with a systematic analysis that relates types of movements to different types of natural beings (simple and complex bodies) and that takes into account their respective differences (although at times, as we will soon see, that concern surfaces).[29] Rather his interest lies in providing a theoretical, general discussion of movements and their causes opening the ground for more specialized studies in other works. Consequently, in *Physics* Aristotle sidesteps the soul as the cause of movement for complex natural bodies and tends to speak of such movements in terms of unidentified agency (*to kinoun*) or the apparatus of causes (final, formal, efficient, and material), without further psychological qualifications.[30] To know systematically about growth, decay, alteration, and locomotion as intrinsic movements of living beings, caused by the soul, we need to revert to *On the Soul*[31] while the actual movements involving animals' body parts and their functions, generation, lives, and activities in the external world are dealt with in a series of treatises conventionally called biological.[32]

This array of related treatments depends on Aristotle's overall systematization of the study of nature into a series of objects to which the soul does not strictly belong inasmuch as it is not material nor moving, but to which it nevertheless pertains on account of being the principle of movement of complex bodies, and in particular animals, besides plants. As for the study of nature, the prologue of *Meteorologics*[33] discloses an inquisitive path that goes from the general to the particular and, so to speak, what is close to us. Aristotle's discourse moves from the rather broad consideration of all motion and first causes in *Physics* to the focused, detailed analysis of the same (and additional) issues in a series of treatises that address specific natural contexts, phenomena, and bodies—the stars with their rotation (and the fifth element) in the superlunary world and the four elements with their mutual transformation in the adjacent sublunary world in *On the Heaven*; growth and decay, and other "operations" (also discussed at a level of generality) in *On Generation and Corruption*;[34] the "meteorological phenomena" involving simple natural bodies and including the formation of comets and shooting stars, rain, and winds down to earthquakes and other earthy phenomena in *Meteorologics*; and finally, animals and plants,[35] which represent the culmination of his study of nature. In the prologue of *Meteorologics*, Aristotle pairs them together and proposes to deal with them generally and separately. That is, generally, inasmuch as both animals and plants share some basic movements (i.e. *auxēsis* and related phenomena), and separately, inasmuch as, given their respective constitution, they each deserve a treatment of its own.[36] Indeed, although Aristotle does not speak in these terms at this point, we have seen in Chapter 1 that animals have a complex *logos* that is foreign to plants.[37] Animals' possession of sensation,[38] along with sexual differentiation[39] and capacity for locomotion (for those who possess it), implies a more articulated (and heart-centered) body and a more diversified range of movements, and hence, of living functions and life than those possessed by plants. Provided of merely the nutritive soul, plants only nourish themselves, grow, and decay.[40]

Now to study animals as creatures of nature requires Aristotle to reassess and refine his conception of natural movements. For in animals those *auxēsis, alloiōsis* and movement across space (*kinēsis kata topon*) which Aristotle discusses in rather general terms in *Physics* constitute modes of life (due to the soul) and are realized in a way that is specific to animals on account of their complex bodies, a fact which legitimates and corroborates their separate treatment in Aristotle's study and their ultimate definition *vis-à-vis* other natural beings.[41] Suffice here to remark that sensation, whose possession defines animals, is a *special* form of *alloiōsis*. Accordingly, this and the next chapters will discuss how natural movements are adjusted to suit and reflect animal capacities and life. Chapter 4 will examine the type of alteration (*alloiōsis*) that is sensation, Chapter 5 will address movement across space in relation to desire and *phantasia* (itself conceived as a movement internal to the body) while the present chapter will proceed to pursue the definition of growth (*auxēsis*) as a phenomenon of the living. Focusing on relevant passages this chapter will follow the shift the notion of *auxēsis* undergoes from physics to zoology and attempt to understand why Aristotle assigns to the nutritive soul a prominent position in his animal-oriented psychological theory. For even if he considers the sensitive soul the essence of animal life he still surprisingly presents the nutritive soul as the first (*prōtē*) and most common (*koinōtatē*) soul calling its function the most natural (*physikotaton*) among those possessed by living creatures. Why is it so? The next sections will help clarify, among other issues, such controversial statements further illuminating animals' relation to natural movements and its relevance for a comprehension of Aristotle's zoocentrism.

3.2.1 *From the coincidence of causes to the definition of growth*

In order to pursue how Aristotle redefines growth let us start by looking at how he conceives the role of the causes in nature and their different involvement in the movements of natural bodies in the general treatment of *Physics* itself. We will next consider Aristotle's technical definition of growth as a phenomenon of complex (hence living) bodies in *On Generation and Corruption* 1.5 and end with a discussion of *On the Soul* 2.4 where growth is an activity due to the nutritive soul and fundamental to understand life. In the methodological discourse of *Parts of Animals*, after discussing whether the final cause or the efficient one is prior, Aristotle notes that necessity (and the material cause) is not present in the same way in all the works of nature (*ta kata physin*). By analogy with the construction of a house, he acknowledges the importance of different materials in the formation of living beings and calls it "conditional necessity" (*ex hypotheseōs*). But he denies matter a formal and efficient role,[42] which he attributes instead to an intrinsic end (the final cause).[43] A passage in *Physics* elucidates in general terms and with the help of examples such diversified involvement of necessity and final cause "in things natural" indicating a radical difference between simple and complex natural bodies (namely, the elements and living beings):

> Why not say, it is asked, that nature acts as Zeus drops the rain, not to make the corn grow, but of necessity (*ex anankhē*) [for the rising vapour must be condensed into water by the cold, and must then descend, and incidentally, when this happens, the corn grow (*auxanein*)], just as, when a man loses his

corn on the threshing-floor, it did not rain on purpose (*ou toutou heneka*) to destroy the crop, but the result was merely incidental to the raining (*symbainein*)? So why should it not be a coincidence that the front teeth come up with an edge, suited to dividing the food, and the back ones flat and good for grinding it, without there being any design in the matter (*epei ou toutou heneka genesthai, alla sympesein*)?[44]

What concerns Aristotle here is to clarify and assess the impact of the final and material causes on phenomena of change that involve complex bodies—those of animals[45]— showing how complex bodies differ from simple bodies in this respect. In doing so, he is replying to the Presocratic philosophers, especially Democritus and Empedocles, who saw in matter—and therefore chance (*tykhē*)—the exclusive cause for the formation of the bodies of living beings.[46] Aristotle deprives matter of such a shaping role, at least in the case of living beings. He resorts to the example of a body of water that evaporates into air and of the new body of air that cools down and becomes water in the form of rain returning to the earth.[47] Water becomes air, and air becomes water again, and, in each case, the transformation of the element is accompanied by a change of location. Both "movements"—one substantial, the other locomotive—are unrelated to the goal of watering the crop to make it grow (*auxēsis*). Finality and the good do not apply to the cycle of water, which evaporates and becomes rain solely on account of material dynamics (under external conditions, namely a change of temperature in the environment); and it is, incidentally, remarkable that of the possible "ends" of the movement of water—whether the very subject of change (i.e. water) or "someone" else,[48] external to it—Aristotle refers to the "someone" else—the crop—thereby showing to exclude the applicability of the end and the good to inanimate beings.[49] The case of water triggers for Aristotle an immediate question, whether every body of nature "moves" like it. More specifically, what prevents the parts of the body of a living being, or its body as a whole, to form and grow in the same way—as connected to material dynamics, but actually disconnected from any specific goal informing such material dynamics? Aristotle asks this question about an animal whose front teeth grow sharp and suitable for biting, while the back teeth are broad and serviceable for chewing.[50] A set of teeth like this cannot form by chance under the shaping force of matter.[51] Rather it is formed this way by the design of nature, that is, to anticipate the foregoing discussion of *On the Soul* in this chapter, to enable the living being to perform complementary activities—biting and chewing—for the overall activity of nutrition, which preserves the animal qua being and makes it grow, and eventually enables it to reproduce itself, attaining "immortality."[52] Thus, in *Physics* itself Aristotle marks a distinction in nature between natural simple (soulless) and complex (ensouled) bodies and, accordingly, in the study of living beings he prioritizes the final cause subordinating to it the material cause.[53] As in the example of the teeth, complex bodies "move" in the sense of increasing in size and taking on a shape functional to their activities, while simple bodies merely move according to material constraints.[54] Finality directs the formation of living beings, and *telos* (end) is form (*eidos*), considered (this latter) as the factor that enables each living being (or thing) to operate and ultimately live.[55] Fundamental in differentiating the movement of ensouled beings from that of soulless beings, this nexus encompassing end/form/activity constitutes a main methodological pillar of Aristotle's discourse on animals and represents the phenomenon of growth conceived of as the touchstone of life.

Besides the teleological, form-bent development intrinsic to complex bodies, just discussed, in *Physics* we find another distinction within the conceptual framework for "things natural" which is appropriate to the study of animals. Aristotle claims the importance for the student of nature (*physikos*) of adopting the four causes, but he also alerts us to the way such causes are often related, remarking on their coincidence:

> Clearly, then, the "becauses" being such and so classified, it behoves the natural philosopher to understand all four, and to be able to indicate, in answer to the question "how and why," the material, the form, the moving force and the goal or purpose, so far as they come within the range of nature. Three of these becauses coincide: for the essential nature of a thing and the purpose for which it is produced are often identical (so that the final cause coincides with the formal) and moreover the efficient cause is the same in form to the effect (so that the efficient cause too must coincide with the formal; for instance, man is begotten by man).[56]

Without being completely explicit, in this passage Aristotle is not addressing the coincidence of final, formal, and efficient causes in reference to all natural bodies. Granted that, as he says, all four causes should be identified in the study of the phenomena of nature, the coincidence of causes presented here applies to the complex bodies of living beings (and things) alone. Aristotle points out that often in nature final and formal causes coincide, and, moreover, that the efficient cause too "is the same in form" to the "effect." The use of "often" restricts this coincidence of causes (final, formal, and efficient) to the generation of living beings understood in terms of an intergenerational phenomenon as illustrated by the example that "man is begotten by man."[57] Thus in the domain of natural complex bodies, that is, in the domain of animal life, formal causality is most crucial.[58] Indirectly, the causal attraction toward form is made understood in terms of transmission and reception, and hence of living beings' conception and growth.

Although in the two passages discussed above Aristotle introduces important distinctions relating to natural bodies' growth and causality, in *Physics auxēsis* remains without further qualification. It is one among the three movements (the others being *alloiōsis* and *kinēsis kata topon*) that characterize natural bodies, both simple and complex.[59] In *On Generation and Corruption*, however, Aristotle attunes physics to living beings and life[60] and redefines *auxēsis* identifying the phenomenon of growth that pertains to living beings as distinct from other natural phenomena that may be labeled as growth, but merely deal in fact with an increase in size of a given (simple) body. In this new theoretical formulation, growth catalyzes the influence of the final cause (and the good) in animal formation and the centrality of form versus matter[61] and in applying to living bodies' orderly increase in size it represents a core feature (to return to the beginning of this chapter) of the animal as such (*toiouto to zōon*).[62] So Aristotle writes,

> It appears that every part (*hotioun morion*) of that which grows has increased, and likewise in diminution every part (*hotioun morion*) has become smaller, and further, that growth occurs when something is added and diminution when something departs.[63]

Not a mere increase in size, as we would expect from his treatment in *Physics*, now *auxēsis* is initially determined by two conditions that define it *vis-à-vis* phenomena that look like growth but are not: 1) every part (*meros*) of the growing entity is growing and 2) growth is enabled by the addition of "something." Qualified as such, growth emerges "as a certain kind of natural phenomenon" that only applies to natural complex bodies.[64] Indeed it does not merely constitute the capacity of a given entity to become bigger, but it is an increase in size that involves a body growing in *every* part (hence a complex body)[65] and an "extraneous additional factor" that alludes to the intake of nourishment (although in *GC* Aristotle speaks in general terms without actually mentioning food or living beings). Without these two criteria (pervasive growth and additional factor) no increase in size can qualify as growth.[66]

There follows in *CG* 1.5 a negative example that confirms the conception of growth along the aforementioned lines and that, at the same time, firmly underscores its identity as a form of change distinct from other processes involving change such as generation and corruption on the one hand, and increase in size by alteration, on the other. Take a body of water that expands and becomes air—says Aristotle. This phenomenon does not pertain to growth, but to generation and corruption. For although in the evaporation of water the size of the body undergoing change has expanded, a new entity (air) has come into being, while water "has perished." To be considered growth the identity of the growing body must remain the same. Or, said in Aristotle's words, "that which grows is preserved and persists."[67] Thus, identity preservation during the process of change constitutes the third (and last) condition to assess the phenomenon of growth (besides the increase in size affecting every part of the body and the addition of something, mentioned earlier). Now, of the bodies that exist by nature only the complex bodies, i.e. animals (and plants), fulfill these three criteria and grow. By contrast, simple bodies, namely the elements, do not properly grow: when not changing identity (like water becoming air), they only increase in size. For neither do they expand in all their parts (they do not have parts) nor are they added something "from the outside."

Given that complex bodies include uniform and nonuniform parts, in *On Generation and Corruption* 1.5 Aristotle proceeds to discuss *auxēsis* in relation to both these mereological constituents, appealing to his theory of hylomorphism.[68] Growth, we learn, impacts the body first at the level of the uniform parts (i.e. flesh, bones, etc.).[69] It is their growth that makes in turn the nonuniform parts (such as hand, arm, head, etc.) grow. The arm, for instance, is a nonuniform part and is composed by uniform parts such as bone and flesh. When the arm grows, it does so on account of an increase of flesh and bone and the other uniform parts that compose it. Yet (and this is where hylomorphism comes into play), the process of growth is no simple addition of matter (*hylē*) but measured "formal assimilation." More specifically, the uniform parts along with "any other things that have form in matter" (such as nonuniform parts and the living being itself) have a double nature.[70] Flesh—says Aristotle—applies both to the matter and to the form of flesh. But flesh in the sense of matter is only potentially flesh.[71] What makes something actually flesh is form (in matter) and the ensuing capacity to perform the function of flesh. So, in Code's eloquent words,

> The material does not literally attach itself to the form of flesh. It is rather the case that matter that is added accedes to every part of the flesh spoken of in respect of the form. That is to say, the matter accedes to it in so far as it is actual, functioning flesh.[72]

Only in this way can a part of the body, or the body itself, preserve its nature and identity (thereby fulfilling one of the criteria for growth). To illustrate the formal nature of growth, Aristotle compares it to,

> That which happens when a man measures (*metroun*) water with the same measure (*metron*), for there is first one portion and then another in constant succession. It is in this way that the matter of the flesh grows; something flows out and something flows in, but there is not an addition made to every particle of it, but to every part of its figure (*skhēma*) and form (*eidos*).[73]

In this ambiguous passage, it is not clear what the analogy with "the man who measures with the same measure" exactly refers to, whether to the "growing" form or to the matter that is assimilated to form in the process of growth involving the bodies of living beings.[74] But whichever the exact reference is, the analogy doubtlessly stresses the proportion (*logos*), which characterizes the growth of animals in all their different parts and contributes to the proportion of size that characterizes their living body as a whole. On this involvement of *logos* more will be said in the next section which, turning to *On the Soul*, discusses the phenomenon of growth as due to the nutritive soul and partaken by all complex bodies, i.e. animals and plants alike.

3.2.2 Animal growth, nutrition, and the soul

While in defining *auxēsis* in *On Generation and Corruption* Aristotle does not mention the soul (and life),[75] we learn from *On the Soul* that the phenomenon of growth so conceived pertains to bodies that possess the nutritive soul and are living. The discussion carried out in this treatise complements that in *On Generation and Corruption*. For not only does Aristotle revert to the question of whether there may be some elements (i.e. simple bodies) that grow. He also casts the phenomenon of growth into the framework of living beings' life by clarifying that the "extraneous element" involved in the definition of growth is food and by discussing the double effect of nutrition on the living body (with growth being one). As to the simple bodies, Aristotle recognizes that among the elements fire could represent an exception and qualify as a natural body that grows (and therefore lives). And if that were true it would contradict the exclusive attribution of growth to natural complex bodies (and hence, implicitly, living beings and things) made in *On Generation and Corruption*. He writes,

> To some the nature of fire seems by itself to be the cause of nutrition and growth; for it alone of all bodies and elements seems to be nourished and grow of itself. Hence one might suppose that it is the operating principle in both plants and animals. It is in a sense a contributory cause, but not absolutely the cause, which is much more properly the soul; for the growth of fire is without limit, as long as there is something to be burned, but of all things naturally composed there is a limit (*peras*) and proportion (*logos*) of size and growth; this is due to the soul, not to fire, and to the essential formula (*logos*), not to matter (with a slight modification).[76]

Aristotle brings fire up with the intention of underscoring the role of the soul as the cause of nutrition and growth against the *endoxa*, which assigned this role to fire, instead. It is difficult to identify the *physiologoi* that Aristotle might have had on mind here, probably they included Heraclitus and Democritus, with the addition perhaps of Hippasus.[77] Aristotle recognizes the importance of fire and heat as essential for the functions of nutrition and growth[78] (as he did elsewhere in his corpus for other processes associated with life)[79] but denies fire the role of principle.[80] For fire keeps growing as long as it has access to "food" (i.e. fuel) and regardless of how big (or small) its body is it would still be fire. In its case size does not matter nor does, we may add, body mereology and articulation (since fire is a simple body).[81] By contrast, all beings that are "put together" (*synistasthai*) by nature—namely animals and plants—have a limit (*peras*) and proportion (*logos*) of their size and growth.[82] So while a forest fire expands wildly, a goat grows "having goat limbs and stature" and an oak tree "expands" producing "oak leaves and branches having a characteristic configuration and size."[83] Living beings (and things) grow until they reach their *telos*, which is *logos*,[84] the organic limit of which is for blooded animals set by the bones.[85] And when an animal (or plant) trespasses the limit intrinsic to its *telos* it misses the *logos* inherent in the fulfillment of its nature by destroying the *logos* (proportion) pertaining to its size and growth. If it grows too much it undergoes a substantial change (and like evaporating water, we may add, it ceases to exist) or, in case of an excessive shrinkage, it presents a defective condition.[86] But it is also equally "destroyed" (changing as a substance), if it grows too much in *some* parts of the body and the proportion regulating the composition of its parts (*symmetria*) is ultimately altered (for this selective growth would contradict one of the principles of growth outlined in *CG* 1.5 discussed earlier).[87] Thus, in the context of animals' (and plants') growth *logos* is mobilized in two distinct, complementary meanings that illustrate its pertinence to the living body and the animal as such. On the one hand, *logos* features in association with *peras* (limit) and applies to both the size of the animal and the relation subsisting among (and regulating) its parts, where this relation is crucial for the functionality of the body (and life). On the other, *logos* features in opposition to matter (*hylē*) and represents the form, that is, the living being's formal nature, which corresponds in turn to its (embodied) *telos*. As remarked in Chapter 1, however, animals' *telos* is more complex than plants', and their *logos*, so to speak, more articulated. For living beings possess living functions that living things do not, first and foremost sensation. And besides, they hold a defined, heart-centered articulation that is foreign to plants.[88]

It is on account of the soul that living beings' growth is contained and structurally bounded. For were internal fire the cause instead, living bodies and their parts would grow forever as long as they have access to food. This role of the soul in regulating body growth is an aspect of its overall causal enmeshment with the living body.[89] In *On the Soul* 2.4 (which is devoted to the nutritive soul and its subfaculties) Aristotle builds on his earlier definition of the "bare core" of life and the bodies in which it can be found[90] to claim that "nothing shrinks nor grows in nature, unless it is fed, and nothing is fed which does not share life"[91] and proceeds to clarify the coextensive relation of growth and nutrition to life, introducing the related phenomenon of reproduction.[92] Now because they grow, Aristotle recognizes, plants too are alive.[93] They grow (and conversely decay) not only up and down, but in every direction, and they continue to live as long as they are able to absorb food.[94] In this *ad hoc* application of the criteria determining growth, the intake of food (*trophē*) is key to explaining growth and self-preservation, and foreshadows the upcoming distinction between growth and properly nutrition as distinct effects of food assimilation. Both growth and nutrition, Aristotle tells us toward the end of this chapter, are capacities of "that which lives" through the possession of the nutritive soul, and each has its own effect: growth makes that which lives grow until it reaches its *telos*, nutrition preserves it as a substance.[95] In absorbing food plants both preserve themselves as plants and grow, changing in respect to size. And so it is for all the animals, but with a crucial difference. While nutrition preserves them as living beings and makes them increase in size,[96] the nutritive soul is not found alone (as in plants) but works, so to speak, in synergy with the sensitive soul. It is at this point that Aristotle's discussion of soul division (whether *simpliciter* or in account) becomes particularly relevant. For the nutritive soul is still the cause of animals' growth but because animals are sentient beings and are defined by the possession of the sensitive soul, their nutritive and sensitive souls are separated only in account.[97] In the framework of the geometrical model of the soul, the sensitive soul includes the nutritive soul, we have seen. The implication of this psychological makeup is that when an animal takes in nourishment, it "senses" with (at least) the sense of touch and that is liable to the feelings of pleasure and pain. Likewise, the process leading to reproduction too (i.e. animal mating) has to do with sensation, pleasure, and pain, and radically differs from plants' capacity to reproduce senselessly as single entities.[98]

3.2.3 Growth, movement, and the origin of animals' life

A touchstone for life, growth is inextricably connected with life's early stages and represents the first mode of existence for all living beings, humans included. Aristotle investigates this aspect in *On the Generation of Animals*, which complements the abstract discussions of growth in *On Generation and Corruption* and its relevance to living beings' life in connection with nutrition as capacities of the nutritive soul in *On the Soul*. In *On the Generation of Animals* Aristotle discusses animals' sexual reproduction and explains how a new organism is formed and develops pinpointing the appropriation of movement which marks the origin of the new life in the encounter that occurs inside the female's womb between the male's semen (concocted residual blood, originating from ingested food) with the material (unconcocted residual blood, also originating from ingested food) supplied by the female itself.[99] The semen imparts movement to the inert material by transmitting to it the sensitive soul and henceforth makes it "come together" (*synistamai*).[100] In *Parts of Animals* this verb appears in the perfect participial form to qualify animals as products of nature that have *logos* (in the sense of *telos*) and ultimately refers to their body structure which, among natural bodies, is exclusive to them enabling them to move and live. With such a semantic field of reference, this verb lies at the core of the notion of the "animal as such." So it is significant (and not surprising) that Aristotle reverts to it in his treatment of animal generation too to indicate the formal articulation of the fetus from the beginning of the movement, imparted by the male, until it reaches completion, i.e. when the sexual differentiation into male or female with the formation of the respective relevant organs has taken place.[101] So, when at first the semen "sets" the material inside the female's womb (*synistēmi*), the bulky portion collects together and the liquid is separated, and due to the heat conveyed by the semen, membranes form around the incipient bodily compound.[102]

At the embryological level, no part of the body becomes formed without the sensitive soul—neither face nor hand nor flesh.[103] Indeed for Aristotle the female's fetation (*kyēma*) only possesses the nutritive soul[104] and, as such, is unable to reproduce an animal by itself. It ultimately lacks "the movement" toward form,[105] which is necessary to animals' generation and which it receives from the male's semen when it conveys to it the sensitive soul.[106] By contrast, and incidentally, plants are formed and take shape by virtue of the nutritive soul alone, which in Aristotle's view suffices to produce the simple articulation of body structure that characterizes them as opposed to animals.[107]

Aristotle illustrates the mechanism by which the sensitive soul triggers the articulation of the embryo through the analogy with automata, those automatic puppets which, once moved by an external agent, are able to move by themselves. Only after the female's fetation receives the movement imparted to it from the male's semen, it becomes the origin of its own movement and hence an autonomous living entity like "a son that has set up a home of his own independently of his father."[108] The movement's *arkhē* is now inside the fetation and results in the systematic, progressive, and measured articulation of body form we have been discussing in this chapter. The first part which develops is the heart. Unique and ambivalent (differently from the other organs, it is both a uniform and nonuniform part),[109] the heart becomes the "organic" origin of the living body's movement and triggers its process of organization (*diakosmēsis*) from the inside[110] and from upward downward.[111] Under the heart-imparted movement, the fetus behaves like a seed sown in the ground: it grows a shoot (the umbilicus) through which it takes in the nourishment (blood) from its "environment"[112] as the root sprung from the seed takes it from the earth.[113] In addition to the heart and umbilicus, also bones and brain[114] are formed during the first stage of animals' constitution (*prōtē systasis*).[115] The lungs, instead, become articulated later on right before the animal starts breathing, an activity which marks the completion (*telos*) of this part of the body.[116] At any rate, starting in the womb (or for the oviparous animals in the egg) and expanding along the acquired *oikeia morphē* (the form proper to each living being),[117] growth will continue to affect the animal born in this world until it reaches the size and proportion inscribed in its *telos*.

No living being is indeed born perfect in size, notes Aristotle.[118] Growth, however, involves animals' bodies differently, both in terms of what parts of the body grow and in relation to time. For instance, in *Parts of Animals* Aristotle remarks that the quadrupeds are born with the lower portion of the body bigger than the upper portion and that as they grow into "adulthood" their upper portion increases in size more than the lower one. Humans, by contrast, are born dwarfs (*nanoi*), and only with time their lower portion, which goes from the anus to the feet, becomes proportionate to the upper one.[119] Besides, in the case of human beings, growth is particularly intensified in the first five years of life during which they grow half of the size they will grow during their entire existence, a biological feature that explains the lack of semen's production in children and that Aristotle takes seriously into consideration when discussing the proper education of citizens from an early age in *Politics*.[120] Inversely related to reproduction, growth stops when the animal is ready to reproduce itself.[121] At that point, nutrition does not make the living being grow anymore, but both preserves and enables it to reproduce by providing male and female of each species with their respective residues (concocted blood/semen and unconcocted blood/menstrual stuff),[122] which are essential for conception.[123]

As for the relation between the nutritive and sensitive soul in the early process of animal formation Aristotle remains vague. For instance, he does not give us a precise temporal framework for the "actuality" of the sensitive soul in respect to the nutritive soul. To put it in questions, when exactly does the (plant-like) absorption of food become a sentient activity accompanied by the sense of touch, pain, and pleasure? To what extent does the physical environment of the mother and the sense organs' incipient formation allow the embryo to sense? Aristotle merely says that "It is while they develop [i.e. the fetations] that they acquire the sentient soul as well, in virtue of which an animal is an animal."[124] Nor does he offer us a systematic explanation of how the nutritive and sensitive souls work together in the formation of the animal. Still he may have not found it necessary to clarify it. The two souls constitute the living being's soul, can only be separated in account, and are entangled with the "organ-based" functions of the body *qua* receptive to its living environment.[125] Overall, it is clear that Aristotle considers the animal soul in relation to the body and in terms of a passage from potentiality to actuality in line with an organic development that unfolds through different stages. The nutritive soul is present in the female's fetation and the male's semen only in potentiality and becomes "actualized" before the sensitive soul[126] at the time in which the fetus starts absorbing food. It is the male's semen that confers the sensitive soul and "activates" the female's material, marking its separation from the body of the mother and the beginning of a self-moving living organism. On the other hand, the actuality of the sensitive soul marks, at least according to *GA*, the beginning of an animal properly as a living being that (actually) feels.[127] Only subsequently does the living being become differentiated into a specific animal whether horse or human being.[128]

On this view, animals' common nature (*koinē physis*), the physical and psychological continuities among different forms of living beings that are at the core of Aristotle's project are grounded in a shared beginning. That is (except for spontaneous generation), all animals partake of the same mechanism for generation via sexual reproduction and of an equal mode of existence that will subsist in the organism along with further formal body articulations and the psychological capacities such articulations entail. In this scheme, by possessing in actuality only the nutritive soul (and an undifferentiated body), for some time immediately after conception inside the female's womb all blooded animals[129] live the life of a plant (*zēn phytōn bion*):[130] they live by nourishing themselves and grow, without sensing and taking on a more complex shape. Consistently, similarly to plants,[131] at this plant-like stage, animals possess a body that potentially has both male and female parts, but whose sex is not determined yet—the embryo being "complete" (and we may add, ready to be born) only when it is formed in respect to the sex (and is potentially able to move).[132] In the course of the embryological development, every new organism starts sensing (the living thing has become a living being), although, as mentioned earlier, in *GA* we do not find clear indications of when exactly the individual senses (and particularly touch)[133] start operating. Regarding the distant senses, those animals that are born with eyes shut will be able to see only once their eyelids become separated since psychological activities become effective capacities when the parts of the body, which are their instruments, become fully formed.[134] At any rate, once born, some animals will sense only with touch, others like dogs and humans with all the (five) senses, each according to its own nature.[135] But despite the differences in their sensorial make-up, they will all live by sharing at least the sense of touch, which is involved in the process of nutrition and, as noted earlier, is accompanied by pain, pleasure, and desire.[136]

3.3 Nutrition, reproduction, and the desire for immortality

In *On the Soul* the nutritive soul (*threptikē* or *threptikon*) holds functions that in current tripartitions and bipartitions were judged by the measure of morality. For instance, in Plato's tripartition of the soul these functions were pinned down to the desire for food, drink, and sex, belonged to the irrational part of the soul, and were considered an impediment to the exercise of reason and the practice of virtue. On the other hand, Aristotle recognized the moral relevance of the pleasures related to the activities of nutrition, i.e. eating, drinking, and sex,[137] even if when he first introduced his model of soul bipartition he considered the nutritive soul irrelevant to a moral discourse.[138] In *On the Soul*, by contrast, the nutritive soul is upfront taken to be as fundamental to life, and as such integrated with the other living functions bestowed by the (composite) soul, sustaining them. Aristotle attributes to the nutritive soul nutrition, growth, and reproduction.[139] Able to subsist alone, no other parts of the soul can exist without it. All living beings partake of the nutritive soul, animals and plants alike, and to emphasize its fundamental role Aristotle's discourse moves from superlative to superlative:

> First then we must speak of food and reproduction; for the nutritive soul belongs to all other living creatures besides man, and it is the first (*protē*) and most widely shared faculty (*koinotatē dynamis*) of the soul, in virtue of which they all have life. Its functions are reproduction and assimilation of food. For this is the most natural of all functions (*physikōtaton tōn ergōn*) among living creatures, provided that they are perfect and not maimed, and do not have spontaneous generation: *viz.*, to reproduce one's kind, an animal producing an animal, and a plant a plant, in order that they may have a share in the immortal (*to aei*) and in the divine (*to theion*) in the only way they can; for every creature strives (*oregesthai*) for this, and for the sake of this performs all its natural functions. "That for the sake of which" (*to d' hou heneka*) has two meanings: (1) that for the purpose of which, and (2) that for the benefit of which. Since, then, they cannot share in the immortal and divine by continuity of existence (*synekheia*) because no perishable thing can remain numerically one and the same, they share in these in the only way they can, some to a greater and some to a lesser extent; what persists is not the individual itself, but something in its image, not identical as a unit, but identical in form.[140]

The nutritive soul is the first (*protē*) and most widely shared faculty (*koinotatē dynamis*). Not only, as we saw earlier, through nutrition every living being/thing starts living autonomously, but it also preserves itself *qua* living and grows.[141] Further, the nutritive soul enables the living being to perform the most natural of all functions (*physikōtaton tōn ergōn*), namely reproduction, on whose greatest degree of naturalness more will be said soon.[142] In this passage Aristotle does not provide an explanation for the association between nutrition and reproduction since he intends to present the power of the nutritive soul in a general way that may apply to all living beings—too many details on the two processes and how they relate would have diverted from it. Besides, he may have written a treatise dedicated only to the function/operation of nutrition, which has not survived down to us.[143] But from *On the Generation of Animals*, discussed in the previous section, we have learned that the semen through which animals reproduce themselves derives from the last residue of concocted nourishment.[144] And in another passage of the same treatise Aristotle argues by deduction that nutrition and reproduction must be operations of the same psychological power because ultimately they work with the same matter (*hylē*) whether this is food or the more refined material from which "the natural object being formed" (*to physei gignomenon*) derives (the semen).[145] In this way, the nutritive function whereby the living organism processes food, preserves itself, and grows,[146] exists for the sake of reproduction,[147] which Aristotle defines as a living being's "production of another such as itself" (*to poiēsai heteron hoion auto*). A plant produces a plant, and a human being another human being, and so with the other living beings. Each one reproduces its kind and significantly not for the perpetuation of the species, as a Darwinian interpretation would have it, but on account of the individual's desire to live forever.

We have confronted the regularity of the process of generation in Chapter 1 when discussing the methodological boundaries Aristotle set forth in *Parts of Animals* to study animals. There genetic correspondence (i.e. a plant reproducing another plant, a human being another human being) spoke against the dominion of chance (*tykhē*) in animals' formation and in favor of an intrinsic, regular form, guiding their physical change toward a complete functional shape. In *On the Soul*, however, the emphasis is different, and the scale of the discussion somehow grander. Here the regularity informing animals' formation is linked with immortality and divinity. It corroborates retroactively—explaining it—the view according to which our philosopher assigned a privileged ontological status to the nutritive soul within the geometrical model he has devised to understand life in its multiplicity.[148] Indeed, not only does the nutritive soul enable basic survival, preservation, and growth, it also enables reproduction. Hence it provides the conditions for the existence of the other types of soul, in both "shorter" (i.e. during one's life) and potentially indefinitely "longer" terms (i.e. through reproduction). In producing another like itself, each living being fulfills its desire to live forever and partakes of the immortal (*to aei*) and the divine (*to theion*) to the extent that it is allowed to.[149] So while it is impossible for a living being/thing which is endowed with a perishable body to become immortal, it has at least the possibility to perpetuate itself through its progeny.

As a number of scholars have remarked, in referring to a universal desire for immortality and presenting reproduction as what allows living beings to attain it Aristotle is echoing Plato's *Symposium* where Diotima explains to a diligent Socrates the extensive power of love.[150] There are indeed some significant linguistic and philosophical overlapping between *On the Soul* and the Platonic text.[151] More however can be done with this intertextuality so as to bring to full focus Aristotle's position on reproduction and the nutritive soul in the context of his study of animals and to illuminate, in particular, his claim that for living beings "to produce another like itself" is the most natural (*physikotaton*) operation with the important theoretical consequences this assertion entails.[152] In Plato's dialogue Diotima does not only link the desire for procreation to that for immortality, making it a feature that involves humans (*anthrōpoi*) and animals (*thēria*) alike.[153] Significantly for our reading of Aristotle, she also questions the cause (*aitia*) for this universal drive relating it to the cognitive differences between humans and animals. After associating reproduction with the care for the offspring, she asks Socrates:

> As for humans (*anthrōpoi*), one might suppose that they do these things on the promptings of reason (*ek logismou*); but what is the cause (*tis aitia*) of this amorous disposition in the animals (*thēria*)?[154]

Socrates remains silent, and it is Diotima herself to provide the answer pinpointing it in "mortal nature" (*thnētē physis*). Inasmuch as they are mortal beings animals and humans possess the same nature. It is this shared possession—not reason (*logismos*)—which makes them equally search for immortality (*athanasia*) and perpetuate themselves. The only way this can happen—we heard it already from Aristotle—is through generation by leaving behind "another new being such as oneself" (*heteron neon hoion auto*). No wonder every "mortal being" (*to pan*) values naturally (*physei*) its offshoot (*apoblastēma*)![155]—concludes Diotima. Literally, "offshoot" signifies everything that grows out of something else, and in Greek literature it will become a term properly used for plants.[156] By adopting this word the wise woman reduces the variety of the phenomena of reproduction across the spectrum of living beings—humans and animals—to the same physical core.

Now, this Platonic subtext enables us to understand better Aristotle's qualification of reproduction as the most natural (*physikotaton*) among living beings' operations and, in turn, to gauge how this take on reproduction impacts his conception of animals and their activities, corroborating the zoocentric perspective we brought up in the preceding chapters. In scholarship there is a tendency to downplay Aristotle's qualification. For instance, Polansky argues that the superlative *physikotaton* does not introduce a comparison with other psychological operations of the sensitive or rational power, operations which are therefore not less natural than reproduction. On his view, reproduction is merely a most natural operation among animals. He further cites Lennox who supports this point by remarking that in claiming for reproduction the highest degree of naturalness Aristotle does not exclude that other operations may be equally supremely natural. *History of Animals*, for instance, features a number of organs that are referred to as being most important for animals.[157] On the other hand, Johansen switches the object of reference for "most natural" from reproduction to the nutritive soul itself taking it to mean that "the nutritive soul is a paradigm example for all the different kinds of soul of how the soul works as a final, formal and efficient cause of the body."[158] In other words, it is the extendibility of the causal role of the nutritive soul that makes Aristotle consider reproduction as the most natural operation. Yet, against these interpretations, one should note that our philosopher is adamant in singling out reproduction as the most natural operation and that he supports this claim by laying out different degrees of naturalness for living beings' operations basing them on the notion of finality: living beings undertake *all* their natural operations for the sake of reproduction,[159] which, in being the end (*telos*), is indeed the most natural.

So how is it that for Aristotle among the other faculties of the soul reproduction is natural to the highest degree, particularly, as it has been noted, in respect to sense perception, which defines all animals, and in respect to reason, which provides the "higher life"?[160] Here I follow Diotima's lead and propose to take *physikotaton* in light of the contrast, pointed out earlier, between *logismos*, calculation, as a faculty that belongs to humans alone, and the mortal nature (*thnetē physis*), which all living beings partake of *qua* living. For Diotima reproduction does not depend on reasoning, but on mortal nature. Along similar lines, in *Politics* 1 Aristotle calls the desire to procreate natural (*physikon*) opposing it to purposeful choice (*proairesis*),[161] while in the methodological discourse on animals, discussed earlier, he underlines the irrelevance of the rational soul to nature (*physis*) claiming that along with its operations it has "nothing" to do with it. The rational soul, let us recall it, does not involve the body, matter, and movement and, so conceived of, is not a proper object of the study of "the animal as such." So to claim that reproduction is the most natural of animals' operations for Aristotle must mean that it is disconnected from the rational soul and that it pertains to all living beings (and even things, i.e. plants) conceived as self-moving, and hence living bodies. But it also implies that inasmuch as reproduction leads to the production of another living being/thing like oneself, it involves the subordination of matter and movement to a finality that reflects *par excellence* nature's "artistic work," its masterly design. Considered from this angle, reproduction does not merely indicate the capacity to reproduce *tout court*. Rather, as both Aristotle and Plato (via Diotima) before him emphasize, it indicates the production of a living being "like oneself," that in terms of Aristotle's outlook on animals means a "finished" living being, composed of the same constituent parts and functions, and hence apt to move, live, and reproduce.[162] In sum, it is the realization of a fully articulate living body, accomplishing its *telos* intended as form, that makes reproduction the most natural (*physikotaton*) of living activities.

Significantly, in the passage quoted above Aristotle clarifies the notion of finality at stake in the operation of reproduction *vis-à-vis* the other operations of the soul. As we saw, he claims that for all living beings (and things) (*panta*) reproduction fulfills the desire for immortality and divinity, which is intrinsic in them all *qua* living and that it is for its sake that they undertake the other natural activities.[163] But he also further acknowledges that "that for the sake of which" (*to d' hou heneka*) has two meanings: (m1) that for the purpose of which, and (m2) that for the benefit of which. In other words, finality can be considered as the goal of an action or for the sake of somebody, as an aim or for a beneficiary.[164] Not merely a footnote, as it has been claimed,[165] this distinction is in fact a cornerstone of Aristotle's teleology and of key importance for understanding his philosophical position on animals and every form of life. It reappears soon after in Chapter 4 of *On the Soul*, but in the general context of the soul—and not the psychological faculty of reproduction—and recurs multiple times in his work.[166] In its application to the faculty of reproduction the distinction between the two meanings of "that for the sake of which" (*to d' hou heneka*) gives theoretical ground to the earlier proposition that every living being engages in how many natural activities it does for the sake of immortality and divinity, resolving any ambiguity as to the purpose of why it does so. More to the point, the distinction between (m1) and (m2) highlights two fundamental aspects of the finality of reproduction: on the one hand, it aims at immortality and, on the other, it is for the sake of the living being who strives for immortality, the subject of life itself. As Johnson argues in the wake of the ancient commentators, both meanings are operative in the finality of nature that pertains to reproduction. A living being operates in this world in order to be living and do so forever; and it does that for itself, not for the species. The end of living beings' various activities (i.e. reproduction) is projected back to the agents of such activities and hence to the very subjects of life, and teleology—as was said before—is immanent.[167] Aristotle's clarification of the expression "that for the sake of which" (*to d'hou heneka*) positions animals' finality in the context to which they belong, that of nature (*physis*) and its working conceived of in relation with, and as intrinsic to, each living being. The array of movements which originate from animals constituting their operations (*auxēsis, alloiōsis, kinēsis kata topon*) and culminating in reproduction are referred back to their "origin," that is, the living being itself, and this interpretive move reflects the proceedings of nature and the *logos* informing its "works." Inasmuch as they are creatures of nature and, as such, composed so as to be the origin of their own movement and rest, all animals exist and operate for the sake of their own lives (and ultimately to live forever) and to benefit themselves.

The identity between the subject of living activities and beneficiary Aristotle attributes to living beings in *On the Soul* seems to also support his discussion in *History of Animals* where animals' actions and lives revolve around nutrition and reproduction (and hence self-preservation and "immortality").[168] This angle stands in open contrast with *Politics*, which centers on humans' supreme good, i.e. life in a political community. In this treatise, Aristotle uses an expression analogous to "that for the sake of which" (*to d'hou heneka*), namely, *to d'hou kharin*, to mean "for the benefit of somebody," but he does not apply it to the very subject of its own life (as it happens in *On the Soul*). For in *Politics* the beneficiary of living beings' (and things') lives is always somebody else in a chain of displaced beneficiaries that culminates in the human being, the ultimate beneficiary of all.[169] In this way, by reducing the meaning of "that for the sake of which" to the beneficiary as one who is external to the subject of its own life Aristotle displaces the end (*telos*) from the living being, to which it belongs as a creature of nature, to situate it outside the living being itself. With this maneuver he suppresses, or at least ignores, living beings' intrinsic individual ends and elects humans to be the end of all other forms of life, which are thereby dispossessed of their own relation to life and presented as merely instrumental to man.[170] In this discourse plants exist for the sake of animals, which feed on them, and animals, both domesticated and wild, for the sake of humans. From animals humans derive food, help, clothes, and other supplies, and so they do also, we may add, from plants. Based on the separation of the "subject of its own life" from "its beneficiary," this vision imposes on all forms of life other than human a teleology external to them and is fundamentally anthropocentric. In *Politics* Aristotle still calls it by nature (*physei*), but from the one-sided perspective that privileges the human being and his good and that ignores, violating it, the teleology immanent in each living being which has in itself the origin of movement and rest. It is the task of his zoological investigation as part of the overall study of nature (*physis*) to pursue this angle and capture such nature: without exception, for each living being life unfolds as the manifestation of its intrinsic, everlasting effort of being alive and continuing to live.[171]

Notes

1 *PA* 1 641a15–8; 641a22–5. Cf. Chapter 1.
2 Aristotle insists on the importance of the *suchness* of living beings earlier in *Parts of Animals* too, when referring to the methodological dilemma of whether to discuss animals' process of formation (*genesis*) or their essence (*ousia*), namely the form (*eidos*), which is achieved at the end of the process of formation. He chooses, we have seen in Chapter 1, the latter since form is the end (*telos*) of the process of formation and directs it (*PA* 1 640a10–640b4).
3 *DA*. 2, 1; see Balme, 1992, 88 and Lennox, 2001, 139.
4 See, respectively, *PA* 1 641a19–21 and 640b30–4, and the discussion in Chapter 1 of the methodological boundary pertaining to the forms of the body parts and body which should be discussed in relation to their activities, hence in movement and in the context of animal life.
5 *PA* 1 641b9; cf. Chapter 1.
6 Balme, 1992, 89.
7 For a definition of this notion, see Introduction note 17.
8 All animals originate at least growth, decay, and sensation (touch), and some others, in addition to these movements, also locomotion in conjunction with the distant senses.
9 *PA* 1 641a33–641b4; see Balme, 1992, 89.
10 See Chapter 1 and discussion below.
11 Aristotle proves the noninvolvement of mind in animals' movement by saying that the nonhuman animals are able to move in space despite their lack of intellect (*PA* 1 641b7). This is, in fact, a remarkable observation because it shows that Aristotle is interested in a "physical" explanation of animals' locomotion regardless of the exclusively human, reason-based motivation that leads the human being to undertake movement across space. In *On the Soul* he attributes the power of locomotion to desire (*orexis*) and the desirous soul (*orektikē*) (*DA* 3, 10). On types of motivations in Aristotle's psychology, see Lorenz, 2006.
12 Mind is both unmixed and separated from the body (*DA* 3 629a18–629b6) and such a noninvolvement with the body implies also a different type of separation of the rational soul from the other parts of the soul (see Polansky, 2010, 169–70; Corcilius and Gregorich, 2010). For a recent assessment of the relationship of mind to the body and the support of a physiological perspective, see Mingucci who distinguishes between "unmixed" and "separated" and argues for a scale of degrees of mind's separability from the body (2015, 157–259).
13 Aristot. *DA* 3 429a24–5; 429b5–6. On this point, see again Corcilius and Gregoric who claim that "the capacity for theorizing" (*nous*) is separable from the body and has independent existence, a qualification that makes it, in Aristotle's words, a "different kind of soul" (*psykhēs genos heteron*, *DA* 2 413b26) (2010, 90–1).
14 *GA* 2 736b27–29.
15 *Sens.* 436a8–11. Now it is true that *On the Soul* (1 408b1–15) discusses thinking (*dianoeisthai*) along with feeling joy, grief, and anger as movement (*kinēseis*), but this

association is brought up in Aristotle's critical review of *endoxa* and is referred to as what "appears" (*dokein*).

16 *DA* 1 407a33–4.

17 *DA* 3 427b29–30; see Chapter 5.

18 Aristot. *DA* 3 429a27–9.

19 Aristotle stresses that to discuss a function of the soul, one should first discuss its activity, and before it its object (*DA* 2 415a14–24).

20 The process of thinking is "not being acted upon" (*apathes*) (as it is, by contrast, the case of "feeling," *to aisthanesthai*), but "being receptive to form" (*dektikon tou eidous*), see *DA* 3 429a10–24.

21 In Aristotle's discussion of the soul a fundamental methodological principle is that of studying the objects of the psychological activities before functions and activities themselves (see *DA* 2 415a14–23 and below Chapter 6). Significantly, in *Metaphysics* Aristotle underlines that the knowledge (*epistēmē*) of universals is the most removed from the senses (*porrotatō tōn aisthēseōn, Met.* 1 962a20). There exists indeed a fundamental difference between the sensible objects (*aisthēta*) and the intellectual ones (*noēta*). While the objects of perception are embodied in matter (*De Cael.* 278a11) and external to the sentient subject, the objects of intellect are "internal" (cf. Everson, 1995, 268).

22 In this respect, it is interesting to note that in *Metaphysics* Aristotle defines physics as a theoretical science which deals with what "can be moved and the substance—namely the soul—as inseparable from matter" (*Met.* 6 1025b25).

23 If we approach Aristotle's exclusion of the intellect and mind from the study of nature by considering the ultimate reason of that exclusion, namely the pursuit of the animal as such (*toiouto to zōon*) (and not merely the human animal along with its sublime intellect), there is no ground, I believe, to challenge Aristotle's choice as Frey has recently done to reintegrate the intellect within the domain of nature (2018, 160–74). Aristotle does recognize the dependency of the intellect on the body and does ascribe it to human nature and soul as realization of such nature, but he excludes it not only because it is idiosyncratic of the human animal and nature, and hence, ultimately irrelevant to pursue what is common (and body-centered) among animals, but also because it is derivative and works with distillations of reality, internal to the human being itself, ultimately disconnecting the human being from nature and its embodiments. For in the end, the enmeshment with matter and movement, which grants a given object a "membership" to nature, does not pertain to the objects of the intellect and its actual working. Significantly, in his animal-bound discussion of life in *On the Soul* Aristotle devotes a small section to the rational soul and the activity of mind (3, 4–7) *vis-à-vis* the embodied activities of the sensitive soul while he pursues the role of *dianoia* (thought) in humans' moral life in the ethical treatises.

24 The extraneousness of reason (*nous*) to the body is featured in *On the Generation of Animals* where reason alone is claimed "to enter the body from outside, as an additional factor, and it alone is divine, because physical activity (*sōmatikē energeia*) has nothing to do with it" (2 736b27–9, transl. by A.L. Peck).

25 For the Presocratics' approach to the study of living beings and life as centering on the body and its interaction with the environment, see Zatta, 2019.

26 *PA* 1 641b1–10; cf. *DA* 1 403a4–29 where Aristotle discusses whether the affections (*pathē*) of the soul are all shared by the body or there may be some that are peculiar to the soul. He claims that most or probably all affections involve the body—anger is, for instance, a surging of the blood and heat around the heart—and that the natural philosopher should inquire into the soul, either generally, or at least in connection with the "movement of the body" (*kinēsis sōmatos*) for which the soul is responsible.

27 See Section 1.1.

28 See Chapter 1, pp. 24–5.

29 See, for instance, below, the discussion of the involvement of different causes in the formation of natural (simple and complex) bodies. Another difference pertains to the agency of the movement. For unlike complex bodies, simple bodies do not move themselves inasmuch as they lack an (internal) agent of movement (*to kinoun*) and are simply moved. This is confirmed by the fact that they are unable to stop themselves (*Phys.* 8 255a7–11); see below n. 76.

30 See, for instance, 8 254b13–33. A veiled reference to the soul features perhaps in Aristotle's label of bodies that move themselves as *empsykha*, living, with an adjective adopted in *On the Soul* for the soul-centered distinction between living bodies (*empsykha*) versus nonliving (*apsykha*) (see *Phys.* 7 244b9–245a1, 8 255a7, 259b2; cf. 265b34 where we find a reference to Plato and the Academics' theory that the soul is self-moving and the cause of movement. Against the first attribution Aristotle firmly argues in *On the Soul* 1 (408b33–409a21). The soul and his parts (sensitive and rational) are instead explicitly referred to in the denial that dispositions are alterations (*alloiōsis*) (see *Phys.* 7 247a1–248a9).

31 In this treatise, as we have seen in Chapter 2, Aristotle focuses on ensouled beings and reverts to the types of movements identified in *Physics* making them coincide with discrete, yet compatible, aspects of life. Growth and decay, alteration, movement across space are now functions of distinct parts of the soul, the differentiated possession of which allows Aristotle to diversify among categories of beings, from plants to stationary animals to animals endowed with locomotion to the human animal who alone among living beings possesses reason. Furthermore, in *On the Soul* there is a systematic conditionality regulating how living beings access movement. For Aristotle further organizes the different kinds of movements living beings possess by creating an architecture of living faculties where the most basic faculty, namely that of nutrition, is presupposed by the sensitive faculty, and the sensitive faculty in turn is presupposed by that of locomotion. In respect to "movement," no animal can move across space if it cannot undergo the physical alterations required for the distant senses, while, in turn, no animal can sense the world unless it has a body that nourishes itself, grows, and decays.

32 These treatises include *Parts of Animals*, *On the Generation of Animals*, *On the Movement of Animals*, *History of Animals*, and the treatises constituting *Parva Naturalia*.

33 Aristot. *Mete.* 1 338a20–339a9 ("We have already dealt with the first causes of nature and with all natural motion; we have dealt also with the ordered movement of the stars in the heaven, and with the number, kinds, and mutual transformation of the four elements, and growth and decay in general. It remains to consider a subdivision of the present inquiry which all our predecessors have called Meteorology. Its province is everything which happens naturally, but with a regularity less than that of the primary element of material things, and which takes place in the region which borders most nearly on the movement of the stars. For instance the milky way, comets, shooting stars, and meteors, all phenomena that may be regarded as common to air and water and the various kinds and parts of the earth and their characteristics ... After we have dealt with all these subjects let us see if we can give some account, on the lines we have laid down, of animals and plants, both in general and particular; for when we have done this we may perhaps claim that the whole investigation which we set before ourselves at the outset has been completed," transl. by H.D.P. Lee); see also Chapter 1.

34 For a list of *GC*'s topics, see Rashed, 2005, CXLI–CXLV.

35 On the study of animals as pertaining to the inquiry into nature (*peri physin historia*) and the student of nature (*physikos*), see, respectively, *PA* 1 639a13–4 and 639b9, 641a32; cf. *DA* 1 402a5–8, discussed at the beginning of Chapter 2, where the discussion of the soul as principle of living beings (*arkhē tōn zōōn*) is said to make a substantial contribution to the study of nature (*pros tēn physin*).

36 While Aristotle claims to devote a special study to plants (see, for instance, *PA* 2 656a4–6), no original work on the subject has reached us. For the circulation of a work on plants attributed to Aristotle as late as the second century AD, see Rashed, 2011. Ps.-Aristotle's *On Plants* is a Greek retroversion of a Latin rendition resulting from a series of translations whose original source was a Greek compendium by Nicolaus Damascenus (see Drossart-Lulof and Poortman, 1989). On Aristotle's references to plants in the extant zoological treatises and the association of animals and plants as objects of a general, unified study of life, required by the theory of scientific explanation in *Posterior Analytics*, and its actual accomplishment, see Falcon, 2015, 75–91.

37 Chapter 1, pp. 28–9. But animals' *logos* is even more foreign to the elements, which lack both body composition, and hence articulation, as well as "external" measure; see discussion of fire in the context of the soul below.

38 See Chapters 1 and 2.

39 For this aspect, see below.

40 *DA* 2 412b1–4; on plants' inability to sense see Section 4.3.1 and on simple *logos* p. 81 above.

41 Some qualifications are here in order. If Aristotle denies plants sensation (a form of *alloiōsis*) and locomotion (which are therefore specific to animals) he still attributes them growth, nutrition, and reproduction. Yet this set of phenomena is realized in a different way than for animals and contributes to plants' *idia physis*. Growth, for instance, is much simpler and at the same time less linear and complete than for animals because for vegetal bodies a minimal articulation into parts is sufficient to fulfill the functions of nutrition and reproduction while within the shape and size proper to it, a plant continues to grow, i.e. new leaves and/or branches form in spring. See Zatta, forthcoming.

42 In this passage Aristotle is referring to the complex bodies of living beings as it is clear from the comparison with the construction of the house, which, like the bodies of living beings, is also a complex product made of different materials (and parts).

43 Aristot. *PA* 1 639b21–3. Unlike his predecessors, who made matter the cause of living beings' formation, Aristotle deprives matter of such a role and distinguishes, instead, between absolute and conditional necessity. Different materials are necessary for the formation of living beings in the same way as different materials go into the construction of a house (i.e. conditional necessity), but the materials themselves and their movements do not make the finished product which rather depends on the final cause (see discussion in Chapter 1).

44 *Phys.* 2 198b18–27, this and upcoming translations of *Physics* are by P. Wicksteed and F.M. Cornford.

45 Needless to say, the mention of teeth's different conformation makes the reference to animals clear.

46 For instance, Empedocles thought that the vertebrae of the backbone were formed in the fetus under the impact of matter breaking the backbone into pieces, while Democritus made inhaled air the cause for the formation of the lungs, see Chapter 2 of Zatta, 2019. For a critique of air as responsible for the articulation of the embryo, see also *GA* 2 741b38–742a16. On the displacement of *tykhē* from the study of animals, see above Chapter 1, pp. 25–6.

47 For a discussion of the same example of evaporating water, see *On the Generation and Corruption* (1, 4 and 5) in which Aristotle points out that this type of change is a substantial change, and not a phenomenon of growth or alteration, inasmuch as the "changing subject" does not keep its identity: water indeed becomes air.

48 In *On the Soul* 2, 4 Aristotle distinguishes two meanings of finality. "That for the sake of which" (*to d'hou heneka*) means: (m1) that for the purpose of which and (m2) that for the benefit of which. This second meaning raises the important question whether the end of a living being's activity and ultimately life is for the sake of the living being performing such an activity or for someone else, external to it. See discussion at the end of Section 3.3.

49 By contrast, as discussed below, the change the living being undergoes fulfilling its nature and realizing its *logos* has a finality intrinsic to the living being itself.

50 On the complementary functions of this combination of teeth in humans and other animals, see *PA* 3 651b1–10.

51 For a preliminary discussion of this example, see Chapter 1, p. 25.

52 See below.

53 Consistently, in *Parts of Animals* Aristotle discusses animals' body parts in terms of their function to which he relates, subordinating it, the type of material a given part is made of. For instance, in some animals teeth discharge only the function of mastication, in others they are also "a means of force" (i.e. sawlike teeth and tusks). It is thus by necessity (*ex anankhēs*) that—Aristotle claims—these parts are made of earthy and solid material (*PA* 2 655b9–13).

54 See n. 81.

55 See *Phys.* 2, 7 and 8. Their reference to the teeth is not casual. For teeth are instrumental to nutrition, which is for the sake of the living being's growth, preservation, and reproduction.

56 *Phys.* 2 198a21–31, with slight modifications.

57 Here Aristotle is concerned with pointing out the "species-related" coincidence of form with regard to the causes involved in sexual generation. In *On the Generation of Animals* he explores aspects related to the coincidence of form across generation such as the resemblance of the offspring to male and female parents of different species (*GA* 2 738b28–38); as for character inheritance in human reproduction, see 4 767a36–768b15).

58 See Polansky, 2010, 208.

59 At *Phys.* 2 192b8–16 and later *Phys.* 3 201a10–9 Aristotle defines *auxēsis* along with its opposite, decay (*phthisis*), as "the actual growth or shrinking of anything capable of expanding or contracting." With this definition he appeals to the same theoretical framework of potentiality and actuality, which pertains to the other types of movements (alteration and locomotion) and shows to conceive of *auxēsis* in a loose sense that includes *any* phenomenon involving an increase in the magnitude of natural bodies, without distinguishing neither among the types of bodies involved nor what causes their change in size. See also *Phys.* 4 214a32–b3, where evaporated air is presented as a phenomenon of *auxēsis*. At the same time, in *Physics* too we can detect some reluctance on Aristotle's part to define the increase of simple bodies' "growth." For instance, when discussing the evaporation of a body of water into air and the expansion that accompanies it Aristotle simply mentions the change in terms of size, from smallness (*ek mikrotētos*) into greatness (*eis megethos*) without labeling it as growth (*auxēsis*). Besides, in the same passage, he observes that in this case matter (*hylē*) changes without taking on anything in addition, a comment that suggests that already in *Physics* he has in mind a defining feature of growth (i.e. the addition of something extraneous). A similar observation is made in reference to the expansion of a body of air: it is the alteration in temperature that determines its increasing, which therefore does not qualify as growth (*Phys.* 4 217a33–217b2).

60 For Burneyat in *On Generation and Corruption* Aristotle anticipates the conceptual needs of more complex areas of his natural philosophy pertaining to living beings and life (2004, 13–24), while Rashed remarks that the physics discussed in *On Generation and Corruption* is "pre-biological" and oriented toward "biology" (2005, xv, cxl–clxxxvi).

61 In resulting from sexual generation, growth "marks" the coincidence of efficient, formal, and final causes, which Aristotle addresses in *Phys.* 198a21–31 (see above).

62 While in *On the Soul* growth represents the minimum requisite for life and is the only living function possessed by plants, the mention of flesh as a uniform part that grows shows that in defining growth in *CG* 1.5 Aristotle has in mind the growth of animals. Indeed, in being defined by the possession of sensation, and having, besides,

the capacity to move across space (for those among them who have it) animals have more complex bodies than plants and manifest a more complex process of growth, which combines the increase in size along the lines discussed below with the capacity of sensation; see also note 66.

63 Aristot. *GC* 1 321a4–5, all translations of *Generation and Corruption* are by E.S. Forster.

64 See Code, 2004, 171. Significantly, Aristotle illustrates growth as a phenomenon affecting all the parts of a growing "body" by the example of flesh, which—he claims—grows in every part (*GC* 1 321a20–1). The reference to flesh indicates that Aristotle is thinking about growth at a mereological level and not at the level of the whole body of a living being—not surprisingly since the first leads to the second; see below. For the technicalities pertaining to the nature of that "something which is added" (whether it is a body or incorporeal) and to other challenging aspects of Aristotle's treatment of growth, see the extensive *ad hoc* discussion of Code, 2004, 179–91; Kupreva, 2005; and Rashed, 2005, xi–clxxxvi.

65 As Kupreva remarks, while the use of "every part" (*hotioun morion*) is quite vague, the expression "perceptible point" (*sēmeion aisthēton*), adopted later (*GC* 1 321b4), indicates that Aristotle is thinking "of the proportionate increase throughout the whole of a bodily structure, in such a way that no part could be left out of the process" (2005, 120).

66 The general description of the conditions for growth makes it applicable to plants as well, but see n. 41 above.

67 Aristot. *CG* 1 321a22–3.

68 Aristotle lays out the fundamentals of his hylomorphic theory to understand change in *Physics*. Everything in nature is a compound of matter and form: form makes matter be the matter of a particular physical object, while matter persists through change and underlies form (1 190a13–191a22, see Shields, 2014, 60–73).

69 *GC* 1 321b17–25.

70 In discussing this passage I follow Code, 2004, 186–90.

71 The most proximate matter for flesh is blood, and before this stage the food the animal ingests.

72 Code, 2004, 188.

73 *GC* 1 321b24–28.

74 Joachim takes the measure analogy as referring to the form as it grows (expands or contracts) (1922, 131); also Giardina remarks on the formal reference of the analogy. That water is added to water indicates that what is added has the same nature of what is "augmented." In her words, "it is matter constituted by a certain matter, namely as possessing a determined form or belonging to a determined species" (2009, 115). On the other hand, Code interprets the measure analogy as the quantity of matter that is constantly changed (and assimilated to form) in the process of growth (2004, 189–90).

75 Interestingly, as in *Physics* in this treatise too Aristotle seems purposely to avoid any reference to the soul and uses the periphrasis *to kinoun* (the motive agent) in order to build a general, open-ended interpretation of growth (and alteration), seemingly unrestricted to the living but subsuming biological phenomena under purely physical ones (see, for instance, *CG* 1 321b8–11). At the same time, as remarked earlier, his analysis of growth in relation to uniform and nonuniform parts and, besides, his mention of flesh clearly, although tacitly, contextualize growth as a phenomenon that pertains to living beings.

76 *DA* 2 416a10–18; on fire in a comparison with "a growing complex body," see also *CG* 1 322a11.

77 The causal role of fire has to be interpreted in the context of the Presocratics' psychological doctrines: Heraclitus identified the soul with fire, while for Democritus the soul was constituted of fiery atoms. In *Metaphysics* Aristotle cites Hippasus in asso-

ciation with Heraclitus for having considered fire as a fundamental principle (*arkhē*) (*Met.* 1 984a7–8); see Movia, 1979, 302–3; Giardina, 2009, 122, n. 146.

78 On heat as producing digestion and its presence in an ensouled body, see also *DA* 2 416b28–30.

79 See, for instance, *GA* 2 732a17–23 on the soul's heat enabling the functioning of large animals (as self-sufficient beings) or 743a19–36 on the agency of heat in the production of sinews and bones. In this last passage we learn that the heat triggering the formation of the bodily parts in the embryo resides in the father's sperm and that its variation in temperature in respect to an optimal standard impacts negatively the process of formation. Also in *PA* 2 652b10–3 heat is recognized as having a major role. Besides nutrition it enables motion and is more generally said to be required for all the activities of the soul.

80 For the denial that fire may be the only cause of formation and growth, see also *GA* 2 734b28–30.

81 In fact, the exceptionality of fire has already been stressed upon in *On Generation and Corruption* for a different reason. While the other simple bodies tend to be borne to their own places (*khōra*) and their shape (*morphē*) and form (*eidos*) consist "in their limits" (*en tois horois*) (namely the limits of the bodies that contain them), fire alone "is of the nature of form (*eidos*) because it naturally tends to be borne towards the limit (*pros to horon*)" (*CG* 2 335a18–20); see also *De Cael.* 3 306a9–11, where simple bodies, especially water and air, are said to acquire shape by adapting to the places that contain them. By contrast, fire has a "formal nature" that the other simple bodies do not have (hence resembling in behavior the natural complex bodies).

82 On living beings' limit (*peras*), see also *PA* 2 646b8.

83 I owe these examples to Polansky, 2010, 212.

84 See Chapter 1.

85 *GA* 2 745a4–6; cf. Giardina, 2009, 123, n. 150.

86 Considered the result of an organic growth and a composite body, also the *polis* has an optimal size. See *Politics* 7 1326a29: "for certainly beauty is usually found in number and magnitude, but there is a due measure of magnitude for a city-state as there also is for all other things—animals, plants, tools; each of these if too small or excessively large will not possess its own proper efficiency (*dynamis*), but in some cases will have entirely lost its true nature (*physis*) and in others will be in a defective condition," transl. by H. Rackham; cf. Carbone, 2010, 100.

87 In this respect, see again *Politics* (5 1302b34–1330a3) where Aristotle uses an illuminating analogy between constitutional revolutions and the disproportionate growth within an animal's body. That is, like the body of an animal, so a political constitution is made of parts and radically changes into another form of constitution when the proportion (*symmetria*) informing the relation of its parts changes on account of a part's disproportionate growth.

88 See n. 41 above and the discussion of the growth of the embryo below; regarding the heart, Sections 1.4.2 and 3.2.3.

89 So Aristotle defines the soul "cause and first principle of the living body" (*aitia kai prōtē arkhē tou zōntos sōmatos*) (*DA* 2 415b8).

90 See *DA* 2 412a13–5 and Section 2.1.

91 *DA* 2 415b27–9, with a slight modification; see also *DA* 2 413b1–2 where the capacity for nutrition is said to be the cause of life for all living beings.

92 It should be remarked that the order of Aristotle's discussion of the subfaculties of the nutritive soul prioritizes reproduction, pairing it with nutrition, and only later deals with growth. Such an order reflects the teleology informing these subfaculties: for the generation of another like oneself is the end of nutrition (and growth); see next section.

93 In fact, Aristotle discusses plants' life in the context of the meanings of life, pinpointing in plants' growth an incontrovertible proof of their living status (and soul posses-

sion) and resolving a dilemma that opens the treatise *On Plants* where we read that "life in plants is hidden and not clear" (Ps.-Arist. *de Plant.* 1 815a11–2).

94 *DA* 2 413a26–31. On Aristotle's critique of Empedocles' confinement of plants' growth to downward and upward and the endorsement of the same standpoint in *On the Soul* 2.2 and 2.4, see Kupreva, 2005, 120–21.

95 *DA* 2 416b13–5; even if not explicitly, the distinction between nutrition proper and growth was already anticipated in *GC* as the addition of something is instrumental to the body's increase (see above); cf. also *GA* 2 744b33–37 where Aristotle distinguishes the nourishment (*trophē*) which properly nourishes (*threptikon*) from that which promotes growth (*auxētikon*).

96 *DA* 2 416b13–18.

97 Let us recall that while the nutritive soul can be simply (*aplōs*) separated from the sensitive soul and as such is found in plants, the sensitive soul can be separated from the nutritive soul only in account (*logōi*), see Corcilius and Gregorich, 2010, 92–5; see Chapter 2.

98 See n. 107.

99 In the discussion that follows I will focus on the process of generation involving the viviparous (blooded) animals and presented in book 2 of *On the Generation of Animals*. In fact, in this treatise, Aristotle identifies a series of types of generation according to animals' specific body temperature, from the hottest (the viviparous) to the coldest (the insects producing larvae) (see *GA* 2 733b1–17 and for the generation of animals other than viviparous book 3). For the sexual reproduction of blooded animals, see *GA* 1 716a3–14; 2 737b27–739a14. It should also be remarked that the female's matter is already "high-level," informed matter and, as such, receptive to the "agency" of the semen (for a recent reassessment of the female's role, see Connell, 2016).

100 *GA* 1 727b15–18.

101 See, for instance, *GA* 1 727b15–18, 729b7, 730a26, 730b3; 2 733b21; consistently, this verb is used for the complex bodies of living beings whose parts Aristotle discusses in *Parts of Animals* (see, for instance, 640a23, 640b4, 644b22, 645a14) and the complex natural bodies (as distinct from simple bodies) in *Physics* (see, for instance, 1 193a36, 2 193b13 4 254b31). On the other hand, *synistēmi* is used in the active to indicate the *dynamis* of the male's semen on the female's "matter" or of some female animals among the birds, which do "set" a fetation, albeit imperfect (see, respectively, *GA* 1 729a12–15, 729a20, 729b32, and 730a32–33).

102 *GA* 2 739b27–34.

103 *GA* 2 741a10–13; see also 1 726b21–24, 2 734a14–15, 734b25–28, and 735a6–8. For some time, however, the sensitive soul is only present in potentiality inasmuch as the embryo lives the life of a plant; see below.

104 *GA* 2 741a25.

105 As proofs for the presence of the nutritive soul (*threptikē*) in the material supplied by the female, Aristotle provides the example of the wind-eggs, which unlike wood and stone are alive, but not in the same sense as fertile eggs are (*GA* 2 741a19–26).

106 *GA* 2 741b6–7. Emblematically, to clarify the male's contribution to generation, Aristotle compares the activity of the male's semen on the female's matter resorting to the analogy of a carpenter, see *GA* 1 730b11–23.

107 Plants are able to reproduce themselves by the nutritive soul alone because for Aristotle they possess a *sui generis* organism. They do not have male and female individuals properly, but only conventionally and by analogy with animals (see *GA* 1 715b17–25 where Aristotle discusses the testaceans in a comparison with plants). In not being differentiated between male and female and encompassing in a single individual both male and female principles, plants produce seeds in which the nutritive soul alone is generally enough to provide the movement necessary for their growth into plants (see n. 131).

108 *GA* 2 740a6–7, transl. by A.L. Peck.
109 The ambivalent status of the heart has likely to do with the intention of explaining the transition of the female's fetation from an inert mass of blood to a "moving," living body involved in a process of articulation via the absorption of blood from the uterus but also of conforming to the doctrine that it is the heart that produces blood from the food, see Section 1.4.2.
110 On the priority of the heart, see *GA* 2 735a23–26, 738b17–9, 740a1–4, 741b15–18, and 743b26–28.
111 *GA* 2 741b25–642b18; cf. Chapter 1.
112 While the analogy with the seed planted in the earth indicates that the environment from which the fetus takes its food is the uterus, Aristotle also presents the possibility that the residual blood of the fetation (the portion not used for its construction, *systasis*) might be used as nourishment (as it happens in the case of seeds of plants) (*GA* 2 740b3–8).
113 *GA* 2 739b34–740b2; see also *GA* 1 728a26–31. Aristotle recurs to the analogy between seed and fetation multiple times in the course of his explanation of the early development of the embryo and stresses the "dependency" of incipient plants and animals from their respective nourishing environments.
114 Also the brain is formed at an early stage of the animal's formation, right after the heart (at least for those animals which have a hotter heart), inasmuch as its role is that of counterbalancing the heat of the heart (*GA* 2 743b28–33).
115 *GA* 2 744b28–30; cf. *GA* 2 745b4–6. As mentioned earlier, the growth of bones sets the limit of animals' growth while the sinews are the bodily part that holds them together (*GA* 2 744b29–745a9).
116 *GA* 2 742a3–7. Aristotle mentions the priority of the articulation of the lung in respect to the activity of breathing to reply to those *physiologoi* (among the Presocratics, Democritus) who thought that the embryo became articulated by inhaling air (cf. also Hippocrates, *Nat. Puer.* 17).
117 *GA* 2 733b22. In his study of *On Generation and Corruption* Rashed distinguishes four different terms for "form" and further qualifies *morphē* as silhouette (2005, cvi). Thus, the phase from conception to the embryo's acquisition of the *oikeia morphē* is not growth (*auxēsis*) as technically defined in *CG* because it implies an articulation of form rather than an increase in size of the body parts that are already formed.
118 *GA* 2 733b2–3.
119 *PA* 4 686b6–17.
120 In reporting this biological feature Aristotle is likely referring to Plato's *Laws* where, however, it is noted that "intense growth" characterizes "the first shoot in every living creature" (*prōtē blastē tou zōou*) and not just humans (*GA* 1 725b23–25 and *Leg.* 7 788d); in *Politics* Aristotle recommends to keep children free from any study and compulsory labors until they have reached five years since these activities would hinder their growth (*Pol.* 7 1336a30).
121 *DA* 2 415a26–9.
122 For once the animal has grown to completion, the excess of the concocted food (blood) which is not used for the animal's nutrition (i.e. preservation) goes into the formation of the male's seed and the female's stuff.
123 On the male's production of semen and the female's production of menstrual stuff as residues from concocted food, see *GA* 1 725a22–727b34, 728a19–22, and discussion above.
124 *GA* 2 736b1–2.
125 In other words, for the soul's functions to be operating, their respective objects must be available.
126 *GA* 2 736a24–736b29. Unlike plants' nutritive soul (see above), the female's nutritive soul is unable to give rise to individual living beings: animals' parts (those, for

instance, enabling the embryo's nutrition) may be analogous to plants' parts, but they still need the sensitive soul in order to be formed and subsist.

127 Once conveyed to the female's matter and setting the fetus, for a definite period of time the sensitive soul is present only in potentiality (see below the reference to living beings' plant-like life). The switch to actuality may arguably coincide with the moment in which the heart becomes the seat of sensation in conjunction with the sense organs' formation. The first to be formed would be flesh, which "sets" early in the embryo under the effect of the cold on the nourishment oozing through the blood-vessels (*GA* 2 743a8–11) and is considered in *Part of Animals* the sense organ for touch, the most corporeal of the sense organs (see *PA* 2 647a14–21). There is, however, some ambiguity in Aristotle's treatment of this issue. In *DA* 2 417b17–24, discussed below at 4.2, Aristotle attributes the actuality of sensation to the moment of birth as if denying that the living being might feel before. This ambiguity is likely due to the fact that in *On the Soul* Aristotle is thinking of sensation as the defining capacity of the *aisthetikon zōon* (i.e. the living being as a fully formed sentient being) and in terms of the objects of sense as located in the outside world; cf. *DA* 2, 6.

128 *GA* 2 736b2–9.

129 Consistently, as soon as the fetus has been set and starts getting nourishment from the womb it is an incomplete (*ateles*) animal (*GA* 2 740a24–5).

130 *GA* 2 736b13–14.

131 Aristotle claims that plants are not separated into males and females, but the two principles are found together. This is the reason why plants are able to produce by themselves their seeds (*spermata*), which are the equivalent of an animal's fetation (*kyēma*) (*GA* 1 731a1–4); on animals as "divided plants" and, conversely, becoming as plants during copulation, see, respectively, *GA* 1 731a21–23 and 731b7–9. This notion that plants were encompassing both sexes and autonomous self-reproducing living beings was already present in Empedocles, whom Aristotle praises in this respect (*GA* 1 731a5–10; DK 31 B 79; cf. Zatta, 2019).

132 See, for instance, *GA* 1 2 740a27–28 (with respect to the capacity of locomotion).

133 See below and n. 127.

134 *GA* 4 765b35–766a10; on this point, see also Mingucci, 2015, 48.

135 On the problem related to the number of senses, see Chapter 4.

136 On touch as the common denominator of all living beings and the basis for the other senses, see *DA* 2 414b4–10 and Chapters 1 and 4; on its relation to pleasure and desire, see Chapter 5.

137 See *NE* 3 1118a2–23 and Chapter 5.

138 See *NE* 1 1102a34–1102b13 and Chapter 2.

139 Nutrition and reproduction are, in fact, already associated in terms of their pleasures and attached to the same "unnamed" (and third) part of the soul by Plato in *Republic* 4 (436b).

140 *DA* 2 415a24–415b9.

141 On self-preservation and growth as distinct effects of nutrition, see *DA* 2 416b13–20.

142 For reproduction and nutrition's origin in the nutritive soul, see *GA* 2 740b30ff.

143 The title was *On Nutrition* or *On Growth and Nutrition*; see Bonitz, *Ind.* 104b 16–28. For a discussion of the passages in Aristotle's work where he refers to this treatise, see Louis, 1952.

144 *GA* 1 724b21–726a29; cf. *PA* 4, 689a5–8. Aristotle conceives a chain of concoctions leading from food to the production of semen. Blood is the result of concocted food while semen derives through concoction from the residue of the blood that has been distributed to the entire body in order for it to grow and/or maintain itself.

145 See *GA* 2 740b30–741a4 and 2 735a16–21 where Aristotle combines growth and reproduction as functions of the nutritive faculty (*threptikon*).

146 See *DA* 2 416b13–20.

147 On the priority of generation over nutrition, see also Louis, who considers food "l'aliment de la génération" and underlines its role in allowing the development of the living being once it has been engendered (1952, 34).

148 For the geometrical model, see Chapter 2.

149 Every living entity is attributed desire for immortality through reproduction but see Philoponos' commentary to *On the Soul* that desire exists only for those living beings capable of perception and since plants do not have perception they do not properly feel desire. He concludes that by desire Aristotle means "natural constitution" (*physikē kataskeuē*) (*In de An.* 270); on "positive" desire shared by all animals, differentiating them from plants, see Section 1.4.2.

150 Hicks, 1965, 340; Hamlyn, 1968, 95; Polansky, 2010, 204; Johansen, 2012, 119.

151 For Diotima "mortal nature," encompassing both humans and animals, achieves immortality "by leaving behind it a new creature in place of the old" (*Smp.* 207D; cf. 208B, *to apion kai palaioumenon heteron neon enkataleipein hoion auto ēn*).

152 So far scholarship has ignored or minimized the force of this superlative. For instance, Hicks (1965) and Hamlyn (1968) do not address it; for other positions, see below.

153 Differently from Aristotle and in line with later Philoponus (n. 149 above), Diotima does not include plants.

154 *Smp.* 207B–C, transl. by W.R.M. Lamb, with a slight modification.

155 *Smp.* 208B; cf. also 207D.

156 See Theophrastus *CP* 1 20 on the small offshoots of trees bearing fruit; for the use of a cognate form (*aploblastanō*) to indicate childbirth, see, for instance, S. *OC.* 533 (where it refers to the birth of Antigone and Ismene from Jocasta).

157 Polansky, 2010, 204 and n. 5.

158 Johansen, 2012, 124.

159 See *DA* 2 415b1: *kakeinou heneka prattei hosa prattei kata physin*.

160 For these lines of consideration, see respectively Johansen and Polansky as in notes 157 and 158.

161 Cf. *Pol.* 1 1252a26, but see also *On the Generation of Animals* where Aristotle uses the "principle of diffusion" to define the nutritive soul as the most natural of the other parts of the soul (2 740b).

162 In this respect, I believe the expression "a man produces a man" is rather eloquent (see above); for it presents reproduction as the production of another similar living being (a man) without considering the fact that the "produced man" has to grow to become such. The focus is on the essence of the subjects involved in this process; cf. the parallel nature/art discussed at Section 1.4.2.

163 Needless to say, in channeling all living operations toward reproduction, Aristotle is considering living beings as fulfillment of their *telos*, namely when their growth has reached the end and they are finally able to reproduce themselves.

164 Wallace distinguishes an end as objective (a point *at which* something ends) and as subjective (a thing or a person *for which* something exists (1832, 234).

165 Hicks, 1965, 240.

166 See, for instance, *Met.* 12 1072b2; *GA* 2 742a22; *Phys.* 2 194a35. For a systematic discussion of this distinction as a cornerstone of Aristotle's teleology, see Johnson, 2008, Chapter 3, and below.

167 See Chapter 1, p. 16. In respect to the immanence of teleology, Polansky forcefully states, "Most natural to mortal things is to seek godlike life for themselves so far as they are able. Aristotle does not say that the individual aims to perpetuate the species, or anything so removed from his topic as that, but rather that the living thing produces another such as itself (*heteron hoion auto*, 415a28). In this way it preserves *its own life* as much as it can. There is no contrasting here of reproductive and sensitive and intellective life—in fact each of these other natural capacities will analogously seek to perpetuate itself—but the generation of offspring is the making of another *self* capable of continuing its very own life and so the extending of nutritive life. Thus aiming

to be a god hardly takes the living thing beyond itself or its own level of being but immortalizes the self through making everlasting the natural life: a plant giving rise to a plant and an animal to an animal" (2010, 205).

168 See Chapter 6.

169 Aristotle writes, "So that clearly we must suppose that nature also provides for them in a similar way when grown up, and that plants exist for the sake of animals and the other animals for the good of man, the domestic species both for his service and for his food, and if not all at all events most of the wild ones for the sake of his food and of his supplies of other kinds, in order that [20] they may furnish him both with clothing and with other appliances. If therefore nature makes nothing without purpose or in vain, it follows that nature has made all the animals for the sake of men," (Arist. *Pol.* 1 1256b15–20, transl. by H. Rackham).

170 For an application of the anthropocentric view voiced in *Politics* to Aristotle's teleology at large, see instead Sedley, 1991.

171 *Physics* 2 194a26–b2 presents an interesting parallel to the notion of finality discussed so far and the question of the identity of the beneficiary. After identifying "that for the sake of which" as the best toward which the continuous movement of living beings is directed, Aristotle claims that humans use all things that exist *as if* they were themselves the end (*telos*). In his words, they treat "other natures like matter reshaping it in accordance with their own ends, as the artist reshapes the matter in accordance with the end of the artifact she intends to create." Thus humans behave like artists in that they take other living beings as matter transcending their natural intrinsic end (*telos)* to attribute them an end, which is instead external to them. See Johnson, 2008, 76–7.

4   The sentient animal

4.1 Setting the problem

In his introduction to the study of the soul Aristotle remarks that "this investigation seems likely to make a substantial contribution to the whole body of truth, and in particular to the study of nature; for the soul is in a sense the principle (*arkhē*) of living beings (*tōn zōōn*)."[1] When we consider that for Aristotle animals are by definition sentient creatures, this remark on the field of relevance of the study of the soul (such field being nature inasmuch as the soul is the principle of living beings) implies that we should take his discussion of psychological capacities as directed to understand animal life, and not merely human life. In other words, for Aristotle it was not the matter of illuminating why "we," the human animals, live, but why animals live, where the category of *zōa* is inclusive of the human being *qua* sentient. The generous attention he devotes to the sensitive soul and its functions before undertaking the analysis of the rational soul (which, among the terrestrial animals, only human beings possess) is in this respect emblematic. Sensation (*aisthēsis*) receives much attention in *On the Soul*, more than other psychological faculties, reason included. Aristotle's treatment extends from book 2, section 5, where he presents a general account of sensation, to book 3, section 2. Here, after having discussed one by one the five senses and the "capacities" that derive to the living being from the senses' synergic grasp of reality [i.e. the awareness of sensing and the simultaneous discrimination of different (special) sensibles],[2] he concludes by stating that the previous discussion should be sufficient to understand "the principle (*arkhē*) in virtue of which we say that the animal is sentient (*to zōon aisthētikon*)."[3] The next sections of book 3 will be respectively devoted to the sensation-based capacity of *phantasia* (sensory imagination), to thought and mind (which only humans possess),[4] and to the power of locomotion, which humans equally share with many creatures of different kinds.

Now, given that animals' soul does not encompass the rational soul and that the sensitive soul, combined with the nutritive soul, provides the psychological capacities of living beings *qua* sentient, it is compelling to ask what type of cognitive experience such capacities enable animals to access. In other words, how does sensation enable animals to feel and understand the world? What does it empower them to know? And more generally, what is, importing a modern expression, "the

DOI: 10.4324/9780367816001-5

philosophy of mind"[5] of the *aistētikon zōon*, considered in the full richness of its sensitive nature and sensation-based capacities, separately from the advent of reason and calculation?[6] On the other hand, Aristotle's systematic conception of the soul also raises a question pertaining to the human being, namely, whether and how the possession of the rational soul and powers conditions, and interacts with, the other psychological parts. For instance, does reason impact sensation and the pleasure and pain that may be associated with it? And if so, how? This and the next chapter will deal with the first set of questions pertaining to animals' cognition and discuss respectively Aristotle's general treatment of sensation along with the awareness it provides the animal with (Chapter 4) while pleasure, pain, desire (also integral to animal sensory awareness), and the sensation-based capacities of *phantasia* and memory (*mnēmē*) will be the focus of Chapter 5. As for the impact of reason on sensation,[7] here suffice to note that Aristotle continues the treatment of man's philosophy of mind in *Nicomachean Ethics*[8] where he addresses the synergy of thought (*dianoia*) and character (*ēthos*) in moral action[9] underscoring in the formation of the latter the greatest importance of pleasure and pain (and relative training). To like and dislike the right things is indeed the *sine qua non* of a virtuous character and life[10] because the desire for pleasure constitutes the trigger that moves a human being to action (thought indeed "does not move anything"). In this way, the biological gives way to the ethical, which is, however, still integrated into nature, albeit as the particular domain pertaining to the fulfillment of the human animal *qua* rational and requiring a study of its own. Insofar as they intersect Aristotle's conception of animal pleasure some of these issues will be addressed in Chapter 5.

This chapter will first look at *DA* 2.5 and discuss Aristotle's preliminary distinction of the different senses in which the living being is said to be sentient showing that the aim of this distinction is to make living beings' life coextensive with sensation. It will then focus on *DA* 2.12 where Aristotle considers sensation a form of receptivity that qualifies the bodies of animals as opposed to other natural bodies (such as plants or the elements) and will subsequently proceed to fathom animals' physical understanding of the world along with the array of cognitive acts it involves.[11] Not only do these two angles of Aristotle's general treatment of sensation, i.e. the sentient being's "definitory multiplicity" and animals' sentience, complement each other,[12] importantly, for the interpretation advanced in this book, they also relate to the notion of *logos*, pursued in Chapter 1. Indeed, each of these angles points to the consideration of animals' body structure in relation to the environment and reveals that it is in the actualization of the structure's inherent potentialities that for Aristotle resides the core of animal living.

4.2   From the dialectics of sensation to a new form of alteration

While the origin of sensation resides in the living being, which possesses in itself the principle of movement and rest, what triggers sensation lies in the world outside. Sensation is a movement (*kinēsis*) that affects the animal body *qua* capable of sensing, but needs an external cause to set it off. So is the nature of fuel—explains

Aristotle—which "does not burn by itself without something to set fire to it."[13] If this were not the case, animals would be sensing continuously, without intermission. In *On the Soul* 2.5 Aristotle proceeds to account for animals' sensitive nature in relation to the world by appealing to a complex interpretive framework that mobilizes the notions of potentiality (*dynamis*) and actuality (*energeia*). He writes,

> But since we speak of perceiving in two senses (for we say that that which has the power of hearing and seeing hears and sees, even if it happens to be asleep, as well as when the faculty is actually operative), so the term sensation must be used in two senses, as potential and actual. Similarly to perceive means both to possess the faculty and exercise it … But we must also distinguish certain senses of potentiality and actuality; for so far we have been using these terms quite generally. One sense of "instructed" is that in which we might call a man instructed because he is one of a class of instructed persons who have knowledge; but there is another sense in which we call instructed a person who knows (say) grammar. Each of these two has capacity, but in a different sense; the former because the class to which he belongs; i.e. his *matter*, is of a certain kind, the latter, because he is capable of exercising his knowledge whenever he likes, provided that external causes do not prevent him. But there is a third kind of instructed person—the man who is already exercising his knowledge; he is in actuality instructed and in the strict sense knows (e.g.) this particular A. The first two men are both only potentially instructed; but whereas the one becomes so in actuality through a qualitative alteration by means of learning, and after frequent changes from a contrary state, the other passes by a different process from the inactive possession of sensation or grammar to its active exercise.[14]

Potentiality merely consists in the possession of sensation, as, for instance, when a living being is said to be seeing or hearing during its sleep; actuality is instead the active exercise of sensation when one actually sees or hears. But as we understand later in the passage, for Aristotle the domain of potentiality has an internal articulation, proper attention to which will reveal a fundamental stance in his study of living beings as creatures of nature. Indeed Aristotle envisions two types of sensing potentialities—in Polansky's words, "undeveloped potentiality" and "developed potentiality"—which he illuminates by analogy with what happens in the field of knowledge (*epistēmē*) and in reference to a knowing subject (*epistēmōn*). One can be said to be capable of knowledge, in two ways. That is, one can know because s/he belongs to the class of human beings, who by definition are able to know as opposed to other living beings. This corresponds to "undeveloped potentiality," when knowledge is taken for granted on the basis of class membership. On the other hand, however, one can also know because as an individual s/he has achieved some knowledge; and this corresponds to "developed potentiality." Knowledge exists in the obscure recesses of a human being's mind, waiting to be actualized.[15] But further, Aristotle contrasts these two modes of potentiality with another capacity for knowledge, that of a "third" knowing subject, who is able to know on account of exercising what s/he has learned, by knowing "this particular A."[16] This capacity to know is different from the previous two because it corresponds to knowing in actuality;[17] it is, so to speak, knowledge applied at a given moment.

Not futile, this analogy with knowledge holds significant repercussions on the very conception of sensation as a phenomenon of the living. As Burneyat has remarked, in the triple scheme encompassing undeveloped and developed potentiality and actuality, the distinction between the two modes of potentiality allows Aristotle to rectify the notion of sensation as ordinary alteration. It is indeed the transition from undeveloped potentiality to developed potentiality that brings about a change that is destructive and alters the original state into a contrary state. But from the (successive) interplay of developed potentiality and actuality, sensation emerges as a peculiar form of qualitative change (*alloiōsis*) that is preservative rather than destructive.[18] To work with Aristotle's analogy with the knowing subject (*epistēmōn*), the change from undeveloped potentiality to developed potentiality destroys the condition of ignorance by realizing one of knowledge: a human being who has the capacity to know inasmuch s/he belongs to the category of human beings—but does not know yet anything—achieves the capacity to know as developed potentiality by actually acquiring knowledge. In this case, the acquisition of developed potentiality destroys the original condition represented by undeveloped potentiality and corresponds to an ordinary form of alteration. On the other hand, in actualizing developed potentiality (that is, in applying to a specific situation what one already knows), the type of *alloiōsis* at work results in a change that preserves the altered condition. Rather than negative, it is a change to a positive state that reinforces the living being's capacity for knowledge fulfilling its nature of human being. Now s/he actively exercises the knowledge s/he already has—rather than undergoing an ordinary alteration that would suppress its original state as in the case of a change from undeveloped to developed potentiality. In sum, Aristotle warns us that,

> There are two senses of alteration, one a change to a negative condition, and the other a change to a positive state, that is, a realization of its nature.[19]

If we apply this triple scheme (involving undeveloped and developed potentiality, and actuality) to the comprehension of the animal qua sentient (*aisthētikon zōon*) and the different senses in which it can be said sentient, sensation emerges as a capacity that is coextensive with animal life embracing it from conception. Significantly, when Aristotle proceeds to apply the relevance of the analogy between knowledge and sensation to the relation of the living being to sensation he appeals to his theory of animal generation and identifies specific stages in the formation of the animal that correspond respectively to its possession of undeveloped and developed potentiality. So he claims,

In sentient creatures (*tou d'aisthetikou*) the first change is caused by the male parent, and at birth the subject has sensation in the sense in which we spoke of the mere possession of knowledge. Again, actual sensation corresponds to the exercise of knowledge; with this difference, that the object of sight and hearing (and similarly those of the other senses), which produce the actuality of sensation are external (*exōthen*). This is because actual sensation is of particulars, whereas knowledge is of universal.[20]

Aristotle pinpoints "the first change of the sentient animal" (*aisthētikou prōtē metabolē*)," which corresponds to the acquisition of sensation as undeveloped potentiality, in the father's transmission of the sensitive soul (and form) to the inert material provided by the mother.[21] It is at this moment that the animal starts existing as an organism that is potentially sentient because it belongs to the category of sentient beings (i.e. *ta zōa*). It will acquire, however, the developed potentiality of sensation and be actually able to exercise it only at birth—when its sense organs are fully formed, according to the particular structure of its body and under the stimuli of the sensible objects that constitute its environment.[22] On the other hand, the interplay of (developed) potentiality and actuality explains the terms of animal sentience in this world. As Aristotle reminds us in *DA* 2.5, once a living being is born it is sentient both when asleep and awake: when asleep because it has the potentiality for sensation, and when awake because it is actually sensing.[23] Animal life and sensation are therefore coextensive.

But besides, the "triple scheme" has other implications that further illuminate the position of zoology within the study of nature as well as his conception of the living being in relation to the environment. In leading to a reassessment of alteration as a nonstandard form of change (preservative rather than destructive) the distinction between undeveloped and developed potentiality allows a re-conceptualization of alteration that runs parallel to the re-conceptualization of growth (*auxēsis*) in biological terms. We have seen in Chapter 3 that in *GC* 1.5 and *DA* 2.4 Aristotle reassesses the phenomenon of growth as a phenomenon of change that pertains exclusively to the bodies of living beings (and things). In *DA* 2.5 it is the turn of the alteration in which *aisthēsis* consists to be redefined. Sensation emerges as an idiosyncratic form of qualitative change (*alloiōsis*) that only applies to animals thereby supporting Aristotle's program of understanding animals on their own terms within the wider realm of nature to which they belong. As for Aristotle's conception of the living being in relation to its environment, the distinction between undeveloped and developed potentiality makes birth mark the coming into being of the subject that possesses sensation as a capacity it can immediately actualize under the stimuli of the external world thereby fulfilling its nature of sentient being. So in *Categories* Aristotle claims that sensation comes into being at the same time as the living being itself (*zōon*).[24] The implication of this statement is that in order to fulfill its sentient nature (and move from sensation's developed potentiality to actuality) the living being needs the world. In *Metaphysics* too sensation is considered an inborn power (*dynamis syngenēs*) but further contrasted with later-blooming powers. Intrinsic to the living being from the time of birth (and potentially from conception), sensation is different from the other capacities that are developed later in life by habit (*ethei*) or instruction (*mathēsei*).[25] Animals do not learn to sense the world, but rather build their knowledge of the world upon sensation[26] and, we may also add, an integrated collection of sensations, a fact that will become particularly relevant in the upcoming discussions of common sense and *phantasia* (imagination).[27] At any rate, and let this be a last point on Aristotle's technical discussion of sensation in *DA* 2.5, the distinction between developed potentiality and actuality with the novel formulation of alteration as a nonstandard form of change implies that the actuality (*energeia*) of sensation is for Aristotle preservation of the living being in respect to the activity that defines it, namely sensing, rather than its destruction. For in actually sensing the world (the modality of which along with the sensible objects will be clarified in the next section) the sentient animal preserves its own life.[28]

4.3 Sensation and logos

Aristotle continues his general discussion of *aisthēsis* in *DA* 2.12 after having examined one by one the five senses and the objects perceived through them, whether directly, "synergetically," or incidentally. The array of sensibles is divided according to an architecture of sentience that encompasses special, common, and incidental sensibles.[29] But while in *DA* 2.5 Aristotle discusses *aisthēsis* in terms of potentiality and actuality, referring it to *to aisthētikon*, "that which feels," and engages with the notion of alteration (*alloiōsis*), in 2.12 he focuses on the capacity to sense (*aisthēsis*) as distinguished from the organ of sense (*aisthētērion*) and appealing to form(s) and matter.[30] In this way, he gives an abstract definition of sense rather than pursuing the terms in which "the sensing subject" experiences sensation in relation to its conception, birth, and the stimuli (or lack thereof) from the world outside.[31] Such a definition applies to all the special senses (touch, taste, smell, hearing, and sight) and hence gives a distillation of what the core living capacity proper to animals consists in. Sense emerges as a form of receptivity that mobilizes yet another meaning of *logos*, and one that, we will see, ultimately depends on animals' constitution as what differentiates them in a fundamental way from the natural bodies of plants and the elements. Aristotle writes,

> We must understand as true generally of every sense that sense is that which is receptive of the form of sensible objects without the matter, just as the wax receives the impression of the signet ring without the iron and the gold, and receives the impressions of the gold or bronze, but not as gold or bronze; so in every case sense is affected by that which has color, flavour, and sound, but by it, not *qua* having a particular identity, but *qua* having a certain quality, and according to *logos*; the sense organ in its primary meaning is that in which this potentiality lies. The organ and the potentiality are identified, but their essential nature is not the same. The sentient subject must be extended, but sensitivity and the sense cannot be extended; they are a kind of ratio (*logos*) and potentiality (*dynamis*) of the said subject.[32]

This is a dense, difficult, and much-discussed passage.[33] The difficulty does not only depend on Aristotle's laconic treatment, but also on the possible different renderings of the syntax.[34] The capacity to sense is defined as the capacity to receive the form without the matter and is explained through an analogy with the wax, which receives the impression of the signet ring without the material of which the ring is made—iron, gold, or bronze. With the image of the signet ring impressing its seal on the wax Aristotle stresses the passivity of the sense, which is liable to be subjected to the action of its proper sensible in the surrounding world, as well as its intrinsic malleability allowing it to be faithfully receptive. Key to understanding sense receptivity and the information it conveys is the notion of *logos*, which Aristotle brings up to qualify the sensible forms and has used earlier in Chapter 11 of *On the Soul* to define the sense itself.[35] The sense is affected—Aristotle tells us—not by the colored or the sounding thing, but by its quality "as such and according to its *logos*."[36] *Logos* in this context means "proportion" (or "ratio") and is indicative of the type of transmission of information that takes place in sense perception. As Ward has explained, each special sensible (color, sound, flavor, etc.) is defined by a *logos* (ratio) between two extremes. So taking up the example of a red rose, when a living being sees it, it experiences that red color, through receiving the determinate ratio (*logos*) of white to black, that makes it (i.e. the color) up.[37] Aristotle has a lot to say about the *logoi* of sensible qualities in *On the Sense*[38] and even though he acknowledges that in some cases the two extremes may not be defined by finite numerical relations at all but are rather characterized by asymmetrical excess or defect (of the two extremes),[39] he still admits that these combinations as well the sensible forms "act in the same way as when they are in harmonious proportions."[40] In sum, when Aristotle makes sense receptive of the sensible forms (*aisthēta eidē*) without matter, each sensible form (as such) according to its *logos* (ratio), he conceives of sense as taking on the *logos* (ratio) of the sensible form and becoming like it.[41] It is then that the sense becomes aware by grasping the formal specificity of the sensible quality.

As for the type of assimilation at stake, Caston has persuasively argued that "it involves a kind of *transduction*, where information is transmitted in a different form."[42] In his thorough study, Caston aims at responding to the long-lasting diatribe between contrasting approaches to Aristotle's theory of perception, which can be pinned down to "literalism" on the one hand, and "spiritualism" on the other. The first approach conceives of the action of the sensible on the sense as making the sense organ take on literally the sensible quality possessed by the sense-object, while the latter discards the possibility for any physiological change in the sense organ to assert that the sense is a "coming to awareness."[43] In considering sense as depending on a transmission of information[44] from the agent (the special sensible) to the patient (the sense) Caston argues instead for a third approach,[45] stressing that what the sense takes on is not a replica of the sensible, but a relevant characteristic of it, conveyed in a "transduced form" that is not enmattered. For instance, in looking at the red rose qua red, the eye jelly does not become literally red but only "analogically" red and in doing this it senses and knows the color red.[46] Thus there is a sharedness of the physical characteristics of the world, of its temperatures and "fabrics," of its tastes and sounds, colors and scents, of which only living beings partake. They become aware of these external features by assimilating them on account of their unique form of receptivity *vis-à-vis* other natural bodies such as elements and plants, on whose incapacity to sense more will be said in the upcoming section. It is animals' capacity for this qualitative sharedness that constitutes the basic building blocks of their life and the foundation for further interactions with, and comprehension of, the world and that ultimately cements their common nature (*koinē physis*).[47]

In fact, the sense holds this capacity because it is itself a form of *logos* (ratio) and potentiality[48] whose actuality consists in a *balanced* relation with the opposites that constitute its sensible object. We understand this feature from the discussion of the sense of touch in *DA* 2.11, in which, differently from *DA* 2.12, the use of the term *aisthēsis* subsumes sense and sense organ. Aristotle states,

> Hence that which an object makes actually like itself is potentially such already. This is why we have no sensation of what is as hot, cold, hard, or soft as we are, but only of what is more so, which implies that the sense is a sort of mean (*mesotēs*) between the relevant sensible extremes. That is how it can discern sensible objects. It is the mean that has the power of discernment; for it becomes an extreme in relation to each of the extremes in turn; and just as that which is to perceive white and black must be actually neither, but potentially both (and similarly with the other senses) so in the case of touch it must be neither hot or cold.[49]

To sense implies a passage from potentiality to actuality and in becoming like its object the sense is already potentially like it. There is no sensation of a sensible that is the same as that which senses, but only of what exceeds (and falls short) of it.[50] So one cannot sense what is as hot or cold or hard or soft as itself, but only what is more (or less) so. Certainly, this claim does not mean that the sentient being is not aware of the sensible (which is as itself), but only that it is not capable to detect the difference in quality with it.[51] Referred to the capacity to sense, in the passage above, *krinein* signifies to discriminate (rather than judge).[52] The sense is capable to detect qualitative differences between itself and the sensible affecting it (by its specific form) because it is a sort of mean (*tis mesotēs*) between the sensible extremes that define the range of its relevant sensible. And in being a midpoint it acts like a standard against which it discriminates the qualitative difference with the sensible object. In sum, sense receptivity is due to sense plasticity. So to illustrate Aristotle's claim, when the sense feels something cold it is because it becomes hot in respect to it and when it feels something hot it is because it becomes cold. Each time the sense is able to detect a specific sensible quality because it acts as an opposite and can do so because it constitutes an optimal midpoint. Holding this standpoint enables the sense to receive the sensible forms without matter, each form as such and according to its *logos* (ratio), and to become itself (the sense) *logos* (ratio).[53] Now in *On the Soul* 2.11 Aristotle attributes *mesotēs* specifically to the sense of touch and extends the midpoint position to the sense of sight (with the reference to black and white) but he also adds that the same applies to the other senses. More "complex" than all,[54] touch takes center stage in Aristotle's discussion because it is essential for the living being to exist. Taste itself is a form of touch while none of the other senses could exist without touch.[55] In this respect, significantly, Aristotle points out that while an excess of the other *sensibles* merely destroys the sense organ, an excess of the tangible, as, for instance, of heat or cold, destroys the animal itself. On this view, it is especially touch that makes sensation coextensive with life.[56] For not only is touch shared by all living beings, it is also coterminous with each living being's body (with the exception of nails, hair, and similarly earthy parts).[57]

In his discussion of sense in *DA* 2.12, quoted above, Aristotle clarifies that the formal aspect of sensation is inseparable from its physiological aspect[58] because sense is a capacity that inheres in the sense organ. There is no sense without sense organ. The two are in fact the same, and yet—he continues—they allow for some distinction because the sense organ is a magnitude while sense is not.[59] Being a magnitude implies having a "body," whose physical make-up allows it to be receptive to the sensible form by working as the middle in respect to the extremes whose mixture determines its sensibles.[60] The abstract capacity to sense is intimately related to the constitution of the sense organ that embodies such a capacity. In *On the Soul* Aristotle is rather reticent to fathom the physiological aspect of sensation (as well as of other body activities enabled by the soul),[61] but in the conclusive chapters of book 3 he returns to the capacity to sense and discusses it in relation to the constitution of bodies that possess it and of the organs of sense. He writes,

> It is obvious that the body of an animal cannot consist of a single element such as fire or air. For without a sense of touch it is impossible to have any other sensation; for every body possessing soul has the faculty of touch, as has been said. Now except for earth, all the other elements would become sense organs (*aisthētēria*), but they all produce sensation by means of something else, that is through media. But touch occurs by direct contact with its objects—and that is why it has its name. The other sense organs perceive by contact too—but through a medium. Touch alone seems to perceive immediately. Thus no one of these elements could compose the animal body. Nor could an earthy one. For touch is a kind of mean (*mesotēs*) between all tangible qualities, and its organ is receptive not only of all the different qualities of earth, but also of hot, cold and all the tangible qualities. And this is why we do not perceive by our bones and hair and such parts of the body, because they are composed of earth.[62]

For Aristotle it is elemental composition that makes animals' bodies sentient, in an implicit contrast with the elements' simple bodies and likely the body of plants. No animal can possess a simple body because a body constituted by a single element cannot sense. Here too Aristotle focuses on touch inasmuch as it is the fundamental life-preserving sense, coterminous with the body. Its receptivity of tangibles depends on a mixture of the four elements. Earth alone would not allow a body to sense (the tangible qualities of the world). That is why—Aristotle adds as a proof—bones, hairs, and similar parts are not loci of sensation. As Shields remarks, in this passage Aristotle's compressed prose relies on the doctrine of *Generation and Corruption* where the elements are characterized each by a couple of qualities—fire is hot and dry, air is hot and moist, water is cold and moist, and earth is cold and dry. For the sense of touch to perceive all these tangibles, it cannot be merely earthy because, constituted as such, it would be able to sense only dry and cold.[63] But, importantly, this passage resonates also with *Parts of Animals* in whose framework the elemental composition of the sense organ of touch pertains to the first level of *synthesis* of animals' bodies and is ultimately inscribed in the notion of *logos* conceived of as animals' artful form and end, we brought forward in Chapter 1. Regarding the senses (*aisthētēria*), in *PA* Aristotle distinguishes them from the instrumental parts (*organika merē*) claiming that they are always uniform in contrast to the latter, which are nonuniform. Hence he shows to isolate the receptive part through which the sense functions[64] and to consider it uniform on the basis of the exclusive relation between the sense and its object with ensuing assimilation. If the sense is potentially what its object is in actuality, then it must be of the same kind. And if the object is single so must the sense be. On this view, the receptive part of the eye, i.e. water, is uniform because it is impacted by color, that of the ear, i.e. air, because of sound. Now also the sense organ of touch (flesh) is uniform but in having an array of sensibles is a compound of the four elements and hence the least simple and the most bodily (*sōmatōdestaton*) of all.[65] Yet even the "simplicity" of the other sense organs requires some artful balance. Sight, for instance, "works" best when the sense organ is midway between a large amount and a small amount of fluid: too little fluid would hamper the movement produced by the colors (by not conveying it at all), while too much fluid would convey it with difficulty (*dyskinēsia*).[66] As surprising (or contradictory) as this elemental view of the sense organs might appear, it shows that Aristotle relies on his doctrine of the different levels of *synthesis* constituting the animals' bodies (the eye is a nonuniform part, but receptiveness happens at the level of the uniform one composing it). The uniform parts have a role of their own besides contributing to the nonuniform parts, which indicates the complex operativity of the body, its parts, and design. But Aristotle's reliance to his doctrine of body *synthesis* also reveals that, in attaching sensation to the uniform parts (whether water, air, or flesh), he follows in principle the Presocratics. For, he notes, only with the sense organs did the *physiologoi* associate them to the elements, saying that one sense organ is air, another fire.[67] Nobody among them, however, seems to have discussed touch[68] and it is Aristotle to make it the primary sense (*prōtē aisthēsis*) that resides in flesh (and its analogous in bloodless animals) and the sensorial basis of animal life.[69]

Aristotle concludes *On the Soul* (3.12–13) with a discussion of sensation in relation to the teleological working of nature, thereby integrating the study of the soul and its capacities with the methodological principles advanced in book 1 of *Parts of Animals*. Now the emphasis is on the capacities animals possess and how, in a systematic relation, they empower them to live. "Everything in nature is for a purpose" (*tou heneka*), he writes in 3.12[70] and proceeds to account for sensation in relation to nutrition-driven locomotion. How could animals find nourishment and nourish themselves if they had locomotion without sensation?[71] They need the senses in order to move successfully across space and find their food. But if the contact senses of touch and taste are indispensable to animals' survival (for how would they seize or avoid something?), the distant senses are instrumental to living well (*tou eu heneka*).[72] The terms of animals' good life are explored at the very end of *On the Soul* (3.13), with a consideration on the purposefulness of each sense. Aristotle writes,

> The animal possesses the other senses, as has been said, not for mere exist-
> ence but for well-being (*ou tou heneka einai alla tou eu*); for instance the
> animal has sight in order that it may see, because it lives in air or water,
> or generally in a transparent medium; and it has taste because of what is
> sweet and bitter, in order that it may perceive these qualities in food, and
> may feel desire and be set in motion; and hearing that it may have signifi-
> cant sounds made to it, and a tongue that it may make significant sounds to
> another animal.[73]

Redundancy and contradiction with the preceding chapter may here be an indication of some editorial mishaps[74] but there is also something new. Without con-
tradicting the living being's overall teleology,[75] Aristotle also stresses, I believe,
the state of well-being that for the living being is inherent in the fulfillment of the
capacities of the distant senses per se. In other words, there is a "good life" for
animals that are able to see and taste and hear. It consists in the actual reception
of the colors, tastes, and sounds with which they come into contact actualizing
their senses and fulfilling their nature of sentient beings (provided that the sensibles are not excessive and hence painful and destructive). Sight is the instrument
that allows animals to see through a transparent medium (whether air or water),
taste is that which enables them to perceive sweet and bitter in food and makes
them desire and move, while hearing is what empowers them to grasp meaningful sounds. Rather than being attached to eating or mating (and hence reductively
and critically to the sense of touch), for Aristotle animals' good life consists in
the actualization of those sensorial potentialities that constitute the fabric of life
beyond touch and the knowledge they provide. More on the pleasure intrinsic
to animals' sentience against a moralizing and reductive view of it will be said
in Chapter 5. Suffice here to remark that in assigning to hearing the function of
receiving purposefully meaningful sounds and to the tongue that of producing
meaningful sounds Aristotle points to animals' communication[76] hence subtract-
ing them from the solipsism of sensation and from being passive vehicles of pain
and pleasure as in *Politics*. Hearing exists also in view of animals' ability to interact and their sociality[77] further contributing to the richness of their life.[78]

4.3.1 On the inability to sense

In contrast to animals, neither elements nor plants are sentient beings.[79] In *On the Soul* 3.12 Aristotle couples them together rather indirectly. He uses a periphrasis claiming that,

> Sensation is not necessary in all living things; indeed, neither those things having a simple body, nor those incapable of receiving the form without the matter can have it.[80]

In being simple, the elements lack a body constitution and articulation viable to sensation. On the other hand, while possessing a "composed body" plants are still insentient because they are not capable to receive sensible forms without matter. The sense Aristotle has in mind when discussing plants is obviously touch. For he claims that plants become warm and cold (without becoming aware of it),[81] and these qualities pertain to the sphere of the tangible.[82] Plants could possess a sense organ of touch in their "flesh" (*sarx*), which *On Plants* by Pseudo-Aristotle mentions among plants' uniform parts in addition to knots and fibers and whose analogous in animals plays the role of organ for the sense of touch.[83] For the Presocratics, we may recall, plants were indeed sentient beings.[84] But Aristotle's soul division in *On the Soul* categorically excludes it: plants only possess the nutritive soul, which can exist *simpliciter*.[85] And as for the connate capacity to sense,[86] also its incapacity has a formal explanation that depends on plants' bodies and is ultimately rooted in their elemental composition and ensuing physiology. Plants lack a mean (*mesotēs*).[87] That is, they lack an "internal" standard against which to measure the opposites determining the specific quality of the tangibles affecting them. Without an "internal mediation with the outside," for Aristotle plants are affected by the sensible forms "with the matter" (*paskhein meta tēs hylēs*). Scholars have interpreted this expression differently but, whether referring to the matter of sensible form[88] or the matter of plants,[89] it points to plants' unsophisticated bodies. Plants miss to grasp the *logoi* (ratios) of the tangibles they come in contact with. Because, as we learn from *On the Soul* 3.13, they are composed "of earth," (*gēs*) and hence not by the mixture of the four elements that make up animals' flesh. This basic, elemental constitution makes them close to the senseless parts of animals' bodies or to being simple bodies,[90] and is incompatible with touch.[91] In having an earthy body, plants lack a sufficient body temperature[92] that might work as a benchmark enabling them to discriminate thermic differences without being affected "with matter." [93]

Aristotle completes the general discussion of sense (*aisthēsis*) in *On the Soul* 2.12 with an important clarification that brings to the foreground the utter passivity that characterizes inanimate bodies as opposed to living beings. After discussing plants' incapacity to sense he asks whether it is possible for other (natural) bodies to be affected by the sensible forms if they are not equipped with the relevant senses.[94] For instance, can a body be affected by smell if it cannot smell? Even the simple bodies of inanimate beings—he observes—can be affected (*paskhein*) and altered (*alloiousthai*). So it is air when it takes on a certain scent, but this affection remains distinct from the act of smelling. Air may well become perceptible to a sentient being, but for air to become scented does not imply that it perceives.[95] In this way, Aristotle makes crystal clear that sensing (*aisthanesthai*) and "undergoing something" (*paskhein ti*) are not coextensive: if sensing is always "undergoing something," it is not conversely true that "undergoing something" (*paskhein*) is sensing (*aisthanesthai*).[96]

4.4 Relating to the world: Sensorial architectures and animal awareness

Sensation establishes a unique relation with the world, a relation whose ultimate referent is the living being itself. In Aristotle's view, there cannot be sensation unless there is an object of sensation,[97] and the physical conditions that allow it to be sensed: the presence of a medium and of a living body so structured as to be receptive of the objects' qualities. The medium receives the sensible forms carrying them to the respective senses. Depending on the kind of sense perception it can be part of the body or external. For touch and taste, the so-called contact senses, the medium is flesh (or its analogous).[98] For sight, hearing, and smell, i.e. the distant senses, it depends on the living environment of the sentient subject and may be air or water.[99] Not all animals possess the distant senses, we have seen, but all have at least the sense of touch.[100]

A fundamental methodological principle that Aristotle endorses in *On the Soul* 2.4 and subsequently applies to his treatment of the soul is that in order to study the soul and its capacities one must start from their relevant objects, whether sensibles or intelligibles.[101] Regarding the sensitive soul in *DA* 2.6 he identifies three types of sensibles, proper (*idia*), common (*koina*), and incidental (*kata symbebēkos*), a distinction that depends on the type of relation subsisting between the sensible object and "that which senses."[102] In the case of the proper sensibles, they hold an exclusive relation with their relevant senses (sight, hearing, smell, taste, and touch).[103] That is, color is the proper sensible of sight, sound of hearing, scent of smell, flavor of taste, and a set of tangibles (hot/cold, soft/harsh, etc.) are the proper objects of the more complex and elusive capacity of touch.[104] There are no more than the five senses, and later in *On the Soul* (3.1) Aristotle goes on at length to explain why this is so, relating the specific capacities to sense to the elemental composition of medium and sense organs,[105] and hence, as we would expect by now, to the structure of animals' bodies considered in relation to the physical constitution of the world.[106]

In taking this approach to sensation Aristotle goes along with the early students of nature. Theophrastus' synopsis of the early theories of sensation tells us that the Presocratics envisioned sensation as happening between likes or unlikes, and in terms of affection and alteration, hence pointing to a physical encounter of the living being with the outside world.[107] Also for Aristotle sensation is an alteration but, as we have seen, of a particular type that preserves the sense by actualizing it rather than destroying it. This process implies first an encounter between unlikes and then between likes:[108] the sense detects the difference with the relevant special sensible (the colors, the sounds, the scents and the hots and the colds, the sweets and bitters, etc.) and tunes into it. By grasping the *logos* constituting the sensible it becomes itself *logos*.[109] To consider sensation as happening solely between likes is problematic for Aristotle because it undermines its epistemological validity: on this principle the sense could be aware of itself with no need for an external, tangible object to be activated.[110] Likewise, conceived as an encounter between likes, sensation is at risk of emerging as a "partial" phenomenon, without any tangible counterpart in the environment.[111] For Aristotle by contrast the special sensibles are real, objective characteristics of the world. They preexist sensation and are ready to be grasped in the direct or mediated encounters with the living beings' senses, in the "living body,"[112] hence offering a firm ground for the perception of common and incidental sensibles as it will be soon clarified. Besides, another and fundamental problem in conceiving sensation as happening solely between likes lies for Aristotle with the difficulty of having a unifying sensation that enables the living being to grasp the entities of the world, for instance, using Aristotle's example discussed below, to see the "son of Diares"[113] or in the case of animals, to perceive their family members, enemies, or natural features such as a hole in a rock, trees, or likable food. This issue was raised early in *On the Soul* when Aristotle replies to those theories that consider the soul consisting of elements in virtue of its capacity to cognize an elemental-constituted world. He writes,

> But it remains to see what is meant by saying that the soul is composed of the elements. This theory is intended to account for the soul's perception (*aisthanesthai*) and cognition (*gnōrizein*) of everything that is, but the theory necessarily involves many impossibilities; its supporters assume that like is recognized by like, as though they thus identified the soul with the things it knows. But these elements are not the only things existing; there are many—to be more exact, infinitely many—other things composed by the elements. Granted that the soul might know and perceive the elements of which each of these is composed; yet by what will it perceive and know a composite whole (*to synolon*)?: e.g. what God or man, flesh or bone is? And similarly any other compound whole?[114]

How is it possible to avoid the fragmentation and crude bareness of an elemental-operating sense perception and reach a sensible understanding of the world? To grasp entities rather than what is merely like to "what senses"?[115] To avoid this difficulty and fully understand sensation and the perceptual range available to the living being Aristotle mobilizes the common (*koina*) and incidental (*kata symbebēkos*) sensibles, which build upon the special sensibles. Among the common sensibles Aristotle lists movement, rest, number, shape and size, which are not peculiar to any one sense, but shared by all senses,[116] while incidental sensibles are about the recognition of particular entities. Aristotle illuminates incidental sensibles with the example relating to the perception of "a white thing" as "the son of Diares" thereby indicating that what propels the apperception of a specific individual (the son of Diares) is a white object *(to leukon)*. For him incidental sensibles imply an instant and spontaneous critical "gesture."[117] For to be perceived as the son of Diares that "white thing" must be discriminated on the basis of the common sensibles characterizing it (shape, size, and even movement and rest, besides number), besides the perception of the special one (color). On this view, sense perception involves an architecture of sensibles which Modrak has effectively spelled out into a built-in cognitive sequence, "because we see colors, we see colored shapes; because we see colored shapes, we see faces and tables."[118] In fact the example "perceiving that 'white thing' as the son of Diares" entails even more than the capacity to apprehend general entities (i.e. faces and tables). For it implies the recognition of specific identities and hence constitutes the empowering capacity that allows the living being to comprehend its environment in its different features (whether other living or natural beings) and to establish interpersonal relations and a relation to place (and even much more).[119] For instance, the continued association between parents and offspring that characterizes some animal species and the choice and the use of a lair, upon which Aristotle remarks in *History of Animals*,[120] are due to the grasp of incidental sensibles (and the sensitive soul).

If the perception of incidental sensibles consists in the immediate discrimination of an association of sensibles (special and common) what type of awareness does the mere perception of common sensibles, of, let's say, a "(white) shape" constitute? Not effecting by itself the recognition of an entity, the grasp of common sensibles corresponds to the undetermined discrimination of bodily shapes and movements as in the sensorial spectacle opening up to a sentient being just born into the world or the bare realization of the present embodiments of reality (on the basis of the perception of size, movement and rest, number and unity). In Aristotle's analysis of the architectural elements of sensation common and special sensibles contribute to the living being's perceptual discrimination at work in the grasp of the incidental sensibles, which, differently from the other sensibles, provide perceptual focus and depth. Indeed, from the example of that "white thing" as the son of Diares we understand that to reveal identities (general or specific) the sensation of incidental sensibles must emerge from the association of the present proper sensible with past perceptions, at once and regardless of memory.[121] Such a process implies that the living being must have at his disposal a storage of interiorized sensations, the source of which scholars have differently identified.[122] Be that as it may, the grasp of incidental sensibles depends on living beings' particular sensorial conformation[123] and set of sensorial experiences. Upon seeing the "white thing," at that very moment s/he recognizes Diares because in the past s/he has been exposed to him. It ensues that in the actual unfolding of life sensing involves the various sensibles in architectures that are unique to each sentient being according to their particular experiences, granted that the building blocks, i.e. the colors, odors, and shapes, are the same for all (who possess the relevant senses to grasp them) in that they embody specific proportions (*logoi*).[124]

Extending beyond the grasp of the sensible objects discussed above (special, common, and incidental) the capacity to sense (*aisthēsis*) is involved in other complex activities such as the awareness of sensing and the simultaneous perception of different special sensibles, with which Aristotle completes his discussion of the living being *qua* sentient (*aisthētikon*) in *DA* 3.2.[125] How is a living being aware of sensing? That is, how is it able to know that it is seeing? Or hearing? And how is it able to discriminate different sensibles at once and perceive, for instance, white as different from sweet (and vice versa)?[126] The explanation for both cognitive operations ultimately lies within the realm of the five senses and the sensitive soul. Admittedly, in *On the Soul* Aristotle presents the awareness of sensation as a problem that he does not fully resolve while he makes perceptual discrimination depending on a single faculty (*to hen*), without further qualifications. In the treatise *On Sleep and Wake*, however, he is adamant that both functions belong to a common (sensorial) power (*koinē dynamis*). Sleep is an incapacitation of the senses, while wake consists in their activity or capacity of becoming active. Sleep is for the sake of wake (since in the sublunar world no activity can be continuous) and both are considered in relation to a perceptual part that subsumes all the senses. When this part shuts down the living being sleeps, when it turns on the living being is awake and perceives.[127] Aristotle subsequently writes,

> Now every sense has both a special function of its own and something shared with the rest. The special function, e.g., of the visual sense is seeing, that of the auditory, hearing, and similarly with the rest; but there is also a common faculty (*koinē dynamis*) associated with them all, whereby one is conscious that one sees and hears (for it is not by *sight* that one is aware that one sees; and one judges and is capable of judging that sweet is different from white not by taste, nor by sight, nor by a combination of the two, but by some part which is common to all the sense organs (*tini koinōi moriōi tōn aisthētēriōn hapantōn*); for there is one sense-faculty, and one paramount sense organ, but the mode of its sensitivity varies with each class of sensible objects, e.g., sound and color); and this is closely connected with the sense of touch (for this is separable from the other sense organs, but the others are inseparable from it. We have discussed this in our speculations *On the Soul*). It is clear then that waking and sleep are affection of this common sense organ. This explains why they are attributes of all animals; for touch alone is common to all animals.[128]

There is a common power which accompanies all the senses and is responsible for a common function which empowers a person (qua sentient being) to be aware of being sensing, of seeing while s/he sees or hearing as s/he hears and which equally confers her the capacity to distinguish between white things and sweet things. As Gregoric remarks, in qualifying this common part as the controlling sensory organ Aristotle is here alluding to the heart or its analogous in bloodless animals.[129] And in adding that it subsists chiefly with the faculty of touch he makes clear not only that sleep and wake pertain to all animals but significantly that the awareness of sensing along with perceptual discrimination are faculties possessed also by the simplest animals, that is, those endowed with merely the contact senses. In this way for Aristotle there is no awareness without self-awareness and each time an animal is actually sensing, whether a color or a sound or merely hot and cold or sweet and bitter, it knows it is sensing.[130] Likewise, as long as an animal senses it is able to detect differences, whether among heterogeneous sensibles (i.e. white and sweet) or homogeneous ones (i.e. hot and cold). The fact that in *On Sleep* Aristotle deploys a common faculty subsuming all the senses a given animal possesses points to his overall effort to explain how animals are able to integrate different sensations and to have available a unified awareness, or said in Gregoric's words, "a single awareness attending to the activity of any number of senses."[131]

On the Soul further illustrates the question of perceptual discrimination with a geometrical model that complements the discussion based *On Sleep and Wake* carried out so far. We read,

> The fact is that just as what some thinkers describe as a point is, as being both one and two, in this sense divisible, so too in so far as the judging faculty is indivisible, it is one and instantaneous in action; but in so far as it is divisible, it uses the same symbol twice at the same time. In so far then, as it treats the limit as two, it passes judgment on two distinct things, as being itself in a sense distinct; but in so far as it judges of it as only one, it judges by one faculty and at one time.[132]

There is no contradiction in claiming that a single power judges opposite sensibles at once. For it acts like a point, which is indivisible in that it is single, but also divisible in that it sets the limit of two (or more) crossing lines.[133] Applied to the conception of "common sense" this model means that at any given moment it (the common sense) is able to discriminate the range of sensibles that form the living being's perceptual field unifying them—proper, common, and incidental sensibles. This is a complex operation made possible by the capacity to sense alone, and of no little consequence. For as Polansky remarks, it implies "a start to perceiving a 'world' of things with various relationships,"[134] according to, we may add, a living being's specific psychological make-up. For the possession of the distant senses implies a depth of awareness of the world.

Discussed so far in abstract terms on the basis of *On the Sense* and *On the Soul*, the capacities inherent in the animal qua sentient (*aisthētikon zōon*) can be illustrated with a scene from *History of Animals* in which the consciousness rising from sensorial architectures inherent in animal life determines animal behavior. Animals' character is a natural capacity (*physikē dynamis*) inscribed among the affections of the soul (*ta tēs psykhēs pathēmata*), by which we need to understand an entity encompassing nutritive, sensitive, and locomotive parts except the rational. After describing life and work of bees and likes, Aristotle portrays the character of the lion "in action," that is facing specific circumstances. He writes,

> In character it [the lion] is not shy nor suspicious of anything, and towards those reared with it and familiar with it is very playful and affectionate. When it is being hunted, even if it is in view it never runs away nor takes cover, but

if it is compelled to withdraw because of the number of hunters it retreats at a walk, step by step, and turning its head at short intervals; nevertheless if it reaches thick cover it runs away fast until it has come into full view; then it again withdraw at a walk. In open country if it has been forced by the crowd to run away into full view, it runs extended and does not leap [...] And what is said of it is true, both that its chief fear is of fire (as in Homer's verses: "and burning torches from which he turns in fear even when he is charging furiously") and that it watches for the man who is shooting and then rushes upon that one. And if one shoots without hurting it, then if it can catch him by springing on him it does not injure him nor wound him with his claws, but after shaking and frightening him he lets him go again. They approach the towns and injure humans mostly after they have grown old, being unable to hunt because of their age and because their teeth have suffered.[135]

Not indiscriminately wild, also the lion has for Aristotle its moments of gentleness. There is certainly more than a tinge of anthropocentrism in this passage, for instance, in the lion's measured, almost heroic-like behavior, who proudly retreats under the eyes of its hunters but runs instead when it is not seen, hidden by the thickets. But despite this humanizing portrayal we can still understand the lion's behavior as a sensible reaction to real-life situations on the basis of what it perceives and recognizes,[136] that is, in light of a perceptive field that emerges from the array of sensibles discussed earlier (special, common, and incidental) integrated by the common sense and interpreted in relation to himself as the subject of sensation. For while sensing the animal knows it is doing so. The lion is gentle and affectionate "with those reared and familiar with it," and this behavior is consequential upon the perception of incidental sensibles, which the animal is able to grasp on account of previous sensory impressions. For the lion recognizes in "those colored things" its companions adopting a different behavior than with animals and people it does not recognize. On the other hand, the lion retreats with a composed gait having sensed the multitude (*plēthos*) of the hunters, who in his psychological theory are incidental sensibles grasped through the common sensibles of seize, figure, movement,[137] and number on the basis of the special sensibles inherent in them, i.e. their colors, sounds, and scents. The lion may not count in numerals, but it still perceives number by perceiving lack of continuity, again by virtue of the special senses,[138] and acts accordingly.[139] Further, the animal recognizes the man who shot it (another instance of incidental sensible).[140] Not fragmented, this recognition of individuals is woven into a more extensive perceptive fabric encompassing the environment with its different features and variables, the awareness of which leads the lion to modulate its behavior. For the animal retreats slowly in the open field but runs fast in the thickets. So while in *On the Soul* Aristotle confines the discussion to the apparatus of sensation in relation to the different sensibles and the higher-end operations it affords the living beings with, in this episode of *History of Animals* he offers us an example of how animals' sense-based knowledge is used to act for the sake of self-preservation and how, at any rate, it directs animal behavior, sensibles, and sensitive soul being assumed.[141]

In asserting that he has finished the treatment of the animal *qua* sentient at the point he does (after discussing the various sensibles, the awareness of sensing, and perceptual discrimination on the one hand, and before undertaking the discussion of *phantasia* on the other), Aristotle shows to be pursuing a systematic treatment of sensation unfolding from the bottom up, i.e. from the special senses under the direct effects of their sensibles to more complex levels of sensorial understanding. But he also shows that the pursuit of the animal *qua* sentient (*zōon aisthētikon*) corresponds to a comprehension of the living being in the immediacy of its existence, that is, as connected with, and aware of, the environment it lives in. There is a presentness intrinsic to the life of animals conceived of as creatures that exist in touch with nature and by being in touch with nature, and it is from this sensorial presentness impacting the living body that more psychological capacities arise.[142] They consist in *phantasia*, which provides the living being with a storage of sensorial representations, and memory (*mnēmē*), which adds the awareness that such representations are connected to the past. With these sensation-stemming capacities new temporal dimensions become accessible to animals. If sensation happens in the moment,[143] *phantasia* is a delay of sensation[144] that allows the living being to project itself into the future while memory builds on *phantasia* and relates it to the past.[145] Conceived of in this way, animals' cognitive life stems and irradiates from the living body and the physical experience it entertains with the world, where fundamental to that experience are the feelings of pleasure, pain, and desire, which along with the sensation-stemming capacities of *phantasia* and memory are the subject of the next chapter.

Notes

1 *DA* 1 402a5–8.
2 See below.
3 *DA* 3 427a16–7.
4 See, respectively, *DA* 3.3 and 4–8; cf. *DA* 3 427b8–15, where he argues for a difference between sensation (*to aisthanesthai*) and thought (*to dianoeisthai*) on the basis of their respective relationship to truth and humans' exclusive possession of *logos* (as reason). So he claims, "For the perception (*aisthēsis*) of proper objects is always true, and is a characteristic of all living creatures, but it is possible to think falsely, and thought belongs to no animal which has no reasoning power (*logos*)" (all translations of *On the Soul* are by W.S. Hett).
5 Here I adopt this expression in a loose way to indicate the cognitive life of animals but noting that for Aristotle only the human being possesses mind (*nous*). This fact requires that we should understand animals' cognitive life in terms of the body and its receptivity as it will be seen in the course of this and the next chapter.
6 It should be remarked at this point that while in *On the Soul* Aristotle devotes a special treatment to the locomotive soul (*DA* 3.9), he brings the capacity of locomotion up in his general discussion of sense in *DA* 2.12 integrating it into the discussion of the distant senses. For he remarks that "the other senses are means to well-being; they do not belong to any class of living creature taken at random, but only to certain ones, e.g. they are essential to the animal which is capable of locomotion; for if it is to survive, not only must it perceive when in contact, but also from a distance." (434b24–27); see section 4.3 below.

7   This aspect is already addressed in *On the Soul*, see, for instance, the reference to the self-controlled human being in which reason subdues his "craves" and desires (*DA* 3 433a6–9, and later 433b5–10, 434a12–4). For a discussion of the impact of the rational soul on the other psychological faculties, see Whiting, 2002, 193–6; for nonrational cognition in human adults, see Lorenz, 2006, 86–201. The possession of mind in human beings has led Aristotle to conceive two forms of *phantasia*, one connected to *logos* and available to humans, while the other is connected to *aisthēsis* and available to animals (but to humans as well *qua* living beings) (*DA* 3 434a9–10); see Chapter 5.

8   In this treatise, Aristotle offers a logocentric partition of the soul that suits his discussion of moral psychology (see *NE* 1 1102a28–1103a11 and Chapter 2).

9   *NE* 6 1139a31–35.

10 See *NE* 10 1172a19–25. In *NE* 6, right before the passage referred to above, Aristotle aligns affirmation (*kataphasis*) and negation (*apophasis*) in the sphere of thought (*dianoia*) to pursuit (*diōxis*) and avoidance (*phygē*) in that of desire (*orexis*). On the cognitive nature of human emotions, see Fortebaugh, 2006, 107–30.

11 More specifically, section 4.4 deals with proper, common, and incidental sensibles, the awareness of sensation and perceptual discrimination.

12 Interestingly, Chapters 5 and 12 of *On the Soul* constitute two general discussions of sensation that encompass in between Aristotle's treatment of the different sensibles and ensuing sensations (extending from Chapters 6 to 11 of book 2 of *On the Soul*). Their complementarity resides in the fact that each discusses sensation as a key phenomenon that relates the living being to the environment in which it lives. But while Chapter 5 looks at this relation, so to speak, diachronically including the origin of sentience along with its "dialectics," Chapter 12 looks at it synchronically and explains the "abstracted mechanism" of sensation; see also n. 31 below.

13 *DA* 2 417a9–10.

14 *DA* 2 417a7–417b1.

15 Here I am pressing the analogy for the sake of clarification knowing well that for Aristotle mind is not a physical entity; see *On the Soul* where he claims that mind is unmixed with the body (3 429a24–5); cf. Chapter 3.

16 *DA* 2 417a30.

17 Aristotle uses the notion of actuality (as *entelekheia*) in conjunction with potentiality (*dynamis*) also when he discusses the soul in 2.1.

18 For the ordinary type of alteration see *DA* 2 417a31–2; 417b 2–3; 15; *Phys.* 5 2, 226b1–8; *GC* 1 7, 324a 5–14; Burneyat, 2002, 28–9.

19 *DA* 2 417b15–7.

20 *DA* 2 417b17–24.

21 *DA* 2 417b17–8; cf. Aristot. *GA* 2 737a8–9. For a discussion of the male contribution—form and movement, with movement being soul—to the process of generation in light of *On the Generation of Animals*, see Balme, 1990, 20–31, who, however, focuses on human generation and the problem enveloping the transmission of *nous* rather than sensation. Like sensation, also the potentiality for *nous* is transmitted by the male at the moment of conception.

22 *DA* 2 417b17–19; cf. *Cat.* 8a8–9, discussed below.

23 *DA* 2 417a10–4.

24 *Cat.* 8a8–9, see n. 97 below. It remains a question, I believe, whether blooded animals could start sensing inside the mother's womb. Made of flesh, they could possess touch since for Aristotle flesh is the medium for touch (*DA* 2 422b35–423b27; cf. *PA* 2 653b19–30); see Section 3.2.3 and n. 127, p. 106.

25 *Met.* 9 1047b32–5; on this contrast, see also Plat. *Tht.* 186C; cf. Hicks, 1965, 357. Cf. also Chapter 5 on human beings' habits as second nature in *Nicomachean Ethics*.

26 On animals' capacity to learn as connected to memory and the sense of hearing, see *Met.* 1 980a28–980b26 where bees are said to be intelligent (*phronima*) but unable to learn (*aneu tou manthanein*) because they do not possess the sense of hearing.

27 See sections 5.3 and 5.3.1.

28 Aristotle uses the notion of *sōteria* (417b3; cf. 416b1–3 and 422b4), which is applied also to the activities of the nutritive soul (*sōzein*) as well as to thinking. All operations that put into actuality the capacities for life preserve the relevant capacities (Polansky, 2010, 235).

29 See section 4.4 below.

30 418a3–6, see Polansky, 2010, 339; also Shield, 2016, 247.

31 For an attempt to reconcile the two discussions, see Modrak, 1987, and Ward, 1988. Modrak makes Aristotle's definition of *aisthēsis* in 2.12 a consequence of the "Actuality Principle," while Ward pinpoints in the sense's discrimination of the difference separating it from the ratio (*logos*) of the sensible; see below. In fact, Aristotle's parallel discussions appear to derive from complementary angles of inquiry and serve different purposes in his treatment of the power of the sensitive soul. *DA* 2.5 ultimately aims at defining the sentient nature of the living being (*zōon*) in light of its exposure to the sensibles and redefining the meaning of "sensitive *alloiōsis*" in terms of *energeia*. On the other hand, *DA* 2.12 gives an abstract definition of sense, valid for all special senses, and with a focus on the transmission of authentic information from the sensibles to the relevant senses.

32 *DA* 2 424a17–29.

33 See Hicks, 1965, 415; cf. Ward, 1988, 219; for different interpretive possibilities of the "reception of forms without matter," see Polansky, 2010, 342–6.

34 For instance, scholars have interpreted differently lines 23–24 (ἀλλ' οὐχ ᾗ ἕκαστον ἐκείνων λέγεται, ἀλλ' ᾗ τοιονδί) giving diverse *relata* for, respectively, ἐκείνων and τοιονδί; for a thorough discussion of the different intepretations, see Caston, 2005, 306, 120n. Here I align with the traditional interpretation voiced by Polansky, taking ἐκείνων as relating to the entities in which the special sensibles mentioned earlier (color, flavor, and sound) inhere, while τοιονδί (as such) relates to the particular quality as depending on *logos* (see below) (Polansky, 2010, 344; Cf. Hicks, 1965, 415).

35 The passage from Chapter 11 of *On the Soul* is relevant for the make-up of the sense organ and is quoted below.

36 On the ambiguity of the expression *kata ton logon*, whose attribution may be either the sensible forms or the sense, depending on the role we assign to the preceding *kai* (explanatory or a conjunction), see Polansky, 2010, 344.

37 See Ward, 1988, 221–2; cf. Modrak who takes the sensible's *logos* between a pair of opposites as indicating "the unique character and unity" of the sensible quality in a way analogous to an arithmetical *logos* which "assigns a single value to the relation between the two numbers" (1987, 56–7). In the same passage, she also notes that in considering sensible qualities and sense perception in light of opposites Aristotle is drawing on the Presocratic philosophers. In this respect, it may be useful to recall Parmenides' fr. 16 which presents the *logos* of the mixture.

38 See, for instance, 439b27, 440a13, 440b19 (about colors deriving from the different proportions of the two extremes white and black), 442a12–17 (about flavors from sweet and bitter), 442b17–19, 448a8–12. The same is true for the other special sensibles, although Aristotle is not explicit about it. Sounds are a mixture composed from acute and grave, odors from sweet and bitter, and touch from hot and cold, hard and soft, moist and dry.

39 Sorabji, 1972, 293–308.

40 *Sens.* 439b29–32. It is also noteworthy that Aristotle considers the attractiveness of a color the sign that such a color consists "in an easily computed number" (*arithmos eulogistos*, 439b32–33).

41 On the sense's reception of the sensible's ratio (*logos*), see Modrak, 1987, 56; Polansky, 2010, 344; Shields, 2016, 249–50. Ward also points out that in his definition of sense in *DA* 2.12 Aristotle mobilizes *logos* in three (complementary) meanings: as an abstract ratio, as the ratio between opposites determining a sensible object, and as the ratio received by the sense and "embodied" in the sense organ. In this case, the author continues, "the sensible ratio as it is received by the sense-organ need not to be embodied by the sense-organ in the same way it is in the sensible object ... Yet, since the *logos*, or abstract ratio is the same in the sense-organ as it is in the sensible object, one may say that special perception apprehends the sensible qualities of things as they are" (1988, 222).

42 Caston, 2005, 303; cf. 1998, 268.

43 For the literalist approach, see Sorabji, 1974, 63–89; 1992, 195–226; 2001, 49–61; Everson, 1997; for the spiritualist, Burnyeat, 1992a, 15–26, 421–434; 2002, 28–90; Johansen, 1998, 252–80.

44 Caston underlines that in the analogy with the wax and the impression of the signet ring the Greek word for the signet ring is *sēmeion*, whose meaning pertains to the field of "signification" (2005, 302).

45 For a similar alternative approach, see, among others, Modrak, 1987, 56–62; Ward, 1988, 217–33; Bradshaw, 1997, 57–113, whose contributions will be addressed in the upcoming discussion of *logos* as a crucial feature of sense perception.

46 On the relation between sensation and knowledge, see below.

47 In Caston's words, "our perceptions have *the backing of the world* and so provide us with a warranted basis from which to operate" (Caston, 2005, 307), where, however that "us" should be extended to encompass all living beings qua sentient. On the special sensibles as real features of external objects, holding causal power over the percipient, see Broadie, 1993, 137–52.

48 See the end of *DA* 2 424a17–29 and n. 41 above.

49 *DA* 2 424a2–11; in *Metereologica* (4 382a16–20) Aristotle discusses touch along the same lines, but with an emphasis on the tactile perception of softness and harshness and the inclusion of the discrimination of what "falls short," in addition to "what exceeds" mentioned in the passage quoted above from *On the Soul*; cf. Bradshaw, 1997, 147.

50 See note above.

51 Besides, given that touch implies an array of sensibles, while the sense cannot discriminate the difference in temperature it is still discriminating the differences in dry and moist (see n. 54).

52 Ebert, 1983, 181–98.

53 On the connection between *mesotēs* and the identification of the sense with *logos*, see Bradshaw, who argues for a dual logos theory in Aristotle's discussion of perception. For this author each sense organ has in fact two pairs of *logos*, one that pertains to its own fixed ratio of opposites and another that fluctuates as it grasps the *logoi* of the sensible forms transmitted by the sense organ (1997, 145–7 and 154–5). More simply, Ward recognizes the "equipoise" of the sense, considered as a form of *logos* in terms of, so to speak, the sympathetic association it engages with by receiving the *logoi* of the sensible forms (1988).

54 In Aristotle's words, the range of sensibles for the sense of touch spans over multiples objects (hot/cold, hard/soft, dry/wet, and all like qualities), see *DA* 2 422b26–28. Hence its organ requires a more complex constitution than the other senses (see below).

55 *DA* 3 435b3–4.

56 *DA* 2 435b2–14. For Polansky touch best illustrates the sense capacity to discriminate sensible differences "by somehow interfacing with and being affected by the differences, and that the sense is an exceptional form of the body" (Polansky, 2010, 333). On the other hand, when discussing the individual senses in *On the Soul*, Aristotle starts with sight, which in the "simplicity" of its organ constitution and overall mech-

anism is emblematic for the other senses (cf. *DA* 3 429a2; for an assessment of sight in respect to the other senses, see Stigen, 1961, 15–41).

57 See below.
58 On this point, see also Caston, 2005, 253–4.
59 *DA* 2 424a25–29.
60 In this respect, Bradshaw further notes that Aristotle uses the term *logos* to cover "both the constitution of the organ and the resulting unique ability (1997, 150).
61 An important exception is the passage in book 1 in which Aristotle asks whether the affections of the soul are all associated to the body or belong exclusively to the soul and claims that the majority of such affections (anger, enthusiasm, desire, and sensation, among others) have a physiological basis. He gives as an example the rise of anger, which in terms of body physiology corresponds to a movement (*kinēsis*) of the body in a particular state, i.e. a surging of the blood and heat around the heart (*DA* 3 403a4–403b1).
62 *DA* 3 435a11–435b1 (with a slight modification).
63 *CG* 2 330a30–330b8, see Shields *ad DA* 3 435a11–b3. In fact, Aristotle recognizes a bigger variety of tangibles but he reduces them to the couple cf opposites hot/cold and dry/moist (see note 54 above).
64 As Lennox remarks, the sense organ as a whole (i.e. the eye or the ear) is still nonuniform but the part receptive to the sensible form is uniform (2001, 182–3).
65 *PA* 2 647a14–21; cf. *DA* 2 423a11–15. This mixture changes from species to species and within the same species. For instance, in comparison to animals', humans' touch is more discriminating (*akribestatē*) on account of the elemental blend constituting their flesh (*DA* 2 421a20–21). For variations in the sense of touch among humans, see *DA* 2 421a23–27 (where Aristotle connects the sense of touch to intelligence); cf. Bradshaw, 1997, 149–50, to whom I owe the example in the upcoming note.
66 *GA* 5 780a22–5; cf. Ward (1988, 224) who further includes the sense-organ for hearing noting that Aristotle's conception of high and low sounds in terms of vibrations presupposes a certain structure of the ear (*DA* 2 420a30–31, b2–3). In considering the sense organ of sight water Aristotle is following Democritus (*Sens.* 438a5–6) whom, however, he criticizes with regard to his theory that vision consists in mirroring. For the association of the sense organs to the elements, see *Sens.* 2 435b16–439a5.
67 *PA* 2 647a9–13. Aristotle criticizes the Presocratics' conception of sensation by likes and conceives of it as a passage of the sense from potentiality to actuality (see above). As for the sense organs constituted by fire and air, he has in mind Empedocles (and Plato) and likely Diogenes of Apollonia, among others. The first thought that the eye was fire (DK 31 A 91), discussed it with the image of the lantern (DK 31 B 84) and attributed animals' capacity to see at night to the fire in their eyes (DK 31 A 86), while the second made cephalic air the "detector" of scents (Thph. *de Sens.* 41).
68 In his discussion of the Presocratics' theory of sensation Theophrastus does not mention their doctrines about touch, noting, for instance, that Empedocles did not treat neither touch nor taste (Thph. *de Sens.* 9 and *passim*).
69 On touch as the primary sense, see *PA* 2 653b19–27; cf. *DA* 2 414b1–7.
70 *DA* 3 434a31–32; cf., for instance, *PA* 2 658a9 (on why the fish does not have eyelids), 3 661b24 (on why no animal has saw-teeth as well as tusks); for emphasis on the connection between *DA* 3 12 and *PA*, see Hicks, 1965, 573.
71 *DA* 3 434a31–434b1.
72 *DA* 3 434b9–30. Distant senses too allow the living being to pursue the means of subsistence and avoid dangers, and hence to preserve its life and reproduce itself.
73 *DA* 3 435b19–26; cf. 435b25–7.
74 On redundancy, see Shields, 2016, 377. As for the contradiction, in the earlier passage addressing the finality of the senses, Aristotle makes taste, besides touch, essential to

the animal, while only the distant senses are for the sake of good life. By contrast, in this passage taste too is recognized to exist for the animal's good life.

75 The living being's overall teleology is that of reproducing itself and attaining the type of immortality that lies within its reach.

76 The order of Aristotle's discussion is, I believe, charged: he first brings up the capacity to hear meaningful sounds and then the capacity to produce meaningful sounds, which (this last) is not itself a sense, but has an impact on another animal's sense.

77 It bears noting that bees are political animals (*Pol.* 1 487b33) without being able to hear (*Met.* 1 980b), and hence learn and communicate.

78 For a discussion of the involvement of *phantasia* in animals' language, see Chapter 5.

79 It is noteworthy that Aristotle proceeds to discuss the lack of sentience of both plants and elements in the same chapter in which he defines sense (*aisthēsis*), showing in this way to underline the uniqueness of animals' capacity and body *vis-à-vis* the other natural bodies, simple and complex.

80 *DA* 3 434a28–30.

81 *DA* 2 424b1–2.

82 *DA* 2 422b25–27, but see also *DA* 2 424a34–b1 where heat is referred to as a perceptible form among the tangibles (*tōn aptōn*); cf. Scaltzas (2010, 203, n. 10) and Murphy, 2005, 299.

83 Ps.-Aristot. *De Plant.* 1 819a30; for a thorough discussion of plants' parts, both uniform and nonuniform, also in relation to animals' parts, see Thph. *HP* 1, 1, 9–2,6.

84 DK 31 A 117/DK 59 A 117; for plants' possession of sensation in Presocratic philosophy and Aristotle's response, see Zatta, 2019, 95, along with note 1, and more extensively her forthcoming article (2022).

85 See *DA* 2 413a20–413b11 and Chapter 2.

86 See above.

87 *DA* 3 424a34–424b3; see above note 56.

88 See Hicks, 1965, 105, 417; Hamlyn, 2002, 43, 114–5.

89 See Polanski, for whom "the plant reenmatters the form of what acts upon it ... undergoing a standard alteration" (2010, 253); similarly, for Scalzas the expression "to be affected with the matter" "points to the vehicle of patience: it is the matter of the plant that suffers the change, by receiving the form" (2010, 204).

90 Regarding plants' closeness to simple bodies, see Themistius (122, 33 H., 225, 2 Sp.); cf. Hicks, 1965, 574–5.

91 *DA* 3 435b26–7.

92 In *Part of Animals* the *dynamis* associated to earth is cold (*PA* 2 646a12–18), while in *On Generation and Corruption* earth is qualified as "cold and dry" (*CG* 330b6).

93 By contrast, animals hold a more active stand and sense because they are so constituted as to be able to produce and maintain their body temperature to a greater extent. For the internal heat-related contrast between animal and plant constitution, see Murphy, 2005, 334–6. Consistently with this line of interpretation, it was pointed out in Chapter 1 that plants do not concoct their food, but merely absorb it ready to be assimilated from the soil (see above n. 157, p. 47). It is noteworthy that, among the Presocratics, Cleidemus seems to have anticipated Aristotle in looking at the difference between animals and plants in terms of body temperature. For he is reported to have claimed that plants "were more removed from being animals, the murkier (*tholerōtera*) and colder (*psykhrotera*) they were" (DK 62.3/Thph. *HP* 3.1.4; see Zatta, 2019, 99).

94 Aristotle's concern with explaining the difference between simple bodies and animals is made more urgent because the receptive parts of all sense organs (but touch) have a uniform elemental basis (i.e. water for sight, air for hearing, cf. *Sens.* 438b17–21), and this fact opens the possibility that simple bodies might be sentient too (cf. *DA* 3 435a11–435b1, quoted above on p. 117). As we have seen in the preceding discussion,

such elemental basis is what enables the sense to act as *mesotēs* and has ultimately to do with the structure informing the elemental basis itself and the overall organ.

95 *DA* 2 424b4–19. Along the same lines, in a complementary passage of *Physics* the inanimate (i.e. the simple bodies) is said to be unaware of being affected by alteration (Aristot. *Phys.* 7.2).

96 See Hicks, 1965, 421.

97 In *Categories* 7 Aristotle mentions perception and its objects—along with knowledge and its objects—as "non-simultaneous relatives" (7b15–8a6), and subsequently clarifies that "perception comes into existence at the same time as what is capable of perceiving—an animal and perception come into existence at the same time—, but the object of perception exists even before perception exists" (8a6–9); cf. also *Met.* 5 1010b35–1011a2, where Aristotle advocates the priority of the sensible object over sensation. Focused on the relation between sentient being and its objects, these discussions are not concerned with the actuality of the sensitive soul; see below.

98 See 4.3 and *DA* 2 422a9–10, 422b18–423a22, where, however, Aristotle raises the difficulty pertaining to the identification of the perceptive part of the tangibles if one considers flesh to be their medium. In other words, in having a medium internal to the body touch cannot be squared with the other senses, and remains obscure (*adēlon*) (for what pertains to its perceptive part).

99 The mention of water as a medium shows that Aristotle intends to address sensation as a phenomenon common to all animals and aims to understand its working in different habitats. As we learn later on, also fish are able to smell thanks to the scent-carrying quality of water (see, for instance, *DA* 2 419a35–419b1).

100 See, for instance, *DA* 3 435a11–20.

101 *DA* 2 415a14–24.

102 Suffice here to note that the relation between these three types of sensibles and the sentient subject is considered in terms of exclusivity and directness. In other words, Aristotle evaluates whether the sensibles are perceived by one or more senses and whether they are perceived per se (*kath'auta*) or incidentally (*kata symbebēkos*).

103 *DA* 2 418a7–26.

104 See Sorabji, 1971, 55–79.

105 *DA* 3 424b22–425a15. In addressing the number of senses Aristotle may in fact likely be responding to Democritus, who according to the tradition attributed a sixth sense to the *aloga zōa*, in addition to wise men and diviners (DK 68 A 116; see Hicks, 1965, 423; on Democritus, DK 68 A 116, see Zatta, 2019, 66). On the targets of Aristotle's argumentation and its systematic unfolding, see, respectively, Gregoric, 2007, 178. and Polansky, 2010, 363.

106 On the relation between Aristotle's psychology and physics, see Giardina, 2009; Polansky, 2010, 223. In *De Caelo* (278a11) Aristotle confirms this relation by claiming that everything that is perceptible subsists in matter.

107 Thph. *de Sens.* 1–2.

108 *DA* 2 417a19–20.

109 See section 4.3.

110 This is the first issue Aristotle raises in his positive discussion of sensation at *DA* 2 417a2–5, see Polansky, 2010, 224.

111 *DA* 3 426a20–22. In this passage Aristotle notes that "the *physiologoi* were at fault in this, supposing that white and black have no existence without vision, nor flavor without taste." He is probably thinking of Empedocles and Democritus who believing in the theory of effluences considered the sensibles as deriving from the encounter between the effluences stemming from the objects and the senses themselves. Outside this encounter, the sensibles would not have a tangible reality (for Empedocles, see DK 31 A 89, A 90, where also eyes' effluences are mentioned; for Democritus, DK 68 A 135 and DK 68 B 9 on the conventional nature of the sensibles such as sweet/bitter, hot/cold, and color; cf. Polansky, 2010, 389). Aristotle corrects this position by distinguishing between potentiality and actuality. When not perceived by living

beings the sensibles exist in potentiality, and in actuality when actually perceived (*DA* 3 426a20–26).

112   On the coincidence of the actualization of the special sensible and ensuing sensation as taking place in the organ of sense, see *DA* 3 425b26–426a19.

113   See below.

114   *DA* 1 409b24–410a1. For the elemental nature of the soul and perception between likes, see also *DA* 1 405b14–17.

115   Besides the difficulty of interpreting identities, the belief in sensation between likes raises other aporias such as the fact that the earthy parts in animals bodies (i.e. hair, nails) do not sense and that the elements' inability to sense what is unlike admits a large margin for ignorance (*DA* 1 409b9–15).

116   For instance, one can perceive movement by touch and sight. *DA* 2 418a18–19. Later on, in *DA* 3 425a27–30, Aristotle will refer to a common sense, (*koinē aisthēsis*), as what senses the common sensibles, although likely in a weak sense as denoting a communality of the five senses (Polansky, 2010, 375) rather than a special faculty emerging from the unification of the five senses (as in Modrak, 1987, 62–71; Everson, 1997, 148–57). There is indeed some controversy as to the involvement of the common sense (intended as a higher-order perceptual power accompanying the special senses) in the perception of the common sensibles; for a full discussion of the problems inherent in this interpretation and an alternative solution that takes "common sense" in a nontechnical way as a perceptual ability shared by each individual special sense, see Gregoric, 2007, 68–82.

117   For the perception of incidental sensibles, see Cashdollar, 1973, 156–75; Modrak, 1987, 69–71; Caston, 1996; Bolton, 2005, 209–24; Scheiter, 2012, 271–2.

118   Modrak, 1987, 55.

119   On the extensiveness of incidental sensibles as pertaining to the range of predicates in Aristotle's *Categories*, see Cashdollar, 1973, 163–7.

120   See Section 6.4.

121   On Aristotle's view, memory builds on *phantasia* (i.e. imagination intended as sensorial representation) and involves the perception of time; see 5.3.1. Perceiving the son of Diares in its immediacy does not imply to contextualize this perception in the past. Rather it is the recognition of the identity of a person in the present.

122   Unfortunately, Aristotle does not elaborate sufficiently on the incidental sensibles for us to understand how incidental sensibles work, but that they allow recognition shows that they bridge a present perception to past perception/s. Given that *phantasia* is defined "as a movement caused by sensation actively operating" (*DA* 3 429a2–3), it is a plausible candidate for the capacity that allows the permanence of sensations in the living being's body allowing it to grasp the incidental sensibles. For the involvement of *phantasia* in the perception of the incidental sensibles; see Nussbaum, 1978, 259; Scheiter, 2012, 268–9; and rather indirectly Frede, 1992, 285 n.18. Also Cashdollar presupposes a storage of images from which the living being draws to perceive the incidental sensibles (1973, 165), but in his intent to explain the incidental sensibles at the level of "pure sensation" he refers to the role of habit, not *phantasia*; cf. Modrak, 1987, 107. On the other hand, Everson forcefully denies the involvement of *phantasia* in the grasp of incidental sensibles on the basis that in *DA* 3 428a12–5 Aristotle mobilizes *phantasia* when sense perception is not fully discriminating (1997, 161–3) and that the common sense holds a variety of perceptual operations (see upcoming discussion), including the sensation of incidental sensibles. At any rate, whatever the explanation, the retention of previous sensory impressions still indicates that for Aristotle perception implies a continuity of the living being's sensorial experience, a continuity that is rooted in the body and related to movement, and hence inherent in the conception of the "animal as such."

123   It bears noting that if in the example of Diares as incidental sensible Aristotle makes it rely on color (and sight), incidental sensibles are grasped through the other primary

senses as well (sound, smell, etc.), a fact that becomes relevant with regard of animals whose sense of smell is more developed than in humans.

124 Aristot. *NE* 7 1141a22–4.
125 *DA* 3 427a16–7.
126 Aristotle addresses these topics respectively at *DA* 3 425b11–25 and 426b8–427a15.
127 *de Somn.* 454a22–454b10.
128 *de Somn.* 445a12–26. In *On the Soul*, however, it is the sense itself responsible for the awareness of sensing (sight for seeing and ear for hearing) (*DA* 3 425b13–26). For different attempts to reconcile this seeming contradiction, see Modrak, 1987, 66–7; Johansen, 2005, 235–76; Gregoric, 2007, 174–92; Polansky, 2010, 386.
129 For the location of the common sense in the heart, see *de Somn.* 456a1–14; cf. *PA* 2 656b24–5. For a thorough discussion of Aristotle's conception of *koinē aisthēsis* and its functions, also in reply to Plato's position in *Theaetetus*, see Gregoric (2007, 52–61, 129–92), who excludes that the common sense is responsible for the perception of common sensibles (193–201) and Polansky, 2010, 386, 393–400.
130 For Aristotle's discussion of the perception of perception (i.e. self-awareness), see *DA* 3 425b12–25 along with Polansky who remarks that it is in taking on the form of its sensible that the sense becomes aware of itself (2010, 380 and 384).
131 Gregoric, 2007, 170.
132 *DA* 3 427a9–14.
133 In respect to the reference to "two" we can here remark with Gregoric the underlying idea that "if there can be no perceptual discrimination between two special perceptibles, neither there can be among three or more" (2007, 145–6).
134 Polansky, 2010, 401.
135 Aristot. *HA* 8 629b8–30, transl. by D.M. Balme.
136 Aristotle's description presumably relies on first-hand accounts by observers and hunters, besides Homer (*Il.* 11 554, 17 663).
137 We may indeed well assume that it is from the advancing movement of the multitude of hunters that the lion senses the danger and undertakes its retreat. In this respect, a passage in *On the Soul* may be illuminating. Aristotle claims that in perceiving a beacon (i.e. a lighted object) a man recognizes that it is fire (incidental sensible), and in perceiving that the fire is moving (common sensible), he understands that it is an enemy (incidental sensible). While Aristotle mentions this example in the context of a discussion of images (*phantasmata*) with regard to the rational soul, it still helps clarify, I believe, how at the level of the sensitive soul, and hence in the sphere of animal life, special, common, and incidental sensibles contribute systematically to animals' awareness (*DA* 3 431b5–7).
138 Number falls under the sensible objects because a single thing is seen as continuous, that is, without breaks or interruptions. Multiple discontinuities between similar sensibles (i.e. colored shapes) indicate multiple units (*DA* 3 425a19–20).
139 Had the lion seen only one "hunter" he would have likely attacked him rather than retreating.
140 It should be stressed that in dealing with sense perception Aristotle is interested in animal awareness in the present rather than of the past, a cognitive capacity corresponding to memory and tied to *phantasia* (see Chapter 5).
141 The soul is mentioned in general terms in *History of Animals* 7 588a19–21; 8 608a10–15 without emphasis on the sensitive soul, but the fact that Aristotle distinguishes animals' capacities by degree and analogy in respect to the human ones indicates that he holds in the background of this discussion the conception of soul partition, presented in *On the Soul*. In not possessing the rational soul, animals have capacities analogous to those of human beings by virtue of the sensitive soul.

142 Significantly, in *Posterior Analytics* Aristotle notes that while in some animals sense perception persists, in others it does not, in which case there is no cognition outside the act of perception. Thus, in *On the Soul* Aristotle concludes his treatment of the animal as sentient at the point he does because he intends to discuss sensation as the common denominator of all animals. As he claims in *Posterior Analytics*, among those animals in which sense perception persists repeatedly, only some are able to draw a coherent impression (*logos*) from it (*Post. An.* 2 99b34–100a3).

143 Aristot. *Post. An.* 2 99b35–100a4; *de Mem.* 449b13–16, 27–28.

144 *Post. An.* 2 99b35–100a4.

145 *de Mem.* 450a23–25.

5   Animal pleasure

From sensation to imagination and beyond

5.1 The questions about pleasure

Sensation implies more than the discrimination and coordination of sensibles along with the subject's awareness of being sensing. For all living beings sensation also brings the feeling of pleasure (*hēdonē*) and pain (*lypē*). Early on in his positive treatment of the soul in *On the Soul* 2, Aristotle writes,

> Now of the faculties of the soul which we have mentioned, some living things, as we have said, have all, others only some, and others again only one. Those which we have mentioned are the faculties for nourishment, for appetite, for sensation, for movement in space, and for thought. Plants have the nutritive faculty only, but other living things have this and the faculty for sensation. But if for sensation (*to aisthētikon*) then also for appetite (*to orektikon*); for appetite (*orexis*) consists of desire (*epithymia*), anger, and wish, and all animals have one of the senses, that of touch (*aphē*); but that which has sensation knows pleasure (*hēdonē*) and pain (*lypē*), the pleasant and the painful, and that which knows these has also desire; for desire (*epithymia*) is an appetite (*orexis*) for what is pleasant (*hēdeos*). Again, they have a sense which perceives food; for touch is the sense that does this. All animals feed on what is dry and wet, hot or cold.[1]

Not all forms of life possess every living function. Aristotle mentions nourishment, appetite, sensation, movement in space, and thought, each deriving from different parts of the soul (the nutritive, sensitive, locomotive, and rational). If humans have all these functions, other forms of life may have some, or only one, i.e. nutrition. Animals share with plants the capacity for nutrition, but in addition they also possess sensation. Sentience implies pleasure and pain, and the feeling of desire (*epithymia*), which Aristotle defines as "appetite for what is pleasant."[2] It ensues that even the simplest animals, namely those that live by nourishing themselves, possess a sensorial constellation made by touch, pleasure, pain, and desire.[3] By apprehending an array of tangibles (such as the opposites of dry and wet, and hot and cold), touch (*aphē*) enables living beings to sense their food (*aisthēsis tēs trophēs*).[4] Hunger and thirst are desire (*epithymia*) for tangible,

DOI: 10.4324/9780367816001-6

pleasurable objects: hunger for what is dry and hot and thirst for what is cold and wet.[5] Thus, we understand, the desire for pleasure represents for Aristotle the subjective side of animal nutrition.[6] It impels animals to eat while it is the actual touching of food that provides pleasure, let alone the discrimination of food.

In *On the Sense* Aristotle complements this account. The focus is now on taste (*geusis*), as a form of touch (*aphē*),[7] here considered instrumental not only to the identification of nourishment and inherent pleasure, but also to animals' movement across space. We read,

> Taste must belong to all animals on account of nutrition, for by it one distinguishes (*diakrinein*) the pleasant and the painful in food, in order to flee the one and pursue the other.[8]

Pleasure indicates the edible food, pain the inedible one: animals are to pursue the first and to flee the second. Crucial in understanding pursuit and flight, the feeling of "focalized" desire (i.e. desire for an object) and its opposite, "focalized" aversion, return again in book 3 of *On the Soul* to explain animals' locomotion and action motivation along with the possession of the locomotive soul, but in more general terms. Animals, Aristotle tells us, move toward the objects, the sense perception of which is associated with pleasure, but move away from those that are associated with pain.[9] While unmentioned, here too nutrition must be at the forefront of Aristotle's thought: food is an object of desire, the fruition of which produces pleasure and animals move across space to access it.[10] But the expression "desire for the pleasant" is general as to encompass other animal activities, such as mating[11] or the search for a more suitable and hence pleasurable environment, all of which involve locomotion and are also objects, along with nutrition, of the ethological discussion of *History of Animals*.[12] More on animal movement across space and desire along with the relevant faculty of *phantasia* will be said later in the chapter. Here let us point out that these references to animal pleasure (and pain) (from, respectively, *On the Soul* 2 and 3) exhibit a context-based ambiguity. For on the one hand (in book 2), animal pleasure is seemingly reduced to the senses involved in nutrition (taste and touch) but on the other (in book 3), what animals desire and flee is left undetermined, indicating that there may be a wider range of pleasurable (and repulsive) objects than food. Further, whether mentioning movement or not, in both passages the internal relation of sensation, pleasure, pain, and desire remains largely unexplored. We are simply told that desire is contingent on pleasure and pain, and these in turn on sensation, or that "desire is for the pleasant." In fact, in his corpus Aristotle is rather reticent to discuss animal pleasure to the extent that Pellegrin has recently argued that nonhuman animals are denied the range of pleasures deriving from sensation available to humans. For this author animals would feel only the pleasure related to touch, hearing or scent (and the other senses) as instrumental to nutrition but would be unable to enjoy colors or sounds or other objects of sense per se as humans do because they lack *logos* and thereby fail to grasp the *logoi* defining the object of sense[13]— hence Aristotle's overall laconism about animal pleasure. But this interpretation is based on the perpetuation of the logocentric, human/animal difference voiced in the ethical treatises and introduces a hierarchy in the very process of sensation, which clashes against Aristotle's pursuit of animals' common nature in the zoological discourse and contradicts his general definition of sense in *DA* 2 12 as a sort of mean (*mesotēs*) between the relevant sensible extremes.[14] Granted, the emphasis on nutrition in animal pleasure is consistent with Aristotle's systemic view of life and the crucial importance of the nutritive process as what constantly maintains the body and its activities,[15] including sensation. Such a view constantly orients his discussion, while pleasure is appealed to as instrumental to animal locomotion, thereby contributing to the comprehension of animals qua creatures that have in themselves the principle of movement and rest. But this approach does not deny that for Aristotle animals may enjoy sensing independently from nutrition or related activities (i.e. reproduction). And indeed, as we will see later in this chapter, the ethological accounts of *HA* alert us to the fact that some animals enjoy musical sounds (reproduced by hunters to capture them) and that they are attracted to scents, at times likely independently of nutrition, as in the case of some fishes.

In sum, if Aristotle has a lot to say about human pleasure and its moral relevance in *Nicomachean Ethics*, in *On the Soul* he still insists on its "coterminosity" with sensation raising the possibility that pleasure and pain might more generally inhere in the living being as sentient (*zōon aisthētikon*) and pertain to any act of sensation intended as actualization of a potentiality beyond the immediate field of reference to touch and nutrition. To put it in questions, how should we understand the relation of pleasure and pain to sensation and life? Is it, so to speak, limited (to touch and taste) or coextensive with sentience and the architecture of sensibles that make up a given living being? And further, how should we understand the internal relation of pleasure, pain, and desire predicated in *DA* 2 and 3?

5.2 Pleasure and pain within and beyond morality

Given the paucity of information in *On the Soul*, to approach these questions we need to revert to *Nicomachean Ethics*, which contains two focused discussions of pleasure in, respectively, books 7 and 10 (the only extensive ones in Aristotle's corpus) besides shorter yet significant references in books 2 and 3. Admittedly, as discussed in Chapter 2, in focusing on human beings *NE* works with an "ordinary," reason-dominated, Platonizing doctrine of the soul, and ultimately offers an evaluative account of pleasure[16] but it still provides us with pertinent elements of reflection to understand the relation of animal pleasure (and pain) to sensation and life on the one hand, and the association of pleasure, pain, and desire, on the other. For while ethics unfolds as the field that has to do with the fulfillment of human (rational) nature and is hence irrelevant to animals, in his treatments of pleasure Aristotle engages with notions that are pertinent to natural philosophy [such as movement (*kinēsis*), generation (*genēsis*), time (*khronos*), and *physis*)] and are therefore relevant to living beings other than humans. Aristotle also considers pleasure in reference to a number of living functions identified in *On the Soul*, namely nutrition, sensation, and thought, and at some points he even addresses pleasure from a biological rather than ethical point of view implicitly pointing to his notion of the "animal as such" or animality, pursued in Chapter 3. For instance, when discussing human pleasure he consistently refers to animals, sometimes to underscore differences (in *NE* 3), but others (of interest to the present project) to highlight a commonality. In *NE* 7, for instance, Aristotle addresses human pleasure considering human beings in terms of their animal nature. We learn that for the ancient *physiologoi* the living being is naturally in pain (*zōon ponei*) because hearing and seeing are inherently painful but with time these "activities" become sensorially neutral on account of the numbing effect of perceptual living.[17] Aristotle does not endorse this position but still acknowledges pain as an ineludible presence in living beings' life with consequences on his conception of human pleasure. Given this interface between his ethical and zoological discourses, the following section will pursue animal pleasure within a discussion of Aristotle's conception of the human good.

5.2.1 From virtue to the naturalness of pleasure

Faithfully to a Platonizing doctrine of the soul, in *Nicomachean Ethics* body and soul are considered in separation rather than in the hylomorphic interdependency of *On the Soul*. Now the parts of the soul should be ideally "in conversation" and subsumed into a logocentric model that enmeshes appetites and desires with the body and attributes them to an irrational part of the soul (*to epithymikon*) capable to listen to the rational one. This shift implies an "allotment" of the living capacities of nutrition and sensation under the umbrella of the appetitive soul and an internal distinction of pleasure into pleasures of the soul (*psykhikai*) and pleasures of the body (*sōmatikai*).[18] The first relate to affections of the soul that do not involve the body, such as courage, love of honor, or the love of knowledge but also the distant senses.[19] The latter are by contrast confined to the contact senses of touch and taste and considered relevant to temperance (*sophrōsynē*), on which more will be said in the upcoming pages.[20] True happiness, the main subject of the treatise, is a soul affair and consists in the active life of virtue, which against traditional understandings is for Aristotle human supreme pleasure.[21] And because virtue depends on feeling pleasure when doing something good and pain when doing something bad (for both bodily and psychological pleasures) pleasure and pain have a crucial relevance to the ethical discourse. Training by habit to feel the right responses becomes key in the acquisition of human virtue/s and happiness and, in the wave of Plato, should be practiced from a very young age.[22]

It is within this general framework that in book 3 Aristotle discusses the pleasures of the body in relation to the five senses[23] articulating the human/animal relation, from the ethical, anthropocentric point of view that has influenced current understandings of animal pleasure.[24] The virtue at stake is temperance (*sophrōsynē*) along with its respective vice, i.e. profligacy (*akolasia*). Aristotle writes,

Temperance has to do with the pleasures of the body. But not with all even of these; for men who delight in the pleasures of the eye, in colors, forms and paintings, are not termed either temperate or profligate, although it would be held that these things also can be enjoyed in the right manner, or too much, or too little. Similarly with the objects of hearing: no one would term profligate those who take an excessive pleasure in music, or the theatre, nor temperate one who enjoys them as is right. Nor yet does temperance apply to the enjoyment of the sense of smell, unless accidentally ... nor do the other animals derive any pleasure from these senses, except accidentally. Hounds do not take pleasure in scenting hares, but in eating them; the scent made them merely aware of the hare. The lion does not care about the lowing of the ox, but about devouring it, though the lowing tells it that the ox is near, and consequently he appears to take pleasure in the sound. Similarly it is not pleased by the sight of a stag or a mountain goat, but by the prospect of a meal. Temperance and profligacy are therefore concerned with the pleasures which man shares with the other animals, and which consequently appear slavish (*andrapodōdeis*) and bestial (*theriōdeis*). These are the pleasures of touch and taste. But even taste appears to play but a small part, if any, in temperance. For taste is concerned with discriminating flavors as is done by wine-tasters and cooks preparing savoring dishes; but it is not exactly the flavors that give pleasure, at least not to the profligate: it is actually enjoying the object that is pleasant, and this is done surely through the sense of touch, alike in drinking and in what are called the pleasures of sex. This is why a certain gourmand wished that his throat might be longer than a crane's, showing that his pleasure lay in the sensation of contact. Hence the sense to which profligacy is related is the most common (*koinonatē*) of the senses.[25]

Temperance is a moral virtue that has to do with observing the mean in the enjoyment of bodily pleasures and properly applies to the sense of touch, the most common (*koinonatē*) of the senses (i.e. all animals partake of it). Despite being a form of touch, taste is by contrast less shared. In *Parts of Animals* Aristotle notes that it is more developed in the human being (than in the other animals)[26] and in the passage above he attributes its enjoyment to the category of wine-tasters and cooks[27] denying that it might play a significant role in the attainment of temperance. The tactile bodily pleasures at stake in this virtue are properly those related to eating, drinking (and sex) and are intended more in terms of quantity, i.e. the body touch-receptive surface, than subtlety as the anecdote of the intemperate gourmand desiring a neck longer than a crane indicates. Human beings that indulge in these pleasures are profligate (*akolastoi*) and show to possess a bestial (*theriōdes*) nature.[28] Aristotle bases this claim on the observation that animals enjoy the distant senses only accidentally. That is, different species find pleasure in specific sights, scents, or "sounds" but only insofar as they signal to them an imminent meal. Animals' real pleasure (i.e. pleasure per se) resides instead in the actual touch that takes place while they eat. We will return to this passage in the course of this chapter when discussing *phantasia* (imagination).[29] Suffice here to remark that this discussion of human temperance agrees with the relevance of animal pleasure to nutrition, hinted at in *On the Soul* 2 and discussed above. Along analogous lines, Aristotle endorses a reductive conception of pleasure and further denies to animals the pleasures of the distant senses in themselves, which he attributes only to humans and discusses in terms of particular "fondnesses." The lover of spectacles finds pleasure in seeing and the lover of melodies in hearing.[30] Animal pleasure instead properly lies in the contact senses, and in particular touch, the mostly shared sensorial capacity, conceived in association with the activity of nutrition,[31] and a human being leads a beastlike life when s/he engages excessively in activities that catalyze the lives of the other animals.[32]

Still, in this discourse the bodily pleasures at stake in temperance are considered natural (the desire for them is called *physikē*)[33] and they play a *positive* ethical role by drawing enjoyment from the sense of touch. Indeed, a man who does not enjoy food and drink (or sex) at all falls into a relevant vice. Significantly, when in *NE* 2 Aristotle addresses the omnipresence of pleasure and its contribution to motivating human choice (*airēsis*)[34] he speaks of a common pleasure (*koinē hēdonē*) that the human being partakes with animals inasmuch as it is an animal itself.[35] In respect to this common pleasure (which relates to the body and involves eating, drinking, and sex), "the good man is likely to be right" because he understands the right mean to abide by.[36] Too much pleasure would make him profligate and like a beast, but no pleasure at all would make him utterly insensible (*anaisthētos*)[37] denying his nature of human being qua sentient living being. So while humans share bodily pleasures with animals because they are animals,[38] not feeling bodily pleasures at all undermines their humanity.[39] Hence significantly for Aristotle humans' animalness is relevant to the attainment of their humanness.

Implied in these early discussions, the naturalness of pleasure emerges from two extensive and complementary treatments in *NE* 7 (Chapters 11–14) and 10 (Chapters 2–5) where transcending easy reductionisms and body and soul dichotomies Aristotle engages with current conceptions of pleasure[40] to pursue an adequate assessment and positive definition of this notion. In book 7 Aristotle defines pleasure as unimpeded activity (*energeia*) of the "natural state" (*kata physin exis*)[41] intending by "natural state" both the healthy condition of the body and the virtuous condition of the soul. In book 10 pleasure is characterized instead as "something that accompanies something else." It perfects the activity.[42] Rather than elaborating on the differences of these definitions, we will focus on their shared points in the effort to obtain a fuller view of animal pleasure. Both definitions stem from the denial that pleasure might be movement, whether referred to as *genesis* in *NE* 7 or *kinēsis* in *NE* 10.[43] Both also engage with the notion of activity (*energeia*) providing a hedonic formula general enough to apply to animals as well. Indeed animals too are concerned with the activities that are proper to them. So if early in *NE* Aristotle presents educated pleasure (of body and soul) as a basis and touchstone of human morality, stressing its instrumental, virtue forming, and revealing value,[44] in these later accounts he changes focus and undertakes a systematic reappraisal of pleasure as a phenomenon of nature. And if it is true that, as scholars have discussed, Aristotle's treatment introduces a break with the earlier discussions in the treatise,[45] it is also true that this break is not as radical as it seems. Aristotle has earlier addressed the role of (body) pleasure as a "natural" phenomenon common to humans and animals alike (the *koinē hēdonē*) and argued more extensively for its overall relevance in the active life of virtue[46] and now proceeds to account for it on a general, all-encompassing basis that appeals to nature, making pleasure intrinsic to life, human and non, within ethics and beyond.

The denial that pleasure might be *genesis* or *kinēsis* (i.e. a movement) aims at undermining the claim that pleasure is not a good,[47] but it does more than this. It rescues pleasure from a confining definition that attaches it merely to the body making it relevant for all the experiences of the soul, whether they involve the body or not. Now pleasure is an activity (*energeia*) and an end (*telos*) (or accompanies an activity) and it can be further distinguished in two subgroups, pleasures that are per se (*aplōs*) and those for the sake of something else (*kata symbebēkos*). Aristotle writes,

> Now the pleasures that restore us to a natural state are only accidentally pleasant, while the activity of desire is the activity of us which has remained in the natural state: for that matter there are some pleasures which do not involve pain or desire at all (for instance, the pleasure of contemplation), being experienced without any deficiency from the normal having occurred.[48]

By "pleasures restoring to a natural state" (*hai kathistasai eis tēn physikēn exin*) Aristotle is here referring to the enjoyments of food and drink, i.e. to the activity of nutrition. Pain emerges from a state of physical lack and triggers desire for the object whose fruition gives pleasure. Hence for Aristotle body pleasures are only incidentally pleasant because they depend on a preexisting condition of deficiency. One feels pleasure as long as it provides a remedy. But in a natural state (i.e. not lacking), one does not enjoy the same things (i.e. food or drink) as in a state of deficiency, and, we may add, there is no pain. On the other hand, pleasures that are per se (i.e. absolutely good) are not conditional upon a bodily lack, but are always enjoyable.[49] In presenting such view, this passage complements the mobilization of pleasure in *On the Soul* and *History of Animals* and further illuminates the nexus of pleasure, pain, and desire. But it also suggests the idea that the animal's life qua sentient is enmeshed with pain: its natural state needs to be restored, with pleasure being constantly dependent on pain and fulfilling a universal natural need.

Later in the same book Aristotle returns to this classification of absolute and incidental pleasures in his response to the claim that pleasure is not a good because both animals and children pursue it.[50] On this view, the (body) pleasure these living beings experience cannot be good because they are deprived of the rational soul.[51] So he argues,

> The arguments that the temperate man avoids pleasure, and that the prudent man pursues freedom from pain and the animals and children pursue

pleasure, are all met by the same reply. It has been explained how some pleasures are absolutely good, and not all pleasures are good. Now it is those pleasures which are not absolutely good that both animals and children pursue, and it is freedom from pain arising from the want of those pleasures that the prudent man pursues: that is, the pleasures that involve desire and pain, namely the bodily pleasures (for these are of that nature), or their excessive forms, in regard to which Profligacy is displayed. That is why the temperate man avoids excessive bodily pleasures: for even the temperate man has pleasures.[52]

Aristotle insists on the association of pleasure with pain to voice its ineludible legitimacy in human (virtuous) life. Some pleasures are absolutely good and others incidentally so. The latter compensate a physical condition of lack and subdue pain, as we have seen earlier. Here, however, we also understand that different virtues involve pleasure and pain differently and that while the temperate man (*sophrōn*) pursues moderate (bodily) pleasures and flees excessive ones, the prudent one (*phronimos*) pursues freedom from pain by enjoying the pleasures of the body (but still positively aims at the pleasures of the soul which grants him a virtuous life).[53]

Yet animals' pleasures are not merely those remedying states of deficiency and revolving around nutrition as the above discussion would lead us to think from a perspective that aims at addressing the human virtues involved with the common pleasure (*koinē hēdonē*) humans partake with animals. Aristotle's definition of pleasure in terms of activity (*energeia*) seems to expand the range of animal hedonic experience transcending the replenishment and tactual pleasure characteristic of nutrition. To fully understand the extensibility of pleasure and how it might rise above nutrition it is now useful to revert to a passage in *NE* 2, which is usually neglected in the commentaries on Aristotle's definitions of pleasure but of critical importance to the present discussion. We read,

Virtue being, as we have seen, of two kinds, intellectual and moral, intellectual virtue is for the most part both produced and increased by instruction, and therefore requires experience and time; whereas moral and ethical virtue is the product of habit (*ethos*), and has indeed derived its name, with a slight variation of form, from that word. And therefore it is clear that none of the moral virtues is engendered in us by nature, for no natural property can be altered by habit. For instance, it is the nature of a stone to move downwards, and it cannot be trained to move upwards, even though you should try to train it to do so by throwing it up into the air ten thousand times; nor can fire be trained to move downwards, nor can anything else that naturally behaves in one way be trained into a habit of behaving in another way. The virtues therefore are engendered in us neither by nature nor yet in violation to nature; nature gives us the capacity to receive them, and this capacity is brought to maturity by habit. Moreover, the faculties given us by nature are bestowed on us first in a potential form; we exhibit their actual exercise (*energeiai*) afterwards. This is clearly so with our senses: we did not acquire the faculty of sight or hearing by repeatedly seeing or repeatedly listening but the other way about—because we had the sense we began to use them, we did not get them by using them. The virtues on the other hand we first acquire by first having actually practiced them, just as we do with the arts. We learn an art or craft by doing the things we shall have to do when we have learnt it: for instance men become builders by building houses, harpers by playing on the harp. Similarly we become just by doing just acts, temperate by doing temperate acts, brave by doing brave acts.[54]

In this passage Aristotle claims that virtues are acquired by habit and supports this claim by contrasting them with the dispositions given by nature. Fire, he tells us, cannot be trained to move downward, but will always go up. Earth always falls down, no matter how many times it is thrown up in the air. You cannot train these natural bodies to behave differently. This feature pertains to everything that is naturally disposed to behave in a certain way, that is, we may add, equipped with a body that makes it so disposed.[55] But the case of (human) virtues and arts is different. While not born strictly just, a human being can become just by acting with justice and s/he will become courageous by acting courageously. Likewise, s/he becomes a harper by playing the harp or a builder building houses. Nature does not make human beings virtuous or skillful in a particular virtue or art (it does so with nonhuman animals though),[56] but gives them the capacity to become such, an intrinsic and wonderful malleability.[57] Aristotle's ethics builds on this aspect of human nature. For mindful of it, Aristotle advocates that human beings should be trained in feeling appropriate pain and pleasure in order to acquire and continue practicing virtue.

   In the above passage Aristotle does not merely mention the elements' behavior. He also appeals to the capacities that are given by nature and refers to the senses, which (unlike virtues or arts) are given in a potential form (ready to be actualized) and not acquired by habit. The human being qua living being (*zōon*) merely activates them by using them and this activation is called *energeia*.[58] So one does not need training to see or hear, but sees and hears immediately, as long as there is a functional organ of sense and a relevant stimulus, although, we may also add, previous perceptions are essential to the identification of an object of sight and hearing (qua incidental sensibles). What seems significant to stress at this point is that activity (*energeia*) is also the notion involved in the definition of pleasure and that, in this hedonic application, it pertains to both nature-given and acquired dispositions, which (these latter) end up to constitute a "second nature."[59] Hence, by being located at this critical intersection (of the capacities given by nature and those provided by habituation), the unimpeded *energeia* that is pleasure overarches nature and ethics (as second nature) and concerns humans and animals alike. So conceived, pleasure is intimately related to the activity of life (i.e. actual living) in the different natural dispositions that make it up, namely the different senses and activities, with nutrition being *only* one of them. "What is natural is pleasant," claims Aristotle in *History of Animals* to comment on animals' dietary inclinations in terms of bodily constitution.[60] And although he obviously does not elaborate on these points in *NE*, we can think at the pleasure inherent in animals' fulfillment of the natural relation to their habitats and diet. A fish cannot be trained to be in the air and fly, nor a terrestrial bird to live in a swamp. And interestingly even within their terrestrial or aquatic habitats, animals find pleasure in specific environments. Horses, for instance, enjoy (*khairein*) meadows and marshes because they like to take baths (*philoloutron*), drink muddy water, and water in general (*philydron*). Again, a carnivorous animal will not become herbivorous, and vice versa, and within the type of diet generally ascribed they will enjoy specific foods.[61] All living beings find pleasure in what is natural to them. And if their nature is that of being sentient they will find pleasure in activating it in the different ways they are naturally predisposed to sentience, distant senses included, beyond taste and nutrition. Significantly, Theophrastus will make explicit what Aristotle left rather understood in this respect, forcefully claiming that "sense perception is in accord with nature, and no such process does violence and brings pain, but rather it has pleasure as its accompaniment."[62] But, besides sense, diet, and locomotion, we can also think at animals' "poietic activities," i.e. at the spiders' webs or at the ants' underground chambers, which are also made on account of nature (and not by art).[63] Pleasure should be implied in all these and like activities that animals accomplish according to nature, and that are not impeded. In sum pleasure is or emerges (depending on which definition we chose, i.e. according to book 7 or 10) from the fulfillment of one's nature, whether animal or human, strictly natural or trained.

And indeed, in book 10 of *NE* Aristotle writes,

> And it is thought that every animal has its own special pleasure, just as it has its own special function: namely, the pleasure of exercising that function. This will also appear if we consider the different animals one by one: the horse, the dog, man, have different pleasures—as Heracleitus says, an ass would prefer chaff to gold, since to asses food gives more pleasure than gold. Different species have different kinds of pleasures. On the other hand it might be supposed that there is no variety among the pleasures of the same species.[64]

This passage is part of a longer chapter (5) devoted to account for the existence of different kinds of pleasures in relation to relevant activities (*energeiai*), which they perfect. The pleasures of the senses are different from the pleasures of the intellect inasmuch as the activities themselves are different. Here Aristotle's outlook is hierarchical and anthropocentric, consistently with the subject of *NE* and the Platonizing approach to the soul.[65] Pleasures are discussed in terms of purity, with those of the intellect at the top and those of touch at the bottom.[66] But casting hedonic hierarchies aside, the difference among pleasures extends to animals as well. Each one among them has its own special pleasure (*oikeia hēdonē*), inasmuch as each one performs an activity (*ergon*) proper to itself and pleasure relates to this kind of activity. Citing Heraclitus, Aristotle mentions the pleasure of a donkey, which differently from a human being enjoys more chaff than gold.[67] Along the same lines, he extends the difference of pleasure to other species: horses, dogs, humans, all have their own.[68] But he also stresses how the human being is the animal that has the most variety of pleasures because the activities s/he undertakes are the most various (i.e. a lover of music would find pleasure in music rather than philosophy). The other animals by contrast seem all to enjoy the same sets of pleasures, without intraspecific variety and certainly with fewer activities than humans. Still Aristotle shows some hesitancy in asserting categorically nonhuman animals' intraspecific homogeneity of pleasures and he makes it clear that even within the same species animal pleasures are plural and transcend mere nutrition.[69]

To elaborate further on animals' plurality of pleasures we may return to *History of Animals* in which the fulfillment of a number of natural dispositions points to the attainment of "good life." In this treatise Aristotle refers to different birds with the adjective *eubiotos* (which translates as "having a good life"). Not occurring alone, this qualifier is accompanied by one or more adjectives, which actually clarify animals' life enjoyment in respect to the activities they engage with. The brinthus is a songster (*ōdikos*), the trochilos ingenious (*tekhnikos*), the swans are good parents (*euteknoi*) and age well (*eugēroi*), while the aigithios has many young (*polyteknos*).[70] Also the elea lives well (*eubiotos*) and,

> Sits in summer where there is breeze and shade, but in winter in a sunny and sheltered place on the reeds round the marshes.[71]

Adapting to the variation of climactic conditions, this bird changes residences and likely enjoys the optimal living space it has reached. From these observations, it seems, animals' good life is made up by a diverse range of activities and dispositions, such as prolificness, parenthood, and resourcefulness, including (with the regard to the *eugēroi* swans) a protracted vitality in the old age. In these cases (as it is for humans), the pleasure is inherent in the activity,[72] does not involve replenishment, and is per se. Further it might also be an incentive to perform a given activity and to do it well.[73]

On the other hand, in *HA* we also learn about a miserable subspecies of eagle whose heavy and clumsy body condemns it to be prey of other animals rather than being an active predator itself. Thus impeded, this bird is characterized by a bad life (*kakobios*),[74] that is, a life of inertia, passivity, and pain.[75] By contrast, animals' life enjoyment and, we could add, quality reside in movement, "agency," and freedom (intended as being in the condition to move and "act" in whatever way it is natural to a given animal). Ultimately, different species possess various and enhanced dispositions to pleasure while some of them are constitutionally more prone to pain. And as for the pleasures of the activities (*energeia*) of the distant senses, Aristotle may have denied them to animals in his discussion of human temperance in *NE* 3 but this denial seems to stem from his overall ethical focus and the concern with establishing criteria about bodily pleasures that might lead to assess and exercise human virtue. If however we pursue his discussion of pleasure as unimpeded activity (or as emerging from an activity) that is natural and hence essential to animals' constitutions (along with Theophrastus' claim cited above) the activity of the distant senses too provides some enjoyment by virtue of being a realization of a living being's nature qua sentient (granted, when the sensible object is not so strong as to cause pain). In fact also the pleasure related to the distant senses may be ultimately differentiated as it depends on animals' specific natural dispositions to sense. Aristotle is not interested in pursuing this aspect in *HA*, but he still tells us that fishermen used to anoint with salt the mouths of the holes in rocks where fish stood to entice them out.[76] But even more emblematically, he also tells us that hunters used to capture the deer by singing and playing the pipes. At hearing the melodies, the animal would feel pleasure (*hēdonē*) and lie down.[77] This last report shows that some animals do feel the pleasure of hearing per se, and not only incidentally as stated in *NE* 3, suggesting that Aristotle's reticence to making strong claims about animals' enjoyment of the distant senses might depend also on the actual lack of evidence, in addition to a systematic view of animal life that turns around the functions of the nutritive soul and the constant effort to preserve one's life.

5.2.2 Life and pleasure

An important doctrine with which Aristotle deals in both his extensive treatments of pleasure in *NE* considers the universal pursuit of pleasure a proof that it is the supreme good. In book 10 this doctrine is attributed to Eudoxus of Cnidus[78] and some points mentioned in the relevant passage have been discussed in the section above, but one in particular is "new" and anticipated in book 7. We read,

> Moreover, that all (*apanta*) animals and all human beings pursue pleasure is some indication that it is in a sense the Supreme Good ... but they do not all (*pantes*) pursue the same pleasure, since the natural state and the best state neither is nor seems to be the same for them all; yet still they all pursue pleasure. Indeed it is possible that in reality they do not pursue the pleasure which they think and they would say they do, but all the same pleasure; for nature has implanted in all things (*panta*) something divine.[79]

All living beings, human and non, pursue pleasure. Pleasures are different because living beings' nature and best disposition are different.[80] All this is said in book 10 as well. But despite epiphenomenal differences for Aristotle all living beings may in fact pursue the *same* pleasure.[81] The reason Aristotle invokes to claim this hedonic identity transcends partisan categories (i.e. the human) and the overall anthropocentric focus of *NE* to embrace all sentient beings. That is, the pleasure is the same because "all living beings possess by nature something divine (*theion*)."[82] This realization is in line with the egalitarian trend of Aristotle's zoology.[83] It echoes his discourse on animals in *Parts of Animals* and his discussion of reproduction in *On the Soul*. For in both passages divinity has been associated to animals. In *PA* 1.5 Aristotle grants legitimacy to the study of animals against potential critiques by appropriating Heraclitus' dictum that even in a humble kitchen there are gods.[84] With it he ultimately refers to the teleological causality, which is inherent in the bodies of animals as creatures of nature and which in the same treatise will be subsequently further defined as having to do with the complex action that is life. For living beings are so constituted by nature (with a set of functional, synergetic parts) in order to live. So ultimately divinity pertains to animals' life. On the other hand, in *DA* 2.4 living beings are presented as aspiring to the immortal (*to aei*) and the divine (*to theion*) which they fulfill within the limits allowed to them by means of the reproduction "of one like itself." In fact, for every creature *all natural functions* are said to be taken in order to partake of the immortal and the divine.[85] In light of this intertextuality, then the hedonic relevance of Aristotle's claim ("all things possess something divine by nature") appears to pertain to living beings' inherent capacity and goal to live but fulfilled in the modalities that are naturally suitable to, and characteristic of, each of them. And in the case of humans, these modalities include moral training and choice.[86]

We find support for this interpretation in a passage of *NE* 10, which reduces the plurality of human pleasures down to one. Aristotle writes,

> It might be held that all men seek to obtain pleasure, because all men desire life. Life is a form of activity, and each man exercises his activity upon those objects and with those faculties he likes the most: for example, the musician exercises his sense of hearing upon musical tunes, the student his intellect upon problems of philosophy, and so on. And the pleasure of these activities perfects the activities, and therefore perfects life, which all men seek. Men have good reason therefore to pursue pleasure, since it perfects for each his life, which is a desirable thing. The question whether we desire life for the sake of pleasure or pleasure for the sake of life, needs to be raised for the present, in any case they appear to be inseparably united; for there is no pleasure without activity, and also no perfect activity without its pleasure.[87]

Whether rightly trained or not, moral or immoral, intellectual or creative,[88] all human beings pursue pleasure because they all desire to live. The particular fondness each of them strives after, fulfilling its nature and disposition, can be reduced to the same universal core. No matter what facet pleasure takes on, each time it fulfills humans' desire to be alive. For life is an activity and pleasure makes them act, and hence live. And while in this passage Aristotle confronts the eclectic plurality of human pleasures, his claim applies also to the other animals, who engage in less numerous pleasurable activities than humans but still partake of that common pleasure (*koinē hēdonē*) which characterizes the life of all living beings.[89] Thus it is significant that early in the same book 10 Aristotle confronts the denial that pleasure might be good and points out its universality by claiming that the detractors of pleasure could be right if pleasure were pursued only by the irrational animals (*aloga zōa*). But as it is, also the intelligent humans (*phronimoi*) run after it.[90] Differently from earlier parallel claims,[91] in this predicament Aristotle introduces the "*logos*-based" difference between humans and animals to denounce powerfully its irrelevance in a discourse that in the end sees pleasure interwoven with life, bridges ethics to nature, and equalizes all animals, humans included. Granted, from an ethical point of view not all pleasures (and hence lives) are equal,[92] but from a biological one all living beings are equal in that they search for pleasure and live on account of pleasure rather than for the sake of pleasure. On this last point let us finally remark that Aristotle's refinement in *NE* 10 of his definition of pleasure as "perfecting the activity" rather than "unimpeded activity" has likely to do with his disavowal of hedonism on all fronts. For if active virtue remains the foundation of his ethical doctrine in his zoological discourse pleasure stands as instrumental to animal life rather than being its end.

5.3   Animals' desire, phantasia, locomotion, and communication

All living beings feel desire,[93] even the most simple as the stationary ones because they have sensation. In *On the Soul* 2.3 Aristotle defines desire (*epithymia*) as "an appetite (*orexis*) for what is pleasant,"[94] and in 3.10 he sees in it the force that explains animals' movement across space (*kinēsis kata topon*). In his view, this kind of movement is always purposive, that is, for the sake of something (*heneka tou*).[95] Hence *to orektikon*, the part of the soul that enables animals to desire, corresponds to the part that moves them (*to kinoun*)[96] in view of an object (*to orekton*).[97] In this respect, we may incidentally note that in defining *epithymia* through *orexis* and using in his discussion of purposive locomotion a word related to the latter (*to orekton*) Aristotle likely wants to indicate the practical and pragmatic side of animals' movement as taking place for the sake of something that has a physical reality. In other words, if in *On the Soul epithymia* denotes animals' intrinsic condition of desire qua sentient beings, *orexis* qualifies that desire in relation to something tangible[98] and such that it both limits and explains animals' movement in space.[99] In contrast to the restricted and predictable locomotion of the natural bodies of the elements, always intrinsically (and unconsciously) directed toward their natural places,[100] animals' locomotion is much more fluid and versatile, but also orderly and purposeful.

When the objects are fully within animals' perceptual field, sensation suffices to trigger desire but if they are removed from it *phantasia* plays a crucial role.[101] In *On the Movement of Animals* Aristotle writes,

> The origin, then, of movement, as has already been said, is the object of pursuit or avoidance in the sphere of action (*en tōi praktōi*), and heat and cold necessarily follow the thought and imagination (*phantasia*) of these objects. For what is painful is avoided, and what is pleasant is pursued. We do not, it is true, notice the effect of this in the minute parts of the body; but practically anything painful or pleasant is accompanied by some degree of chilling or heating. This is clear from the effects produced. Reckless daring, terrors,

sexual emotions and other bodily affections, both painful and pleasant, are
accompanied by heating and chilling, either local or throughout the body.
Memories too and anticipations, employing as it were, the images of such
feelings, are to a greater or less degree the cause of the same effects ... For
the affections suitably prepare the organic parts, the desire prepares the affections, and imagination prepares the desire while the imagination (*phantasia*)
is due to thought or sensation.[102]

Essential to animals' practical life (*en tōi praktōi*), *phantasia* prepares desire
by enabling the living being to imagine what is to be pursued or, conversely,
fled. Desire in turn sets the affections (confidence, fear, sexual attraction, etc.),
whose concomitant heating or chilling make ready the specific body parts that
are involved in the actual locomotion.[103] Internal heating accompanies the representation of the object to be pursued, cooling that of which is to be avoided. Here
Aristotle reminds us that *phantasia* either relates to sensation or to thought, a
distinction which is further clarified in *On the Soul*: *aisthētikē phantasia* belongs
to living beings qua sentient, while *bouleutikē phantasia* is exclusive to human
beings qua rational (*logistika*) and indispensable to their attainment of rational
desire and will.[104] As the qualifier *aisthētikē* indicates, animals' *phantasia* stems
from sensation. In *On the Soul* 3.3 Aristotle pinpoints this causal relation defining *phantasia* as "a movement produced by sensation actively operating" (*kinēsis
hypo tēs aisthēseōs tēs kat'energeian gignomenēn*)[105] and refers to this "practical,"
"action-oriented" use of *phantasia* in living beings' life (*zōa*) through a contrast
between animals (in this case, called *thēria*) and humans (*anthrōpoi*). He claims,

> Because imaginations (*phantasiai*) persist and resemble sensations living
> beings (*zōa*) do (*prattei*) many things according to them, some, *viz.* the ani-
> mals, because they have no mind, and some, *viz.*, men, because the mind is
> temporarily clouded over by emotion, or disease, or sleep.[106]

Phantasiai ("imaginations" that depend on *phantasia*) allow nonhuman animals
devoid of mind (*nous*) to carry out their activities and are hence considered in
parallel with mind and the role that mind plays for the human beings,[107] although
aisthētikē phantasia and its products are still relevant also for those human beings
whose rational power is temporarily unavailable as when overcome by emotions
or sleeping or in the grip of disease. Thus considered in the context of animal
life, *phantasia* contributes to animals' locomotion enabling them to transcend the
presentness of sensation[108] and to project themselves into the future,[109] allowing
order and consequentiality in their actions and lives. For by envisioning prospects
and moving purposefully across space to attain them, animals carry out their life-
centered activities, from food and mate acquisition to parenting and migration.

It is a question for Aristotle whether all animals have the capacity for *phantasia*.[110] His position seems to oscillate in this respect, and the lack of a defini-
tive claim has given rise in the secondary literature to divergent opinions, which
depend on the very interpretation of what *phantasia* consists in.[111] Early on in *On
the Soul* (2.2) Aristotle says that wherever there is sensation there is *phantasia*,[112]

but in 3.3 that not every animal has *phantasia*.[113] In *On the Soul* 3.9 he returns to this question, asking whether those animals, which are capable of locomotion but only possess the sense of touch have *phantasia*. As Philoponus clarifies, Aristotle is here thinking about the *zōophyta*, the "animal plants," such as corals, sea anemones, sponges, and some kinds of mollusks.[114] Do these creatures, which are devoid of eyes, ears, and nose with relative senses, have *phantasia* and desire? Aristotle admits that they have desire because they are liable to pain and pleasure, and about imagination he allows the possibility that they might have it only in an indeterminate way (*aoristōs*), that is, analogously to the kind of locomotion they possess.[115] This is an interesting clarification, for more than one reason. For not only does it stress *phantasia*'s relevance and extensibility to animals endowed with locomotion, hence excluding those creature that feel, but do not move.[116] It also reveals that *phantasia* with due qualifications might be available to those animals that are equipped with only the contact senses triggering their indeterminate movement. But it further indicates that if a bare sensitive apparatus consisting in the sense of touch is enough for animals to have *phantasia* (provided that they are endowed with locomotion) the power of *phantasia* relies on the sense perceptions of all the distant senses,[117] and not merely sight as one would mistakenly assume by considering a number of factors—*phantasia*'s etymology in *DA* 3.3 (from *phaos*, light)[118] along with its modern translation as "imagination." In fact, inasmuch as in *DA* 3.3 *phantasia* is defined as the product of actualized sensation without further qualification, it must relate to all the senses that are naturally within an animal's reach, i.e. proper, common, and incidental.[119]

Conceived of in terms of the wider perceptual network that makes up living beings' life, *phantasia* relates to the perceptual fabric of a given animal and is fundamentally synesthetic. It obviously pertains, for instance, to animals whose senses of hearing and smell are more acute than sight, or operative, we may assume, in conditions of sight disability. In this case, olfactory or acoustic "imaginations" (*phantasiai*) or "appearances" (*phantasmata*) are more involved in representing the object of desire (and locomotion-producing) than visual ones, and integrated in specific combinations that reflect the sensorial experiences constituting the life of a given animal. It is a pity that Aristotle does not give us more information in this respect. In *On the Soul* 3.3 he is admittedly more interested in the theoretical identification of *phantasia vis-à-vis* other psychological powers (sensation and thought)[120] and in its functional, "sight-based" and hence human-centered definition and relevance (given that sight is the most important sense for the human being)[121] along with the emergence of error.[122] But the passage from *Nicomachean Ethics* mentioned in the preceding section about the lion's pleasure in hearing the ox offers us an interesting script. The lion hears the lowing ox and rejoices at the prospect of eating it (anticipated pleasure), and, completing Aristotle's account, we may well assume that it moves toward its prey to have its meal. In this scenario an incidental sensible (i.e. the "ox" voice), grasped via a simple sensible (i.e. the vocal sound consisting in a specific ratio between high and low), makes the predator know about the prey's existence and location and triggers the *phantasia* of the ox as an object of desire (*orekton*) making the animal move and act.[123] When conceived as a "positive" notion stemming from animals' previous sensations,[124] *phantasia* constitutes a synthetizing, critical capacity that allows each living being that possesses it to understand reality in a temporal and spatial continuity beyond the circumscription and, so to speak, solipsism of sensation and to "act"—whether pursuing pleasure or fleeing pain.[125] So too Beare had eloquently claimed that,

> Without its [*phantasia*'s] aid sense-perception would be confined to momentary *energeiai*, lacking in continuity, unassociated, incapable of forming a basis of *empeiria*.[126]

Phantasia may be triggered by a collateral perception and is itself caused by perception/s, but it is ultimately conceived, at least for animals devoid of reason, as the cognitive capacity that empowers them to act on the basis of the embodied awareness they have acquired through their sentient experience of the world. In sum, *phantasia* is key to understanding how animals are the origin of their own movement across space (and rest) in an existence shaped by purposeful and life-bent activities.

Not only involved in animal movement, for Aristotle *phantasia* is also the faculty that allows animals to communicate. This role has been differently interpreted[127] but what can be safely inferred is that in this case too *phantasia* breaks the solipsism of sensory *energeiai* enabling animals to reproduce and convey through vocal sounds their sensorial experiences. Not random utterances, animals' cries carry a meaning that is rooted in their physical contact with reality. In *Politics*, let us recall, Aristotle contrasts human language (*logos*) with animal voice (*phonē*): the first conveys right and wrong, while the second is limited to the expression of pain and pleasure.[128] So as it was for the practical sphere (*en tōi praktōi*),[129] in the case of communication too *phantasia* ultimately holds for creatures devoid of reason a function parallel to that plaid by *logos* for the rational human beings. If human language works with "disembodied" intelligibles (*noēta*) and enables its possessors to express right and wrong, *phantasia* stems from the *aisthēta*[130] and holds a representational role that, tied to the living being's body and its particular experiences, constitutes an integral part of that animality (i.e. the animal as such), which is at the center of Aristotle's zoological study.[131] As for the representational nature inherent in animal voices, it seems to go beyond the mere expression of pain and pleasure, claimed in *Politics*, and reach at least for some species a certain complexity, as we understand from a curious event that happened in Aristotle's lifetime (in 394 or 404 BC), reported in *History of Animals*. Numerous mercenaries went to help Larissa's dynast Medius in his fight against Lykophron of Pherae, but they were horribly massacred at Pharsalus, and lay in heaps on the open field. Concurrent to this event, all the ravens inhabiting the regions of Attica and Peloponnese deserted their places and, to fill Aristotle's account, convened in Pharsalus to prey on the corpses.[132] Aristotle interprets this behavior with some caution but still admits that it was "as if" the ravens "had some perception of 'communication' (*aisthēsis dēlōseōs*) from each other."[133] In some (for us and Aristotle) obscure ways the ravens from Pharsalus must have informed those of Attica and Peloponnese who left their areas for a specific destination.

Hence *phantasia* emerges not only as the capacity that enables animals to possess a sensorial-based awareness of situations and surroundings and to act purposefully or to express pain and pleasure, but also what empowers them (admittedly, in some cases) to reveal such awareness to other conspecifics. Receptive to other ravens' meaningful voices, the ones from Southern Greece imagine pleasurable prospects and, in this anecdote, fly North to pursue their place-bound desires.

5.3.1 Dreams, memory, and the physiology of phantasia

A child of sensation, *phantasia* "works" with living beings' previous sensory impressions. In addition to being involved in animal locomotion and language, its products, called appearances (*phantasmata*) or "imaginations" (*phantasiai*)[134] further constitute the building blocks of dream-production and memory, discussed in *Parva Naturalia*. This title—"little natural treatises"—labels a collection of short studies that presuppose the psychological partition of *On the Soul* and delve into actions (*praxeis*) that are common to body and soul[135] and ultimately pertain to the "animal as such." Here too the search for what animals share is a driving force of Aristotle's methodological procedure.[136] We have touched upon *On the Sense* in Chapter 4 with regard to the mechanism of sensation. In *On Dreams* Aristotle clarifies that dreaming depends on the sensitive part of the soul (*to aisthētikon*) qua imaginative (*to phantastikon*) and considers sleep and dreams interrelated characteristics of the living being (*zōon*).[137] As in the case of sensation, also *phantasia* depends on animals' constitution. In *On the Soul* Aristotle merely defines it as a motion (*kinēsis*) originating from actualized sensation, but in *On Dreams* he discusses the nature of *phantasia*'s movements in terms of body physiology and illustrates how rising from sensation such movements contribute to dream-production and sensory illusions. Aristotle writes,

> The sensation still remains perceptible even after the external object perceived has gone, and moreover ... we are easily deceived about our perceptions when we are in emotional states, some in one state and others in another; e.g. the coward in his fear, the lover in his love; so that even from a faint resemblance (*apo mikras homoiotētos*) the coward thinks that he sees the enemy, and the lover his loved one; and in proportion to his excitement, his imagination is stimulated by more remote resemblance (*ap' elassonos homoiotētos*) ... So men in fever sometimes think they see animals on the walls from the slight resemblance of marks in a pattern. Sometimes the illusion corresponds to the degree of emotion, so that those who are not very ill are aware that the impression is false, but if the malady is more severe, they move themselves in accordance with what they think they see. The reason why this happens is that the controlling sense does not judge these things by the same faculty as that by which sense image occurs ... It is evident from the foregoing that stimuli arising from sense-impressions, both those which have their origin from without and those which have origin within the body, occur not only when we are awake, but also when the affection we

call sleep supervenes, and even more at that time. In the daytime, when the senses and the mind are active, they are thrust aside or obscured in the same way as a small fire is obscured by a greater, and small pains and pleasures by great, although when the latter have ceased even the small ones come to the surface; but at night, because the particular senses are at rest and cannot function, owing to the heat's reversing its flow and passing from the outside to the inside, these stimuli reach the starting-point of sensation and become noticeable, as the bustle subsides. One may suppose that like the eddies often seen in rivers, each movement takes place continuously, often with unchanging patterns (*homoiōs*), but often again dividing into other shapes (*eis alla skhēmata*) owing to some obstruction. For this reason no dreams occur after food or to the very young as infants; for the movement is considerable owing to the heat arising from the food. Hence, just as in a liquid, if one disturbs it violently, sometimes no image (*eidōlon*) appears (*phainetai*), and sometimes it appears but it is completely distorted, so that it seems quite different from what it really is, although when the movement has ceased, the reflections are clear and plain; so also in sleep, the images or residuary movements that arise from the sense-impressions are altogether obscured owing to the aforesaid movement when it is too great, and sometimes the visions appear confused and monstrous, and the dreams are morbid, as occurs with the melancholic, feverish, and the intoxicated; for all these affections being spirituous, produce much movement and confusion. In animals that have blood, as the blood becomes quiet and its purer elements separate, the persistence of the sensory stimulus derived from each of the sense organs makes the dreams healthy, and causes an image to appear, and the dreamer to think, because of the data supplied by the organs of sight and hearing, that he really sees and hears. The same is true for the other sense organs."[138]

Sensation *per se* may be "ephemeral" and dependent on an outside object, but its effects continue into *phantastikai kinēseis* ("imaginative movements") that remain stored in the living being's body after the actual sensations have ceased and emerge into the "consciousness" through the blood in specific conditions. In normal (nonpathological) circumstances, the movements of *phantasia* disappear during the day under the actuality of sensations but they appear at night when the senses are inactive. For as the body heat withdraws during sleep, the blood carries the *phantastikai kineseis* from the peripheral regions of the body (i.e. the organs of sense) to the center, i.e. the heart, which is the origin of sensation. Dreams form at this time in the wake of the body's different internal motions, which Aristotle illustrates by means of an analogy. As with the propagation of circular movements in water, the *phantastikai kinēseis* may reproduce faithfully the "original" sensory pattern but also distort it if something intervenes to block them. In the first case, dreams will be coherent and we may add, life-like; in the second, warped and confused. Overall, it is the hematic quality and state that make the difference. A pure, quiet blood allows for the emergence of healthy, reality-adherent, well-structured dreams but a turbid and turbulent one will lead to erratic visions. When, however, the movement (and heat) internal to the body is too strong no dreams are produced as in the case of sleep after food (due to the process of digestion) or in that of (growing) infants.

While producing dreams at night, the *phantastikai kinēseis* may surface to consciousness, also during the day taking over the process of sensation and distorting the reception of reality. Aristotle mentions human beings that are prey of the emotions (*pathē*) of fear and love, and superimpose their internal affections to the actual sensations, mistakenly seeing in somebody else their enemy and lover. But illusions arise equally from other pathological states such as diseases as when fevering people see animals on a wall on the basis of a few marks on its surface. In all these instances, an object of perception presents small similarities with the appearances (*phantasmata*) stored inside the body, triggering the deceitful sensory illusions. Ultimately, we understand, the seeming veridicity of the illusions depends on the incapacity of the common sense to discriminate the appearances based on what produces them, the sensitive soul qua imaginative. As in a short circuit, instead, the common sense processes the *phantasmata* as if they were deriving from the sensitive soul qua sensitive and hence in terms of the sensible objects. The more severe the pathological condition the more indomitable and real-looking the products of *phantasia*. In extreme cases, the *phantastikai kinēseis* originate from the living body itself detached from its own surrounding.

In Aristotle's psychological system *phantasia* also supports the capacity for memory (*mnēmē*), which is discussed along with recollection (*anamnēsis*) in the homonymous treatise. While recollection is exclusive to human beings because it implies reasoning (*logismos*),[139] many living beings besides humans are able to remember.[140] According to *Metaphysics* it is indeed memory that enables animals to be intelligent and learn.[141] By relying on *phantasia* memory too stems from sensation, and it differs from either of them because it involves the consciousness (*prosaisthanesthai*) to have seen, heard, or learned something before (*proteron*).[142] In *On Memory and Recollection* Aristotle elaborates on the relation of memory to *phantasia* and perception, and proceeds to discuss its physiology. He writes,

It is obvious, then, that memory belongs to that part of the soul to which imagination belongs; all things which are imaginable (*phantasta*) are essentially objects of memory, and those which necessarily involve imagination are objects of memory only incidentally. The question might be asked how one can remember something which is not present, since it is only the affection that is present, and the fact is not. For it is obvious that one must consider the affection which is produced by sensation in the soul, and in that part of the body which contains the soul—the affection, the lasting state of which we call memory—as a kind of picture (*zōgraphēma*); for the stimulus produced impresses (*ensēmainetai*) a sort of likeness of the percept (*hoion typon tou aisthēmatos*), just as when men seal with the signet rings. Hence in some people, though disability (*pathos*) or age (*hēlikia*), memory does not occur even under a strong stimulus, as though the stimulus were applied to running water; while in others owing to detrition like that of old walls in buildings, or to the hardness of the receiving surface, the impression (*typos*) does not penetrate. For this reason the very young and the very old have poor memories; they are in a state of flux, the young because of their growth, the old because of their decay ... Now if memory really occurs in this way, is what one remembers the present affection, or the original from which it arose? If the former, then we could not remember anything in its absence; if the latter, how can we, by perceiving the affection, remember the absent fact which we do not perceive? If there is in us an impression (*typos*) or picture (*graphē*), why should the perception of just this be memory of something else and not of itself? For when one exercises his memory this affection is what he considers and perceives. How does he remember what is not present? This would imply that one can also see and hear what is not present. But surely in a sense this can and does occur. Just as the picture on the panel is at once a picture (*zōon*) and a portrait (*eikōn*), and though one and the same, is both, yet the essence of the two is not the same, and it is possible to think of it both as a picture and as a portrait, so in the same way we must regard the mental picture within us both as an object of contemplation in itself and a mental picture of something else.[143]

Memory belongs to the same part of the soul as imagination, i.e. to the sensitive part (*aisthētikon*) qua imaginative (*phantastikon*).[144] All objects that are liable to be imagined (*ta phantasta*, the array of "enduring" sensibles) are essentially objects of memory,[145] at the condition that the living being possesses the perception of time (*aisthēsis tou kronou*).[146] Unmentioned in *On the Soul*, time must be included among the common sensibles, like size and movement, and is due to the common sense.[147] Early in the treatise Aristotle remarks that "when one is exercising his memory he always says in his soul that he has heard or felt this before"—a claim that, as has been observed, is difficult to square with the attribution of memory to animals.[148] But more apparent than real, this seeming contradiction implies that even without discursive *logos* animals are susceptible to the awareness of having being the subjects of sensations in the past. Aristotle does not explain how such awareness emerges but the fact that memory depends on the sensitive soul as imaginative and that it is a disposition (*exis*) of sensation,[149] rooted in the body, makes it arguably available to many animals, as he states. And if sensation implies the awareness of being sensing[150] and memory relies on sensation via *phantasia*, then those animals that are so constituted as to possess memory by the perception of time must be aware of having felt determinate sensations, and hence capable to connect their memories (i.e. *phantasmata* related to the past) to themselves.[151]

As for memory sensorial reliance, Aristotle clarifies that even though sensation along with its relevant stimulus is gone, the body is still able to retain it (i.e. to have the *exis*) because it functions in the same way as wax which carries impressed the mark of a signet ring. This same model was used also in *On the Soul* 2 to illustrate the capacity to sense and stress its passivity, receptivity, and faithfulness to the stimuli of the environment,[152] and in the discussion of memory it exemplifies the subsequent capacity to retain the sensory impression. Here too the emphasis is on the meaning that sensation makes available to the body by imprinting it (*ensēmainesthai*). Significantly, Aristotle first characterizes the sensation's enduring impression as "a kind of picture" (*zōgraphēma*),[153] but his next qualification "a sort of likeness of the percept" (*hoion typon tou aisthēmatos*) shows that, in tune with the sensorial extensiveness of *phantasia*, the use of "picture" does not mean that all impressions are visual. Rather they relate to the range of sensations an animal possesses and "present" something in their absence. It is up to the living being to take these impressions in themselves (qua pictures, *zōa*) or as referring to an external reality (qua portraits, *eikōnes*). Conceived in this way, in optimally constituted animals sensorial impressions lie at the cusp between *phantasia* and memory: they become objects of the second when the living being consciously relates them to a past experience.

Like sensation, *phantasia*, and dream-production, for Aristotle also memory has a physiological basis. It works well in a body constitution that has a good balance between softness and hardness (and hence an elemental composition) so as to be apt to accept and retain the sensory impressions (*typoi*). Neither young people nor old ones remember well if they do it at all, the first because their body is too much in flux (and hence excessively soft), the second because it is too decrepit and hard. In either case the sensorial impressions cannot last because they either dissipate or do not penetrate.[154] Similarly, also people affected by some (body) disabilities possess an impaired memory. While these examples illustrate human memory, the physiology at stake pertains also to other living beings that are so constituted as to possess the perception of time (*aisthēsis tou khronou*). Animals with excessively soft (moist) or hard (dry) bodies are not apt to remember, but those with a balanced constitution are.

Significantly, by appealing to such a physiology Aristotle attributes memory's frailty and ultimate absence to a failing of *phantasia*, that is, to a failure of the sense organs to retain the sensory impressions from the objects of sense. He could have appealed to a shortcoming of the perception of time and the common sense instead. The fact that he does not proceed so shows his commitment to understand living beings and their life in terms of body receptivity (and hence ultimately *logos* in the terms discussed in Chapter 1 as origin/end and hence structure) and to articulate a systematic psychological architecture, which rooted in the living body as receptive to its environment, goes from sensation to *phantasia* to memory (the first enabling animals to have a "synthetic view" of their environments and engage in purposeful movements and actions in view of a prospect, the second to connect their present to their individual past, even if indeterminately,[155] and to learn). Without exposure to the world and a body sensitive to that exposure and capable of retaining its imprints, there cannot be neither *phantasia* nor memory.

Certainly, we would like to know more about how (and whether) Aristotle conceived of the inner life of nonhuman animals.[156] This is our postmodern twist. But in discussing animals' "psychology" and in particular the workings of *phantasia* in relation to sensation and the perception of time Aristotle remains faithful to his overall conception of, and interest in, the "meanings" of life spelled out in *On the Soul* 2.2. There, let us recall it, he distinguishes life as nutrition, growth and decay, sensation, locomotion, and reason,[157] with the aim of understanding living beings/things in an integrated system based on a common core.[158] Hence, caused by sensations and the living beings' interactions with reality, the (internal) body movements that constitute *phantasia* are relevant for Aristotle to explain animals' more tangible (and life-constitutive) movements across space (in addition to explaining the margins of error in the perception of reality).[159] Through *phantasia* animals envision an object of desire and aversion, of approach and avoidance, and are able to nourish themselves and reproduce.[160] Likewise through *phantasia* they form a sensical representation of the surrounding reality and are empowered to act and respond (or simply exist) holding an awareness that has progressively ripened under previous sensory encounters and that in the particularity of the individual experience is unique to each of them. Moreover, for Aristotle *phantasia* is also relevant for contributing to reason and to a rational understanding of the world via the intelligibles (*noēta*), to which only humans have access.[161] "There is not thought without *phantasiai*," he claims.[162] And it is because of this outreaching role, of its bridging sensation to the other capacities of the soul (for those animals that possess them) that in *On the Soul* Aristotle discusses *phantasia* at the point he does, after sensation *and* before thought (and locomotion).

5.3.2 Body, sensation, and knowledge: In response to the Presocratics

As scholars have recognized, Aristotle's conception of *phantasia* results from engaging with both the Presocratics and Plato. The first have stimulated him to inquire about living beings' cognitive limitations and intrinsic liability to error.[163] The second has provided him with the suitable notion of an affection that complements sensation, endures after sensation is gone, and is indeed liable to cause error.[164] While important, however, these considerations do not exhaust the extent of Aristotle's engagement with his predecessors. For they fail to recognize how with *phantasia* Aristotle responds to the Presocratics' understanding of cognition in light of (his model of) soul partition and ultimately succeeds in explaining animal cognitive life in terms of the sensitive soul and inherent body dynamics. It was pointed out earlier that in *On the Soul* 3.3 Aristotle appeals to *phantasia* in parallel with mind (*nous*) to explain animals' field of action. That is, if humans act purposefully because of mind, animals do so on account of *phantasia*. Not an isolated parallel, these notions are discussed often together also in *On the Movement of Animals*.[165] And they return in a striking identification in *On the Soul* 3.9 where Aristotle bridges their distance to present the possibility that *phantasia* might be indeed a form of thinking process (*noēsis*).[166] Thus, so conceived, in Aristotle's psychological architecture *phantasia* is not only the sensation-related capacity that for constitutionally apt living beings provides the repertoire of impressions indispensable to dream and remember and that enables nonhuman animals to live intelligently. From the point of view of Aristotle's response to the Presocratics, it is also the key capacity that allows him to save mind (*nous*) [and the intelligibles (*noēta*)] for humans and still attribute a form of "natural," i.e. body-rooted, intelligence to animals. But let us proceed in order.

Aristotle opens his discussion of *phantasia* in *DA* 3.3 rehearsing a point he has made earlier in book 1 when discussing his predecessors' psychological theories. There are two special characteristics that distinguish the soul: movement across space on the one side, and perceiving, judging, and thinking, on the other. The distinction between these two defining blocks is made on the basis of internal compatibility. Whether it is animal flying, swimming, or walking, the relevant activity is locomotion, and whether it is animal perceiving or thinking (or judging), "the soul judges (*krinein*) and has cognizance (*gnōrizein*) of something which is."[167] In fact, it is on the basis of the critical element partaken by sensation and thought that, Aristotle let us understand, the older philosophers (*hoi arkhaioi*) considered thinking[168] and sensing the same (*tauton*). And this identity was ultimately conceivable because in being unaware of soul partition and oblivious of the different degrees of body enmeshment in living beings' cognitive processes they took thinking and perceiving to be a *physical* process (*sōmatikon*).[169] Like sensation, thinking also depended on the living being's exposure to the world and happened by the contact between likes (*ta homoia*). Empedocles, for instance, had said that,

> For it is by earth that we see earth, by water water, by aether divine aether, and by fire destructive fire, and fondness by fondness and strife by baleful strife.[170]

> Understanding grows with a human being according to what appears (*pareon*) to him.

> When it befalls them ever to think (*phronein*) different thoughts (*alloia*).[171]

Theories like this were problematic for Aristotle. For in the first place they were neither concerned with, nor apt, to explaining error: all the phenomena known by means of "likes" were (considered to be) true. And if error was to be explained by the contact with the unlike, then it coincided with the knowledge of the unlike.[172] On either view it remained an elusive notion. But such theories were also problematic because of the nature of thought they conveyed. For conceived like sensation a process of physical alteration (*alloiōsis*), thought provided the living being with an awareness of the surrounding world which was overall relative to its body and living circumstances.[173] Doomed to be environment-dependent, and changeable, thought was inapt to hold the stable epistemological grasp of reality Aristotle held it to possess.[174]

Our philosopher undertakes a different path. In *On the Soul* he distinguishes sensation and thought assigning them distinct objects and allocating them to different parts of the soul: the sensitive and the rational.[175] If in his psychological doctrine thinking (*noein*) is shaped on the model of perceiving and considered in terms of receptivity,[176] its objects (*ta noēta*) are the intelligibles. Abstract and universal, once achieved they have nothing to do with the physiology of the body and its relation to the environment, and they are an exclusive possession of the human animal, the only one among the creatures of this world to be endowed with the rational soul.[177] So Aristotle does not only need *phantasia* to provide living beings with a critical capacity (and ensuing knowledge) that stemming from sensation continues to be anchored in the body, and is thereby fallible.[178] Importantly, he also needs it because as a positive notion *phantasia* represents the body-related, thinking-like capacity available to nonhuman animals, which lacking the rational soul and hence (human) thinking (*noēsis*) and calculation (*logismos*) are still endowed with a body so structured as to retain the variety of sensory impressions they have been exposed to, enabling them to understand their environment (natural and interpersonal) and act on it.[179] The kind of *phantasia* animals possess, the *aithētikē*, is deeply constitutive of the animal as such (*toiouto to zōon*), i.e. of that animality Aristotle is pursuing in his study of animals.[180] True, in *On the Soul* our philosopher is rather interested in carving out for *phantasia* a legitimate place *vis-à-vis* other psychological capacities,[181] while in *Parva Naturalia* he limits himself to consider its physiology and relevance in dream-production and memory.[182] But the involvement of *phantasia* in animals' purposeful action and its compatibility with *noēsis* reveal that it is ultimately this capacity that allows animals to possess and live by the analogues of human wisdom (*sophia*), art (*tekhnē*), and comprehension (*synesis*),[183] mentioned in *History of Animals* and there accounted for with regard to animal behavior.[184] For Aristotle there is no need to attribute nonhuman animals *nous* or *logos* as the Presocratics did,[185] nor is there ground to be puzzled at animals' resourcefulness and skills.[186] Sensation and its derivatives suffice to empower them to live in the knowledgeable way that affords them to live. *Phantasiai* (imaginations) and memories (imaginations connected to the past) are effective enough to provide them with some experience (*empeiria*) (i.e. a distillation of previous sensorial encounters).[187] And it is through *phantasia* that, we may well assume, animals possess a power of forethought (*dynamis pronoētikē*) regarding their own lives (*ton hautōn bion*), granting them in *Nicomachean Ethics* the title of *phronima* (intelligent).[188] Be that as it may, if in *On the Soul* Aristotle claims that *phantasia* is "thought" or might be a form of "thinking process,"[189] he firmly considers it in parallel with, and hence as distinct from, mind because he is working with a Platonizing definition of mind that is incompatible with perceptual *phantasia*. Although relying on the information provided by the body and still dependent on (*logistikē*) *phantasia*,[190] for Aristotle mind is crucially disembodied and works with universals. It affords human beings with a degree of detachment from, and "manipulation" of, the physical reality that lies outside animals' grasp.[191] For physically in touch with their environments and faithful to the stored representations of past sensory experiences, animals act, move, and live holding on to a "body continuity" with their space of existence[192] that in humans qua rational beings is (passible to be) loosened and fades away through the activity of all-making mind.[193]

Notes

1 *DA* 2 414a33–414b8; cf. *DA* 413b24–25 "for, where there is sensation, there is also pain (*lypē*) and pleasure (*hēdonē*), and where these are there must be also desire (*epithymia*)."
2 It should be remarked that in the *On the Soul* passage cited at the beginning of this chapter, besides the reference to the use of touch as the sense for food also the strict association of desire to pleasure and pain suggests that Aristotle is thinking of the pleasure connected to nutrition. For, significantly, in his discussion of pleasure in *NE* 7, desire (*epithymia*) is said to accompany the pleasures mixed with pain, i.e. those pertaining to the replenishing of the natural state and involving eating and drinking (see *NE* 7 1152b36–1153a3 and below).
3 In this respect, see also *DA* 3 433b32–434a3, where this constellation is explicitly attributed to the "imperfect" animals endowed with the mere sense of touch.
4 *DA* 2 414b7–9.
5 *DA* 2 414b12–14.
6 For the process in itself, see Chapter 3.
7 *Sens.* 441a2–3; accordingly, an object of taste (*geuston*) is an object of touch (*apton*) (*DA* 2 422a9)
8 *Sens.* 436b15. On the sensation of the pleasure (*aisthēsis hēdonēs*) deriving from food and the desire for what is pleasant (*epithymia tou hēdeos*) as features shared by all animals, see also *PA* 2 661a8–9.
9 For the involvement of *phantasia* ("imagination"), see *DA* 3 431a9–14 and especially 432b29–30 and section 5.3 below.
10 For a discussion of animals' diversified diet and food acquisition along with intraspecific and interspecific relations in their space of existence, see *HA* 7 and Chapter 6.
11 On the pleasure of mating and how it differs among animals, see *HA* 7 588b27–30.
12 For all these aspects of animal life, see Chapter 6.
13 Pellegrin, 2019, 155; for the capacity to sense as a form of receptivity defining the *logoi* of the sensibles (colors, scents, sounds, etc.), see above section 4.3.
14 For Pellegrin there would be a human process of sensation based on *logos* and a nonhuman one without it, but Aristotle discusses sensation in a general way, without this sort of distinction and, moreover, at the end of his extensive treatment of sensation underscores he has discussed the animal qua sentient (*aisthētikon zōon*); see Chapter 4. *Logos* can certainly contribute to human perceptive pleasure as, for instance, a painter enjoying the mixture of his colors, but is far from dividing the sentient living world. Note that when discussing the parts of the soul in *DA* 3.9 (432a30–432b1) Aristotle claims that the sensitive soul cannot easily be classed as rational or irrational, an ambiguity that does not overlap a supposed distinction between human and nonhuman animals, but has rather to do with the sophistication of the process of sensation as such, i.e. as acting as a sort of mean.
15 See, respectively, Chapters 2 and 3.
16 Taylor, 2003, 1.
17 *NE* 7 1154b7–10. For a discussion of this passage in relation to Aspasius' commentary and the pertinence to Anaxagoras, see Warren, 2007, 20–5.
18 The distinction between the pleasures of the soul and those of the body is made explicit at *NE* 3 1117a27–28, but it underlies Aristotle's discussion elsewhere in the treatise, for instance, his endorsement of the distinction between the goods of the body and those of the soul (1 1098b14–6), the claim that human goodness has to do with the excellence of the soul, not of the body (1 1102a16–7), and his conception of temperance intended as control over bodily pleasures (*NE* 2 1104b5–7), or the distinction between necessary pleasures and "desiderable in themselves" (7 1147b23–31). For the sake of clarity, let here point out that such a distinction between pleasures overlaps the model of soul partition voiced in this treatise. Aristotle divides the soul

in a rational and irrational part, with the irrational encompassing in turn a vegetative and appetitive part (see Chapter 2). It is the appetitive part (of the irrational soul) to converse with the rational one leading to the achievement of moral virtues, such as liberality (*eleutheriotēs*) and temperance (*sophrosynē*), while the rational part has virtues of its own, such as wisdom (*sophia*), understanding (*synesis*), and prudence (*phronēsis*). In this scheme both the pleasures of the body and the pleasures of the soul that are involved in moral virtues are under the jurisdiction of the appetitive soul (*NE* 1 1103a5–11; cf. *NE* 3 1117b24–5). By contrast love of knowledge, mentioned below, pertains to the rational soul alone.

19 Thus from this perspective, the pleasures of the soul include the particular fondnesses that Aristotle discusses in book 1 to argue that happiness consists in the active life of virtue. Pleasure, Aristotle tells us, ordinarily emerges from the satisfaction of people's particular fondnesses. A man that is fond of horses will find pleasure in possessing a horse, while one that loves theatrical plays will feel pleasure in going to the theater and watch a play. But these are particular and "idiosyncratic" objects of pleasure and ultimately in conflict with one another ("horses" against "theatrical plays") inasmuch as pleasure does not belong to them essentially (*kath'auto*). Only virtue is qualified as having intrinsic pleasure and hence only the life of virtue is essentially pleasant and provides happiness, considered to consist in the active exercise of virtue (*NE* 1 1098b 30–1099a20).

20 *NE* 3 1117b29–1118b9; on this passage, see also the upcoming discussion. Devotion to bodily pleasures sustains the first of the three types of life Aristotle acknowledges, i.e. the life of enjoyment, of politics, and contemplation (*NE* 1 1095b5–1096a6). It should be stressed, however, that all pleasures (whether bodily or psychological) relate to the soul (see *NE* 1 1099a7–8 (*to hēdesthai tōn psykhikōn*), 10 1173b9–14).

21 Aristotle remarks that happiness is at once "the best, the noblest, and the pleasantest of things" against, for instance, the opinion conveyed by an inscription in Delos (likely a kernel of Apollinean wisdom) allotting these different qualifications to different subjects ("Justice is noblest, and health is best, but the heart's desire is the pleasantest"), *NE* 1 1099a24–9. On Aristotle's distinction of pleasures, see also *NE* 10, 5, and discussion below.

22 *NE* 2 1104b4–29. In this section of *NE* Aristotle provides a long series of considerations to prove the crucial relevance of pleasure (and pain) to virtue.

23 Interestingly, different virtues relate differently to pleasure and pain. For instance, Aristotle observes that, when compared to courage, temperance is more concerned with pleasure and involves pains in lesser and different degrees (*NE* 3 1117b25–6).

24 See Pellegrin, 2019, discussed above.

25 Aristot. *NE* 3 1118a2–23, all translations of *Nicomachean Ethics* are by H. Rackham with slight modifications.

26 Among (the land and by extension all) animals the human being is the most apt in discriminating flavors (and hence in enjoying them) because of the particular conformation of the tongue the double-function of which is that of discriminating flavors and articulating sounds. Aristotle writes, "Mankind has the most detached, softest, and broadest tongue, so that it may be useful for both its operations. The soft, broad tongue is useful for both the perception of flavors (for mankind is the most keenly perceptive of animals, and his tongue is soft, for it is most tactile, and taste is a sort of touch)" (*PA* 2 660a17–23, transl. by J. Lennox). Animals with a small tongue or without (like fishes or the crocodile and the bloodless animals) are least or not receptive to flavors, and their pleasure of eating and drinking depends on touch (see *PA* 2 660b12–25; 4 690b18–691a6). In sum, in Aristotle's view land-animals (*peza* or *khersaia*) are more receptive to flavors than water-animals (*enydra*), and among the first group the human being most of them all. On the other hand, Aristotle distinguishes lizards, snakes, and seals, which in having a forked tongue (*dikroa glōtta*) are particularly gluttonous (*PA* 4 691a6–9; on the double pleasure of snakes, see also 2 660b3–11).

27 The fact that these categories of people find pleasure in tasting wine and food make them skilled in what they do according to Aristotle's principle that pleasures inherent in a given activity make one better in performing it (see *NE* 7 1153a 31–4, 10 1175a12–6, and 1175a30–1175b1).

28 *NE* 3 1118a24–1118b9.

29 See 5.3 below.

30 Human beings who do find pleasure in these senses would not be criticized for profligacy if they were to feel an excessive pleasure in them. On humans' enjoyment of sights and sounds, besides the passage quoted above, see also Aristotle's discussion of "particular fondnesses" at *NE* 1 1099a7–10; cf. note 19 above. While having to do with the senses, these pleasures are not considered "of the body," but "of the soul."

31 The emphasis in the passage quoted above is on nutrition, but as Aristotle makes clear later on (*NE* 3 1118b8–12) the pleasures of the body at stake in human temperance (and shared by animals) are also those that relate to sex. The emphasis on nutrition has likely to do with the fact that this pleasure is a continuous factor in human (and animal) life (living beings need to eat and drink in order to survive) while sexual pleasure is discontinuous (it characterizes youths). Besides, for many animals sexual pleasure is not a constant but relates to mating seasons (see *HA* 5 542a20–544b12).

32 See *NE* 3 1118b 2–5 and 25–7.

33 *NE* 3 1118b10–14. Aristotle remarks that few men err in respect to these pleasures by excess, i.e. eating or drinking more than the body's lack of food and drink calls for.

34 In this passage, Aristotle identifies three motives of human choice, the noble (*to kalon*), the expedient (*to sympheron*), and the pleasant (*to hēdy*) adding that "what is pleasant" is nevertheless concomitant with all objects of choice since both the noble and the expedient are chosen because they are pleasant (*NE* 2 1104b30–1105a2).

35 *NE* 2 1104b4–5; on bodily pleasures humans share with (*koinōnein*) the other animals, see also *NE* 3 1118a 25 and on the sense of touch as belonging to the human beings qua animals (*zōa*), *NE* 3 1118b2–3.

36 *NE* 2 1104b30–1105a2.

37 The terms of the failures to achieve the means by excess and defect correspond to the vices of profligacy and insensibility and are discussed in book 3 in the segue of the passage quoted above (*NE* 3 1118b15–1119a12). On the insensible man, see also *NE* 2 1104a24–5; on the pleasures of the temperate man, *NE* 7 1153b35.

38 *NE* 3 1118b2–3.

39 *NE* 3 1119a6–12.

40 For a list of the kinds of critiques moved against pleasure and Aristotle's systematic reply as well as the overlapping and differences with book 10, see Gosling and Taylor, 1982, 199–203.

41 "*anenpodiston energeian tēs kata physin exeōs*" (*NE* 7 1153a7).

42 See, respectively, Chapters 10–14 of book 7 and 1–5 of book 10. As for their respective approaches, book 7 presents a defense of pleasure as a good, while book 10 argues for it as being the good. One approach is negative, the other positive. For a discussion of these two accounts, their common points, and how book 10 complements book 7, see Kraut's article in Stanford Encyclopedia of Philosophy under "pleasure," https://plato.stanford.edu/entries/aristotle-ethics/#Plea. On the compatibility of the two definitions, see Gosling and Taylor, 1982; Rapp, 2009, 221–2; and Kraut. Recently, Wolfsdort has argued that the two definitions of pleasure in *NE* reflect a change in Aristotle's conception of pleasure and that the early definition in book 7 originally constituted *Eudemian Ethics* 6 (2013, 105–43).

43 With this claim Aristotle is firmly responding to Plato's theory of pleasure in *Philebus*, see Aristot. *NE* 7 1153a9–12 and Plat. *Phil.* 53c.

44 See *NE* 2 1104b4–29 and discussion above.

45 See Festugière, 1936, XXV; Frede, 2009, 183.

46 See, especially, *NE* 2 1104b4–29.

47 The list of points presented in *NE* 7 that pleasure is not a good starts with the argument that in being a process (*genēsis*) pleasure is not the end, and hence (from a teleological point of view) not a good (see *NE* 7 1152b13–5). This same critique supports the denial that pleasure is not the supreme good (*NE* 7 1152b23).

48 *NE* 7 1152b35–1153a5.

49 In the wave of Plato, accidental pleasures are also called necessary (*anankaiai*) in contrast to pleasures that are good in themselves such as "victory, honor, wealth and the other good and pleasant things of the same sort" (see *NE* 7 1147b23–31 and 1252b36–1253a2); to these we should add the pleasures of the particular fondnesses and of the distant senses discussed respectively in book 1 and 3, see above. On the influence of Plato's conception of restorative pleasures on Aristotle, see also Wolfsdorf, 2013, 106–14.

50 *NE* 7 1152b30–1.

51 This claim betrays a reductive view of pleasure as having to do exclusively with the body. As mentioned earlier, in *NE* Aristotle consistently rectifies this circumscription. Let us remind that Aristotle's hedonic classification includes moral and intellectual psychological pleasures (*psykhologikai*) in addition to the bodily pleasures here ascribed to animals and children; see above. Interestingly, a long passage in *NE* is dedicated to understanding the reasons why bodily pleasures appear to be more desidable than others (1154a26–1154b16).

52 *NE* 7 1153a28–35.

53 On this last point, see Steward, 1892, vol. 2, 227.

54 *NE* 2 1103a14–1103b2.

55 In respect to animals' natural dispositions, see below.

56 See Chapters 1, 6, and below.

57 At *NE* 6 1144b1–16 Aristotle speaks of virtues of character in a complementary, yet seemingly discordant, way. For he distinguishes between natural virtues and virtues in the strict sense. It is the second that is not present at human birth but has to be acquired by habituation (using intelligence). The first are by contrast encoded in character dispositions and present from birth in animals and children alike (who will eventually grow into being strictly virtuous); see Lennox, 1999a, and the discussion of animal characters in Section 6.3.

58 For a discussion of the meaning of *energeia*, see Wolfsdorf, who stressing its relation to the verb *energein* (make use) defines it as "the condition of being in use, deployed, or being exercised, or at work" (2013, 115). He also later remarks on the range of activities to which the conception of pleasure as *energeia* applies (120).

59 On acquired dispositions as second nature, see Randall, 1960, 254.

60 *HA* 7 589a9 (*to de kata physin hēdy*).

61 On animals' habitats and diets, see Chapter 6.2 and 6.2.1.

62 Theophrastus' claim is inserted in a reply to Anaxagoras' theory that by occurring between "unlikes" all sensations imply pain (*de Sens.* 31). As noted above, Aristotle too refers to this theory in *NE* letting transpire his interest in considering human pleasure in the context of nature.

63 For spiders and ants and all the other animals (and plants) come into being, exist, and act by nature; see Aristot. *Phys.* 2, 8, 199a20–9, and Chapter 1. At the same time, however, Aristotle also recognizes animals' capacity to learn (*mathēsis*); see Chapter 6.

64 *NE* 10 1176a3–10.

65 Labeling Aristotle's view of the soul as Platonizing is here directed to stress the centrality of reason and its activity characteristic of his ethics (and firmly advocated by Plato) as opposed to the life-oriented doctrine of *On the Soul*, not to undervalue Aristotle's belief in the hierarchy of soul functions in *NE* 10, with contemplation as the best activity (1177b1–5). Accordingly, in *On the Soul* too Mind alone is considered to be immortal and everlasting (*DA* 3 430a23–4) but, as discussed in this book, remains overall irrelevant to understand animal life. With regard to pleasure, still in

the ethics, I believe, a form of life-oriented animal equality emerges inasmuch as all living beings (tend to) find pleasure in the activities that are natural to them and these activities express their desire to live, granted that Aristotle's empirical gaze identifies also animals which by nature do not live well (*kakobia*), like the clumsy eagle (see below).

66 See *NE* 10 1175a23–1176a3, and in particular, for the hierarchy of pleasures, 1175b37–1176a3.

67 DK 22 B 9.

68 In this respect, in *History of Animals*, Aristotle also notes that fish derive pleasure, each species from its proper food, a fact that shows they possess the sense of taste (*HA* 4 533a31–34).

69 Indeed, in the passage quoted above Aristotle claims that there may not be variety *among the pleasures of the same species* (italics by the author).

70 See respectively *HA* 8 615a16 (*ōdikos*), 615a18 (*tekhnikos*), 616b10 (*polytekhnos*), and 619b23 (*eutekhnos*).

71 *HA* 8 616b13–6, transl. by D. Balme. Admittedly, while not implying body replenishment also the pleasure deriving from these environmental conditions is incidental as it depends on the season; but see next note (72).

72 In the case of the elea such an activity would consist in the perception of its living environment in relation to the current climatic conditions.

73 For pleasure as activity enhancer, see *NE* 10 11751a29–1175b1.

74 Aristot. *HA* 8 618b32–619a4; cf. 616b31–3.

75 For a discussion of these terms, see also Chapter 6.

76 *HA* 4 534a16–19. This piece of information emerges from the discussion of fishes' sense of smell as revealed by fishermen's fishing techniques. Granted that smells reveal to them the presence of food, this fact does not impede that fish may enjoy the scents per se, especially if we consider that they flee "as if disgusted" (*hōs osphrain-omenoi*) from fish-washings and bilge on account of their odor (534a27–30).

77 See *HA* 8 611b26–7. Flute-playing had also an entrancing effect on horses which would stand still and droop their heads upon hearing the music, an affection considered a disease and called "nymphing" (*HA* 7 604b10–15).

78 See *NE* 10 1097b9–15.

79 Aristot. *NE* 7 1153b25–32.

80 For Aristotle nature and best disposition explain respectively food and drink preferences (i.e. bodily pleasures) and the sort of virtue one finds pleasure in. In this respect, if to be just is the best disposition of a human being, s/he will find pleasure in exercising justice, if courageous in acting courageously. The distinction between *physis* and "best disposition" here echoes Eudoxus' doctrine, and particularly his parallel (and hence distinction) between the pursuit of one's own particular good and the pursuit of one's own food (*NE* 10 1172b14–5).

81 Rapp gives two alternative readings of this identity of pleasure: either A) all animals seek "a kind of pleasure" or B) "they pursue the pleasure that is intrinsically involved in the *best state for them*" (2009, 227).

82 For the range of interpretations given to this line, see Rapp, 2009, 228–9. Among them, Stewart pursues the universalism predicated by "*panta*" in terms of the organizing principle of nature.

83 For this notion as supporting Aristotle's zoological project, see Chapter 1.

84 *PA* 1 645a19–24; see Zatta, 2019, 154–5, and Chapter 1 above.

85 *DA* 2 415a27–415b2; for a discussion of this passage in relation to the nutritive soul, see Chapter 3.

86 *Pace* Rapp (2009, 228). For the desire of divinity discussed in *DA* pertains to each individual living being (see Chapter 3), and not to the species as this author claims. Hence it is compatible with Aristotle's view that living beings' plurality of pleasures may be in fact seen in terms of identity.

87 *NE* 10 1175a11–22.
88 Aristotle refers, for instance, to the pleasure of "studying" geometry, architecture, music, and other arts (*NE* 10 1175a32–6; see also 1175a11–17).
89 See, for instance, *PA* 2 661a8–9.
90 *NE* 10 1173a3–6.
91 See 5.1 above.
92 *NE* 10 1173b28–32, 1175a23–1176a4.
93 *DA* 2 413b24–5.
94 *DA* 2 414b2–7, see above. From the definition in this passage *epithymia* appears to be unqualified desire while *orexis* is qualified desire, and both belong to the sensitive soul. Here, as Polansky remarks, pleasure and the pleasant precede pain because desire is primarily appetite for the pleasant (2007, 181).
95 *DA* 3 432b15–6; see also *MA* 701b34–7 (*heneka tinos*); cf. *NE* 3 1111b9–14, where Aristotle claims that "the other animals" are capable of voluntary actions and that they feel desire (*epithymia*) and passion (*thymos*).
96 *DA* 3 433b11–2. a21–2.
97 *DA* 3 433a10–433b12 and 433b28–31. In the wake of Plato, Aristotle distinguishes three kinds of desire (*orexis*), namely anger (*thymos*), will (*boulēsis*), and *epithymia* (*DA* 2 414b3–4; cf. Plat. *Resp.* 9 580d–581c). Of these, will is restricted to rational beings and so is anger, which is a longing for revenge (Aristot. *Rhet.* 2 1378a; cf. Polansky, 2007, 190; although see *HA* which attributes to animals a number of emotions, including *thymos*). Such a plurality of desires may create conflict when, for instance, desire for bodily pleasures overcomes reason (*logos*) leading one to be incontinent (*DA* 3 433a14–433b10; 433a1–3; 434a12–5; cf. *MA* 700b17–25).
98 In fact, in *On the Soul* we can detect an effort toward a "systematization" of the notions of *epithymia* and *orexis*. This effort has likely to do with Aristotle's critical reception of Plato's soul tripartition and its mereological labels (*logistikon*, *thymikon*, and *epithymikon*, see Chapter 2) in the context of his explanation of animal locomotion as a psychological activity. In *Parts of Animals*, on the other hand, it is *epithymia* to be straightly directed to "what is pleasant," without any reference to *orexis* (*PA* 2 661a8).
99 In this respect, see *MA* 700b15–18 where Aristotle claims that in being always moved for the sake of some object, animals' movement has a limit (*peras*) (cf. 700b33–4). For a "narrow" sense of desire, see also *DA* 433a6–14, in which Aristotle speaks of self-controlled people (*enkrateis*) as following reason instead of *orexis*.
100 Aristot. *Phys.* 4 208b9. Besides, as Bénatouïl remarks on the basis of *Phys.* 8 255b29–31, the movement of natural bodies is characterized by a condition of passivity. For he claims, "the thing [the elemental body] contains within itself the source of motion—not of moving something or of causing motion, but of suffering it" (2019, 8).
101 Conventionally translated as "imagination," *phantasia* is a key notion in Aristotle's psychological apparatus, but of difficult interpretation. The literature is vast, and scholarly perspectives divergent. Among the different interpretations, suffice it to mention that Beare distinguishes for the "normal side" of *phantasia* a faculty of *presentation* and one of "*representation* or reproductive imagination," noting that (on its abnormal side) *phantasia* can also be the source of illusions and hence pertains to mental pathology rather than psychology (1906, 290); Nussbaum considers *phantasia* to provide perceptual interpretation and action motivation (1978, 221–61); Wedin makes it provide other faculties with mental images but denies it the full status of psychological capacity (1988, 52–61); Schofield circumscribes *phantasia* to "non-paradigmatic sensorial experiences," such as illusions, dreams, and memories (1992, 252); Frede sees it as the capacity to synthetize the individual perceptions offering "a panoramic view of a whole situation" (1992, 284–7). On the other hand, focusing on the causality of *phantasia*, Everson sees it as a psychological capacity internal to the animal organism and providing an explanation for those "quasi-perceptual states,"

which are not accountable by "mere" perception. For this author, coexisting with this technical conception of *phantasia* there is also a general one, which applies to all perceptions and implies a "state of being appeared to," beyond the consideration of causality (1997, 172–81 and n. 96); more recently, interpreting Aristotle's *phantasia* in light of Plato's *Theaetetus* Scheiter considers it an image-producing faculty that plays an essential role in the understanding of the perception of incidental sensibles and the insurgence of error (2012).

102 Aristot. *MA* 701b32–702a19, transl. by A.L. Peck with slight modifications; cf. *DA* 3 430a20–1 (*"phantasia* too when it starts movement never does so without appetite, *aneu orexeōs"*).

103 For the involvement of *phantasia* in desire-formation and animals' locomotion, along with its range of representational possibilities, from perceptual features to objects to situations including "performing an action or enjoying an experience," see Lorenz whose approach I follow here (2006, 119–20, 124–37). On *phantasia* as a selective capacity that allows the living being to focus on a specific object of desire, see also Modrak, 1987, 95–6; on desire as the efficient cause of voluntary animal motion, see also Modrak, 1987, 95.

104 *DA* 3 434a6–9. As Beare remarks, in human beings *phantasia* "cooperates with rational deliberation" and possesses rationality only by accident (*kata symbebēkos*), namely, in virtue of such cooperation (1906, 297–8). On the need of human thought for *phantasia*, see *DA* 1 403a8–10, 3 431a17–18, and *de Mem.* 449b30–450a6, where Aristotle illustrated it with geometrical reasoning. For a discussion of the dependency of mind on *phantasia* providing it with "a permanence in the sensible beings," see White, 1985, 498–505. On the other hand, human beings still resort to *aisthētikē phantasia* under the effects of emotions, disease, and sleep (*DA* 3 429a5–9, quoted below).

105 *DA* 3 429a1–2.

106 *DA* 3 429a5–8.

107 On this point, see also Thomas Aquinas' commentary (*Sententia De anima* 2.1 6 n. 4): "imagination is not the same for all animals. But there are certain animals, which live by this alone, lacking the intellect and being directed in their operations by imagination, just as we are directed by the intellect" transl. by R.A. Kocourek. More on this functional equivalence between *phantasia* and mind will be said in the last section of this chapter, which contextualizes Aristotle's notion of *phantasia* with respect to Presocratic doctrines.

108 See *de Mem.* 449b13–5 and Chapter 4.

109 For attention on this point, see Rabinoff, 2018, 37–9.

110 *DA* 2 414b17–8.

111 So, for instance, while Lorenz has attributed *phantasia* only to the animals provided with locomotion (2006, 138–47), Nussbaum (1978) and Caston (1996) have attached it to all animals.

112 *DA* 2 413b13–25. Interestingly, in this passage Aristotle is discussing the problem of soul partition and whether the soul divides only in thought or also in space. He cites the example of insects claiming that when they are separated the dismembered parts of their bodies move in space and have sensation, and that if they have sensation, they also have *phantasia* and desire (*orexis*). It is clear that here Aristotle has in mind the definition of imagination as caused by sensation (see above).

113 *DA* 3 428a25 ("some wild animals (*thēria*) have imagination").

114 Philop. *Aristotelis de Anima*, 15, 592, 26–9; see Papachristou, 2013, 33 n. 55. Whether attached to the soil or simply stationary, these creatures have the sense of touch and respond to their tangibles by moving. To this group belongs the testaceans (*ostrakoderma*), which Alexander of Aphrodisia singles out as animals endowed with perception but without *phantasia* (*DA* 67, 2–3). In fact, according to Aristotle, while some of these animals are stationary (and utterly devoid of *phantasia*) others move (*kinētika*) and hence, on the basis of *DA* 3 11, should possibly have indeterminate *phantasia*

(see *HA* 7 590a19–590b4, and on the *zōophyta*, *DA* 7 588b12–21). Regarding the sea anemone, Aristotle further notes that "the limpets detach themselves and go about and feed" (*HA* 7 590b33–4).

115 *DA* 3 433b32–434a7. In other words, in being deprived of the distant senses (sight, hearing, and smell) the imperfect animals move without a sense of direction, but indeterminately.

116 In this respect, see *Posterior Analytics* where Aristotle distinguishes between animals whose sensory impressions persist (*engignetai monē aisthēmatos*) and those in which they do not (*Post. An.* 2 19 99b36–100a1). As we will see in the next section, the persistence of sensory impressions is due to *phantasia* and corresponds to the production of *phantasmata*; on animals without *phantasia*, see also Lorenz 2006, 133 and 138–41.

117 On the derivation of *phantasia* from all the senses, and not merely sight, see Aristotle's discussion of the *phantastikai kinēseis* involved in dream-production, *de Insomn.* 461a26–9; cf. Beare, 1906, 299; Everson, 1997; Scheiter, 2012.

118 "Since sight is the chief sense, the name *phantasia* (imagination) is derived from *phaos* (light), because without light it is impossible to see," *DA* 3 429a4–5; on the popular origin of this etymology, see Beare, 1906, 299.

119 For this point, see Everson, 1997.

120 See below.

121 In this respect, see Aristot. *Met.* 1 980a21–3.

122 On the human-bent treatment of *phantasia* in *On the Soul*, see White, 1985, 483–4, with whom, however, it is not possible to agree that in this treatise Aristotle shares with Plato the concern with the human soul as the most perfect and complex of all. Rather in opposition to Plato, Aristotle seems concerned to show the continuity among forms of life and how the life of reason is based on the life of the senses; see introduction and the discussion of the geometrical model of the soul in Chapter 2.

123 *NE* 3 1118a20–24 and Lorenz, 2006, 131, 149–50, who, more generally, accounts for animals' liability to envisage prospects on the basis of *de Memoria* and *de Insomnis* and argues that for Aristotle "*phantasiai* form ordered sequences of indeterminate duration and complexity" (153).

124 I mean "positive" here in contrast to both the "indetermined" *phantasia* possessed by animals without the distant senses, discussed earlier, and to the error-bringing, deceitful side of *phantasia*, on which Aristotle elaborates in his general discussion in *DA* 3.3 and in *de Memoria* when he mentions Antipheron of Oreus and other lunatics (*existamenoi*). These people, he tells us, "spoke of their *phantasmata* as if they had actually taken place, and as if they actually remembered them" (*de Mem.* 451a9–11). For a reconstruction of the working of *phantasia* in terms of the perceptual apparatus and as the capacity that providing animals with "orderly sequences of sensory impressions" allows them to envision prospects, form purposes and engage in locomotion conferring order to their mental lives, see Lorenz, 2006, 148–73.

125 Filling the gap in Aristotle's discussion, it should be observed that in respect to the prey, physically absent from the lion's immediate surrounding, the *phantasmata* of common and incidental sensibles depend on the particular sensory apparatus, capacities, and experiences of a given predator. For instance, the same kind of prey will appear different to a raptor, endowed with acute sight, than it does to a lion, for which arguably the sense of smell is stronger. On the recognition of intraspecific enemies in the animal kingdom (and, arguably, animals' inherent capacity for *phantasia*), see *History of Animals* 8 (608b19–610a30). Food provision, we learn, causes animal conflict, involving particular species against others and hence triggers (even though Aristotle does not say so) behaviors based on *phantasia*.

126 Beare, 1906, 293–4; cf. Frede, 1992, 184, who stresses the synthetic operation of *phantasia*.

127   *DA* 2 420b29–421a1; for instance, for Beare *phantasia* attributes meanings to sounds (1906, 294); for Nussbaum it makes possible to convey meaning through voice by providing the awareness of "an object or state of affair," hence of incidental sensibles (1978, 259–60); Everson, on the other hand, expands the awareness of *phantasia* to include all sensibles (1997, 178–9, 181).

128   Aristot. *Pol.* 1 1253a8–18.

129   See above.

130   Labarrière discusses the difference between animal and human communication and defines the first as "semantic," the second "symbolic" (1984, 34–40), but the episode of the ravens considered below must involve for Aristotle a degree of symbolism in animal language too, at least for some species.

131   See Chapter 3.

132   Aristot. *HA* 8 618b10–7, transl. by D.M. Balme.

133   In this passage *aisthēsis dēlōseōs* echoes Plato's *Minos* where the senses are discussed in terms of their actual role in showing the sensibles, i.e. sight shows (*dēloun*) the things seen by means of the eyes, hearing the things heard by means of the ears (314a–b). Each *aisthēsis* is a *dēlōsis* (an act of showing), internal to the individual. But used in the account about the ravens of Attica and Peloponnese, the expression *aisthēsis dēlōseōs* indicates the birds' capacity to grasp the meanings conveyed to them through their conspecifics' voices hence subsuming animal communication into the realm of sensation (and sensation-derived *phantasia*).

134   On the nature of *phantasia* as internal representation, see Modrak, 1987, 83–107.

135   *Sens.* 436a6–8.

136   More specifically, Aristotle claims to investigate animals and all beings possessing life (i.e. plants), what actions are unique of them (*idiai praxeis*) and what are common (*koinai*) (*Sens.* 436a1–5.).

137   *de Insomn.* 659a12–5.

138   *de Insomn.* 460b2–461a30; for the physiology involved in the movements of *phantasia*, see, among others, Beare, 1906, 292–95; Lorenz, 2006, 148–74; Scheiter, 2012.

139   Recollection requires an active (and hence rational) search for an optimal starting point from which to retrieve the stored information.

140   *de Mem.* 1 450a18.

141   See *Met.* 1 980a20–980b ["now animals are by nature born with the power of sensation, and from this some acquire the faculty of memory (*mnēmē*), whereas others do not. Accordingly the former are more intelligent and capable of learning than those which cannot remember," transl. by H. Tredennick]. There is a question whether *mnēmē* cited in this passage might be in fact *phantasia* (see Scheiter, 2012, 262).

142   *de Mem.* 1 450a19–24; cf. 451a30–451b1.

143   *de Mem.* 1 450a23–450b26, transl. by W.S. Hett.

144   Cf. *de Somn.* 459a12–23.

145   On the other hand, the claim that those objects the memory of which involves by necessity *phantasia* are objects of memory only incidentally refers to thought's dependency on *phantasia* and the fact that because the human mind must have a *phantasma* to work with (and the intelligibles per se are not objects of sensation and *phantasia*) it is possible to remember the intelligibles only through the representational item/s involved in thinking. See, for instance, *DA* 3 427b17–8, 427b29–31, 431a17–8. *On Memory and Recollection* is particularly eloquent in this respect. In geometrical reasonings, for instance, when drawing a geometrical figure one draws it with a specific size even though magnitude is in fact irrelevant. Likewise, when a human being is thinking, she puts before her eyes (*tithetai pro ommatōn*) a finite magnitude even if the object of thought does not involve the representation of this common sensible (see *de Mem.* 449b31–450a9; Beare, 1906, 297; Modrak, 1987, 84).

146   *de Mem.* 1 449b29–30.

147 See *de Mem.* 1 450a9–14 and Hicks, 1965, 427; cf. Lorenz, 2006, 160–2. For Bloch Aristotle does not include time in the list of common sensibles neither in *On the Soul* (see, for instance, 2 418a16–20) nor in *On the Sense* (437a8–9) probably because in these treatises he is concerned with the immediacy of perception (2007, 60–1 and n. 27); for the involvement of the common sense, see Kahn, 1974/79, 8, and n. 23.

148 1 449b22–3 and 1 450a13–19; see Sorabji, 1972, 10.

149 See passage quoted above and 1 451a15–17, where memory is defined as an *exis phantasmatos*. There is a scholarly question on how to render *exis*, with consequences for our interpretation of Aristotle's conception of memory. In his translation, Hett translates *exis* "state" aligning with Beare who, in addition to "state," further qualifies it as "relative state" and "relation" based on its definition at 5 *Met.* 1022b10 (1906, 313, n. 1). On the other hand, Sorabji translates *exis* as "having" (1972) and Bloch, combining both interpretations, as "the state of having something" (2007, 81–2). Against this rendition, recently Strevell has argued that Aristotle's conception of memory is ambivalent and includes the disposition of retention of memories that allows a living being to enter into the "acts of remembering" and the actual act of remembering (2016, 8 and Chapter 2). Here I also take *exis* as "disposition" further noting that in *Met.* 5 1022b10 "disposition" is also related to the order (*taxis*) of that which has parts (*merē*) and, so conceived, can suggestively apply to the composite nature of the living being's body.

150 See Chapter 4.

151 See Strevell's definition of remembering as a perception of the animal's cognitive activity "that includes the awareness that what appears is the animal's own activity and that it is prior" (2742016, 10 and, more extensively, Chapter 2). Bloch however seems to deny that for Aristotle animals might have "personal memory" and to rather attribute them "indeterminate memory" (intended by him as a disjunction between the sentient subject and its objects of memory) on the basis of *de Mem.* 1 452b29–453a24 where Aristotle distinguishes between "indeterminate and determinate remembering" (2007, 83). But in this passage Aristotle is concerned with distinguishing memory based on the exactness of the perception of time (i.e. as something that happened the day before yesterday or, by contrast, at one point in the past), not with the relation of the sentient subject with its past objects of sensation. Still, these two forms of memory may well be reflecting Aristotle's position on human and animal memory, the one being "determinate" in measuring time, the other "indeterminate" because the awareness of the past does not involve an exact reckoning—granted that human beings too, as Aristotle recognizes, may often remember indeterminately. For Aristotle's memory as expanded self-awareness, see Zatta 2022.

152 *DA* 2 424a17–21; see Chapter 4.

153 This is the work of *phantasia*; cf. Caston, 1996, 257–60.

154 Aristotle refers to age categories also when addressing the incapacity to dream. Infants, for instance, do not dream because the intense nutrition-related movements affecting their bodies dissipate their sensory impressions (*de Insomn.* 461a12).

155 See n. 151 above.

156 For instance, can animals freely "imagine," that is, have "internal appearances" disconnected from a present sensation and without further engaging in action, and if so, to what degree can they? The fact that Aristotle attributes blooded animals the capacity to dream suggests that he would attribute them also the capacity to "imagine" in a sense of the term closer to our use. In the case of human beings, Aristotle illustrates the illusory side of *phantasia* and its consequent impact on memory by remarking on the fictitious memories of people like Antipheron of Oreus and other lunatics (*existamenoi*) who "spoke of their *phantasmata* as if they had actually taken place, and as if they actually remembered them" (*de Mem.* 1 451a9–11).

157 See *DA* 2 413a20–6.

158 See Chapter 2.
159 In discussing error Aristotle gives the example of a human being who looking at the sun is mistaken about its size (*DA* 3 428b3–5), but error pertains in fact to all living beings (*DA* 3 427b2).
160 On this important point, see Modrak, 1987, 97.
161 On *logistikē phantasia*, see above.
162 *DA* 3 427b15–6.
163 In this respect, see Caston who claims that Aristotle needs *phantasia* to explain how "the content of mental states could ever *diverge* from what is actually in the world" (1996, 21). Both Presocratics and Aristotle had failed to account for error: the one made sensation and thought depend on the encounter between likes (presupposing these encounters to be always true), while the other had given a basic account for sensation and thought. On error as the link of Aristotle's *phantasia* with Presocratics' theory of cognition based on "likes," see also Scheiter, 2012, 264–5.
164 Scholarly discussion of Aristotle's relation to Plato with respect to *phantasia* varies according to the role they assign to *phantasia* in Aristotle's psychological system. Beare sees Plato's legacy in Aristotle's use of imagination (*phantasia*) as the faculty by which something, either illusory or not, appears to the mind through perception. In other words, Aristotle's use of *phantasia* for the subjective side of a phenomenon of perception is of Platonic derivation. More recently, Lorenz elaborates on the importance of Plato's memory (*mnēmē*), conceived of as the capacity that preserves a living being's sensory impressions and puts it "in cognitive contact with the appropriate replenishing processes" so as to form desire) (see *Phil.* 34A10–35D6). Further, this author considers the low-forms of thought and belief associated with appetitive desire as Platonic antecedents (*Resp.* 9 571D1–5; 10 603A1–2) (2006, 3–5; 158); also for Scheiter Aristotle works with Plato's *mnēmē*, but in its application to the problem of error characteristic of *Theaetetus* rather than *Philebus* (2012, 266–70). By contrast, Schofield considers Aristotle's imagination a discovery independent from Plato's use of *phantazesthai* (see especially *Phil.* 38A–40E and n. 154 below, besides 263D–264B) to which he may have himself contributed as a member of the Academy (1992, 250, and n. 1). Here I agree with Lorenz and consider Plato's *mnēmē* an antecedent of Aristotle's *phantasia* qua involved in animal locomotion and purposeful action (*Phil.* 34A is eloquent: *mnēmē* is defined as *sōtēria aisthēseōs*). But I further stress that in the same *Philebus* (38A–40E) *mnēmē* holds also a more subjective value that involves the permanence of human discursive interpretation of the data appearing to perception and that, considered from this angle, Plato also contributes to Aristotle's conception of *phantasia* qua accounting for error.
165 Besides the passage quoted above, see, for instance, *MA* 700b19–20, where perception and *phantasia* are said to hold the same place as mind (*nous*), 701a29–31, 34–8, 701b16–20 (on sense-perceptions, imaginations, and ideas all causing the alterations leading animals to move), 701b 32–3 [on thought (*noēsis*) and *phantasia* presenting the object of pursuit and avoidance in the sphere of action].
166 *DA* 3 433a9–14; in *DA* 3 427b17–8 Aristotle claims that *phantasia* is not the same *noēsis* as judgment (*hypolēpsis*) showing to consider two forms of *noēsis*, a nondiscursive one that is compatible with *phantasia* and a discursive one that relates to calculation (*logismos*). On *phantasia*'s serving the same functions as thought, see also Lorenz, 2006, 121, 133–5.
167 *DA* 3 427a18–22.
168 Aristotle refers to *phronein* and *noein*, namely practical and theoretical thought, which he eventually discusses separately to stress their difference from sensation (*DA* 3 427b7–15).
169 *DA* 3 427a22–3; 27–8.
170 DK 31 B 109 quoted in *DA* 1 404b13–5, transl. by Laks and Most.
171 See, respectively, DK 31 B 106 and DK 31 B 108, cited in *Metaphysics* (*Met.* 4 1009b); besides Empedocles, Aristotle quotes also Homer, who seems to have made the human mind, so to speak, circumstances-dependent (*Od.* 18, 136: "such is the human mind").

172 *DA* 3 427b3–7.
173 In this respect, see *Met.* 4 1009b, where besides Empedocles Aristotle quotes also Parmenides (B16 "For as is at any moment the mixture [*krasis*] of the much-wandering limbs, so mind *noos* is present to human beings, for them in each and all, that which thinks [*phronei*] is the same thing, the substance of their limbs, for that of which there is more is thought [*noēma*]"). As for the physiology involved in this process, from Theophrastus' introduction to the fragment we learn that for the philosopher of Elea there existed only two elements, the hot and the cold, and that knowledge (*gnōsis*) depended on which one of the two prevailed—the hot providing a better and purer thought (*dianoia*). For the physical basis and the overall continuity between thought and sensation in Presocratics' theories of cognition, see Zatta, 2019, 75–94.
174 That these aspects were at the core of Aristotle's reception of the Preosocratics becomes especially apparent if we read *On the Soul* (427a18–427b7) along with *Metaphysics* (cited above), in which Aristotle addresses the identity of thought and sensation as well as their physical nature.
175 See Chapter 2.
176 *DA* 3 429a10–429b10.
177 See especially *DA* 3 429a19–30, where Aristotle endorses Anaxagoras' conception of mind as being unmixed (*amigēs*), and applies it to its relation to the body. Yet, even thinking needs to recur to (*logistikē*) *phantasia* (see above), and can at any rate be wrong (see *DA* 3 427b9–11) while scientific knowledge itself derives from the crystallization of memories into a single experience, in a cognitive chain that starts from sensation (*Post. An.* 2, 19); on the perceptive basis of rational capacities, see Frede, 1992, 157–73.
178 *Phantasia* is liable to error because it supplements the weakness of sensation (*DA* 3 428a10–15; cf. Plat. *Phil.* 38C).
179 In this respect, note Aristotle's claim in *Posterior Analytics* 2 19 that animals that do not possess "the persistence of sensation" (*monē aisthēmatos*, i.e. *phantasia*), do not know anything outside the immediate act of sensation.
180 See Chapter 3.
181 In *DA* 3.3 Aristotle distinguishes *phantasia* from sensation, thought (*dianoia*), judgment (*hypolepsis*), knowledge (*epistēmē*), mind (*nous*), and opinion (*doxa*), while remarking that *phantasia* is caused by sensation and that like judgment it is a form of *noēsis*.
182 See above.
183 *HA* 7 588a28–31.
184 See Chapter 6.
185 See DK 28 A 45, "Parmenides, Empedocles and Democritus say that the mind (*nous*) and the soul are the same thing; in their view no animal would be altogether lacking in reason (*alogon*)," transl. by C.C.W. Taylor.
186 See Aristot. *Phys.* 2 199a10–19 along with the discussion of this passage in Chapter 1.
187 See *Met.* 1 980b–981a, "The other animals ... have but a small share of experience;" transl. by H. Tredennick; cf. *Post. An.* 2 100a3–9.
188 See *NE* 6 1141a27–9; Hicks, 1965, 455. In *On the Soul* Aristotle admits that some animals share in *to phronein* (practical thinking or in Shields' translation "understanding") (*DA* 3 427b6–7), distinguishing *to phronein* from *to dianoeisthai* ("discursive" thinking, hence reasoning). Some animals including man share the first, while the second is an exclusive possession of the human being.
189 See n. 166.
190 See, for instance, *Met.* 1 980a22–982a3 and above.
191 Lorenz further elaborates on why for Aristotle *phantasia*, despite its cognitive import and closeness to thought is not a form of (practical) thought. It lacks awareness of both the relation between means to end and of any number of possible alternatives to achieve a prospected end (2006, 148, 175–85).

192 Migratory animals represent an interesting case in that they can move across large distances and rhythmically settle in different environments, a behavior that raises the question of their relationship to the space traveled through and to the "spaces" inhabited and traveled to.

193 See *DA* 3 430a14–16 (*panta poiein*). I am here referring to adult humans, who are unimpaired by disease or a weakness of the mind and whose rational nature is fully accomplished. In not actually possessing yet the rational soul children are for Aristotle close to animals, that is, incapable of moral choice and acting upon desire according to whether an object is painful or pleasurable [see *GA* 2 736b27–9; *NE* 3 1111a22–1111b; cf. *DA* 1 404b1–8, where, in response to Anaxagoras, Aristotle recognizes that mind (*nous*) in the sense of practical intelligence (*phronēsis*) does not even appear to belong to all human beings, let alone all animals]. In these cases mind is absent or "temporarily clouded over by feeling or disease or sleep" (*DA* 3 429a5–9).

6 The lives of animals

6.1 The *History of Animals* in Aristotle's zoology

In Chapter 1 we discussed how for Aristotle the finality of body composition (*synthesis*) extends into animals' mode and space of existence. That is, each part in the complex architecture of an animal body has a function that contributes to the multiform range of movements and activities in which the animal engages in order to live. Such a perspective contributes to shed light on the composition of *History of Animals* in which Aristotle turns from a systematic description of animals' body parts, including the sense organs (1–4), to their modes of reproduction (books 5–6)[1] to a discussion of their lives (*bioi*) and the activities (*praxeis*) that relate to food acquisition, migrations, and hidings (book 7);[2] and looks at their characters (*ēthē*) and "psychological capacities," enmities and friendships (book 8).[3] Compartmentalized in such broad terms, Aristotle's discussion still allows some overlapping among the different books as well as the consideration, under each thematic head, of subcategories. For instance, as he discusses animals' mating habits in book 5, he also accounts for their characters and care for the young.[4] In book 8, animals' interspecific enmities are related to their quest for nutrition, which is treated not only in terms of food, i.e. "preys," but also in terms of animals' body parts. When Aristotle accounts for the meat-based diet of rapacious birds he also mentions their curved beaks and claws thereby covering a subject he has discussed systematically in the earlier books.

The diverse range of topics, in addition to the presence of internal discrepancies, redundancies, and bizarre-sounding pieces of information have led scholars to doubt the authenticity of books 7–9[5] and to undermine the overall thematic and conceptual coherence of the treatise. Depending on more or less generous judgments, *HA* has been considered a work of natural history, a collection of notebooks, or an "ill-digested, ill-compiled mass of details." It has been the merit of Balme to acknowledge the authenticity of books 7–9[6] and argue in a number of publications for the theoretical nature of the treatise,[7] a position which Gotthelf and Lennox have supported throughout the years with new elements of considerations. Drawing on Aristotle's theory of scientific method, these authors have revealed that *HA* deals with animal differences conceived at the higher level of generality[8] and is to be located at least didactically, before other etiologically

DOI: 10.4324/9780367816001-7

oriented works. Indeed, as Aristotle himself reveals, *HA* provides the facts (*hoti*), of which the explanation-driven treatises (such as *On Parts of Animals* and *On the Generation of Animals*, but also likely *On the Soul*) discuss the causes (*dioti*).[9]

Yet in its present format *History of Animals* relates to the explanation-driven treatises for an additional reason that lies at the heart of Aristotle's animal philosophy, mentioned at the beginning of this chapter. That is, the diverse facts collected at the highest level of generality and waiting to be explained elsewhere reflect Aristotle's understanding of life as founded on, and deriving from, a set of fundamental and complementary functions and activities that are rooted in the living body, relate to movement, and ultimately express animals' intrinsic desire to live, fulfilling it.[10] Let me further qualify this statement. In *PA* Aristotle offers us a teleological understanding of animal bodies at a mereological level for the overall *telos* of life and in *GA* he discusses the modes of animal reproduction (relevant organs and parts included) according to different degrees of perfection with an emphasis on the efficient cause (the male parent's contribution to the process of reproduction). In *On the Soul* he deals with the soul as the first principle and cause of the living body in relation to the vital activities of nutrition, growth (and decay), sensation, locomotion, and thought, among which only the first three contributes to the animal as such (*toiouto to zōon*).[11] Following upon his psychological discussion, the treatises of *Parva Naturalia* explore in turn elements of animals' physiology among which, importantly, is the notion of connaturate heat (*emphyton thermon*).[12] Now in its conception and from the point of view of Aristotle's scientific method *History of Animals* may well precede the aforementioned treatises because it collects the facts to be explained, but from the point of view of Aristotle's animal philosophy it also complements them in that, after describing animals' body parts and physical appearance, it accounts for an array of behaviors that involve animals' movement across space (*kinēsis kata topon*) and the activities of reproduction and nutrition—[13] from the search of food to its production and storage to migrations and the care of the offspring in their specific natural environments. In other words, in *HA* we witness animals' "active" lives as a phenomenon whose rich diversity stems from a diversely constituted body and fulfills an equally present but differently satisfied desire to live. Ultimately, we see, for Aristotle animal behavior forms a continuum with the body and "ethology" is connected to a body-oriented psychology.[14] Indeed, even the political nature of nonhuman political animals (*politika zōa*) has such a basis. In *HA* Aristotle pinpoints it in the collective dimension of activities such as food provision or migrations thereby providing us with some subtext to understand the notorious as well as elusive statement in *Politics* that the human being is more political than some nonhuman animals.[15] On this more will be said toward the end of this chapter. Here let us stress that for Aristotle a physiological, body-rooted explanation accounts not only for animals' modes of nutrition, reproduction, or migration, but also for their characters, emotions, and emotional dispositions along with the distinction between the temperament of males and females (discussed in book 8 of *HA*). Indeed, these aspects of animals' lives too can be seen in relation to their body parts and overall constitution. Admittedly not addressed in

HA (conforming with the descriptive nature of the treatise), such an explanation has to do with the quality of animals' blood (and hence, ultimately, the body's elemental composition) and features in book 2 of *PA* in the course of an analysis of the body uniform parts.[16]

So when considered in this light, namely *vis-à-vis* the areas and issues which are developed in the other zoological treatises and *On the Soul*, *HA* complements them in that like them, it offers a discourse on animal lives which is ultimately rooted in the body[17] but unlike them it further situates this discourse in animals' variegated space of existence, following their activities in a way that ultimately reflects the meanings of life identified in *On the Soul* along with their order of priority.[18] Indeed, as was argued in Chapter 3, in this treatise the nutritive soul is considered "the first and most natural" (*prote kai physikotate*) because animals' life gravitates around nutrition and reproduction as capacities intrinsically connected to the senses, accompanied by the feelings of pleasure and pain, and enabled (for the animals that possess it) by locomotion. Much of *HA* corroborates this view inasmuch as it consists in a description of the activities related to the nutritive part of the soul (in conjunction with the sensitive soul) and hence pursues mating habits and food acquisition,[19] in a discussion systematically organized according to animals' groups and their living environments.

6.1.1 The articulation of differences and sameness

If methodologically the discussion of *HA* remains at a level of generality[20] and hence conforms to a programmatic predicament of *Parts of Animals*, it also presents a significant "shift" of perspective. In *PA*, let us recall it, Aristotle asks whether to address animals species by species or by those phenomena that happen in common (*ta koinēi symbebēkota*) to different groups of animals. He opts for the common traits articulated in terms of degree (*kata genos*) and by analogy (*kat' analogian*) while he addresses the traits of species that are one of a kind, like that of man, as "specific" (*kat' eidos*).[21] The focus is on commonalities and ultimately contributes to elucidate animals' common nature (*koinē physis*).[22] In *History of Animals*, however, Aristotle takes another path. For he claims he will be looking at animals' differences (*diaphorai*) in relation to four aspects—their characters (*ēthē*), actions (*praxeis*), and lives (*bioi*), besides body parts (*merē*).[23] Hence it is important to assess whether this approach clashes against his discourse on method in *PA* or may instead be compatible with it and even a faithful instantiation. To put it into questions, does the focus on animals' differences[24] in *HA* undermine the notion that they share a common nature? Or can also the differences in the aforementioned aspects of animals' existence be taken as a predicament of their common nature?

Evidence internal to the treatise seems to confirm the second alternative: animals' body differences are ultimately conceived in terms of "identity" (*tauto*), thereby articulating animals' commonality. In a long introductory passage Aristotle writes,

With regards to animals, some have their parts *mutually identical (tauta)*, some have parts of a *different character* (*hetera*). Some parts are specifically identical (*tauta ta men eidei*), for example, one man's nose and eye, one's nose with another's nose and eye, one's flesh with another's flesh, one's bone with another bone; and the same applies to the parts of a horse, and such other animals as we consider to be *specifically identical*; for as the whole is to the whole, so every part is to every part. In other cases, they are, it is true, *identical (ta de tauta)*, but they differ in respect to excess and defect: this applies to those whose genus is the same (*to genos esti tauton*); and by genus I mean, for example, bird and fish: each of these exhibits difference with respect to genus ... Some animals, however, have parts which are *not specifically identical*, nor differing merely by excess and defect: these parts are *the same (tauta)* only "by analogy": of which an example is the correspondence between bone and fishspine, nail and hoof, hand and claw, feather and scale: what is the feather in a bird, this is the scale in a fish.[25]

Often neglected in the relevant scholarship or even "mistranslated," this passage discusses the differences among body parts in terms of "kinds" of sameness and shows that for Aristotle (in the same way as they do in *PA* 1) differences (*diaphorai*) do articulate sameness (*to auto*). In other words, Aristotle recognizes that body parts may be identical specifically (i.e. a horse's nose is the same as another's horse nose); within a kind (i.e. an alcyon's wing is the same as a duck's, it only differs by excess and defect); or across kinds (i.e. a bird's feather is the same as a fish's scale). This third kind of sameness is "by analogy" and it is incidentally noteworthy that the examples provided echo in part Empedocles' fragment 82, preserved in *Meteorologica*.[26] But while Empedocles had built an overarching metaphor identifying leaves, hairs, feathers, and scales, collapsing body analogies into identity,[27] Aristotle underscores the analogies, relating them to a "third kind" of identity—in addition to the identity within species and within kind. Analogies articulate the sameness of body parts across animal kinds. True, Aristotle here predicates identity at an empirical level, but in order to predicate it he presupposes a causal-based interpretative level: for how can a fish spine be the same as a quadruped's bone? For Aristotle parts across animals' kinds are the same because they ultimately play the same functions and entertain the same relation in respect to the whole they belong to. Bone corresponds to fish spine, nail to hoof, hand to claw, feather to scale, and vice versa. In this way, even the more acute differences across kinds are absorbed into a discourse that refers to identity of functions (i.e. the body-supporting role of bone and spine) and that implicitly builds on animals' body structure and life as expression of *logos* in the sense brought forward in Chapter 1.[28] And, true, in the passage above Aristotle speaks of sameness only when discussing the differences of body parts, but it is still legitimate to extend it also to the differences of animals' activities (*praxeis*) and "lives" (*bioi*)[29] in that, as we will soon see, they equally fulfill (or contribute to fulfill) for each animal the same fundamental functions (nutrition and reproduction) and entertain the same relation in respect to the "whole" phenomenon of life.

In *HA* Aristotle mentions about 500 types of animals, from elephants and camels to crested larks and migrating cranes to those insects whose life span coincides with the arc of a day, hence remaining faithful to another predicament of his discourse of method in *PA* 1 according to which species' "nobility" becomes irrelevant when considered against the acquisition of knowledge—granted, *HA* presents facts, but still in view of assessing their causes. With the distinction between worthier (*timiōtera*) and less worthy (*atimotera*) animals eclipsed,[30] Aristotle gathers a wealth of information[31] from multiple sources, from farmers, beekeepers and herdsmen, fishermen and hunters,[32] and from his own observations.[33] In addition to aspects like animals' habitats and interactions, life expectancy, fertile periods and diet,[34] some of which will be discussed more in depth in the upcoming sections, we also learn about human practices and impact on animals' lives, and what, for instance, fishing strategies revealed about animals' "mysterious" perceptive capacities along with their resourcefulness and "instinct" of survival. So despite being without visible organs for hearing,[35] dolphins were still able to sense sounds as the loud noises produced by fishermen in the open sea made them flee toward the land into fatal nets.[36] When grasped, sponges reacted, holding tightly down to their surfaces,[37] a behavior which manifested their sense of touch. On the other hand, the scent of oregano and brimstone sprinkled on the entrances of ants' nests caused them to abandon their dwelling, revealing that they possessed the sense of smell and a susceptibility to unpleasant odors.[38]

6.2 Body constitution, habitats, and life

In pursuing animal differences Aristotle moves with nonchalance among different animal groups, variously identified in terms of body components (i.e. blooded and bloodless animals), modes of reproduction (i.e. viviparous and oviparous), or idiosyncratic features (i.e. the trunk of the elephant), from the most general to the most specific. Yet there is one way of identifying animals based on their relation to their living environments and the distinction of "habitats"[39] that is overall characteristic of *History of Animals* as it inheres in the observation of animals' activities and lives which is at the core of this treatise. In approaching the relation between animals and their living environments Aristotle seems both to echo the Presocratics but also to depart from them. For besides referring to animals' physical (e.g. elemental) constitution he further looks at their nonuniform parts, hence applying his conception of multilayered *synthesis* that gives rise to their bodies. To understand Aristotle on this issue, it is illuminating to start with the evidence relating to his predecessors (specifically Anaximander, Empedocles, and Democritus) and then turn to *History of Animals* 1 integrating the material provided by this treatise with passages from *On Respiration* that are relevant to animals' physiology in relation to their habitats. This strategy will allow us to trace Aristotle's departure from the Presocratics (and Plato) and to deepen our understanding of animals' unique physiology in respect to plants, while ultimately pursuing the centrality of *logos* in his discourse on animals. But it will also help us clarify how the notion of animals' life (*bios*) at stake in *History of Animals* is inextricably linked to the ("biological") conception of pleasure (*hēdonē*) we have been discussing in Chapters 2 and 5.

If we believe the doxography, Anaximander was the first student of nature (*physiologos*) to have discussed animals in terms of physical constitution and the environment. He claimed that animals were born by evaporation from the effects of the sun, pinpointing the origin of life in water. On this view, human beings were at first similar to another living being, namely a fish,[40] and life inception seems to have been inextricably connected to, and dependent on, cosmological dynamics and conditions. Only successively, the dry land became accessible and hospitable to life, a fact which determined a change of animals' bodies. For the first living beings had bodies enveloped by a thorny bark; but increasing in age, they moved to a drier environment where the bark burst open and they changed their way of life (*metabioun*) in a short time.[41] It is difficult to envision what for Anaximander animals' original bodies looked like and how they developed. Other scholars have attempted to do so.[42] What is interesting to note here is that Anaximander saw a correlation between living beings' physical structures and the environments they inhabited. The thorny bark was conducive to life in water as well as in drier lands, but in this last case to a lesser extent inasmuch as it eventually broke giving rise to other forms of life. Further, the extant evidence reveals that also Anaximander paid attention to animals' nutrition (and hence, even if indirectly, to the availability of food in a given environment). For he observed that, unlike the other animals, human beings are not self-sufficient to provide for their food in the early stage of life but require a longer time of nursing than the other animals. It was based on this observation that he claimed that humans were born from a different species.[43] Be that as it may, the evidence suggests that already from the time of Anaximander the correlation between animals' bodies and living environments was thought in terms of the body "elemental" composition and structure.[44] For that in a drier environment (the dry land) the bark burst open implies an adjustment of animals' bodies to their surroundings: the harder part that had formed at the origin under the effects of the sun in water and sustained life in the aquatic environment broke and disappeared.

It is, however, with Empedocles and then Democritus that we have an extensive correlation between animals' bodies and their environments. In the long passage where he discusses Empedocles' zoogonies, Aëtius reports that,

> The species of all the animals became distinguished by virtue of the variety of their mixtures: some had a drive toward water, with which they had a greater affinity; others flew up into the air until they possessed more of the fiery element; heavier ones [scil. went] onto the ground; and those whose mixture was in equilibrium with their †whole chest made sounds†.[45]

This testimony shows that Empedocles considered living beings in terms of physical constitutions, divided into "watery," "fiery," and "terrestrial." Not all animals represented the same combination of the four elements, but different mixtures gave rise to distinct animal types, which in turn inhabited environments compatible with their nature. In this vision, animals' bodies were suited to their natural habitats. A heavier, earthier mixture gave rise to creatures inhabiting the earth (including plants). Those creatures whose bodily constitution was watery found their natural environment in the liquid element, while birds flew in the air because their bodies contained a higher measure of fire. A similar account is preserved for Democritus.[46] After animals were generated from membranes in the muddy earth heated by the sun, those that contained more warmth (and hence fire) "rose up toward the higher regions and became winged," the constitutionally earthier ones became reptiles and other terrestrial animals while those that had a moist nature moved to the region that was akin to them (*homogenē topon*, i.e. water) and became fish. For both Empedocles and Democritus life in different habitats was determined by, and explained through, the material relation between the bodies of the animals themselves and the environments they inhabited. The appeal to the elemental compatibility sufficed, and no explicit attempt seems to have been made to explore further animals' exclusive relation to their habitats based on their organic constitution, as Aristotle will do.[47]

Now in *History of Animals* Aristotle too considers animals in terms of the environments they inhabit—as observed earlier, this is a distinctive trait of the treatise, called for by the discussion of animals' activities (*praxeis*) and lives (*bioi*). Aristotle speaks of creatures of the earth and water (*enydra, khersaia*), but for him these labels do not merely imply (as in Empedocles or Democritus, for instance) a homology or compatibility between the elements composing the living beings' bodies, on the one hand, and those making up their habitats, on the other. As he insists in *Parts of Animals*, criticizing the early Greek philosophers, one should look at the parts of the body, and take account of the "flesh, bone, blood, in fact of all the uniform parts, and indeed of the nonuniform parts too, viz. face, hand, foot,"[48] and not only of the elements. Elemental composition is important, and Aristotle refers to it when discussing, for instance, living beings' structure in relation to locomotion[49] and even diet,[50] but it is one of the layers of body composition[51] and certainly not enough to understand the complex relation living beings entertain with their environments. So in *On Respiration* Aristotle criticizes Empedocles for having claimed that fishes were originally land animals that "escaped" to a new life in water because their body constitution was warm (too much fire inside!) and needed constant cooling off.[52] By contrast, he sees animals' relation to habitats in terms of a physiology that, building on the elemental composition, involves their nonuniform parts and works toward maintaining an optimal body temperature. Animal physiology, we learn, is essential in order for life to continue to exist and has ultimately repercussions on specific behaviors and habits, as we will see in the next pages.

For Aristotle every living body along with its parts possesses some connate heat (*emphyton thermon*). He remarks that "when alive, animals are perceptibly warm, but when dead and deprived of life the opposite of it." In blooded animals the source of this heat is the heart, while in bloodless ones an analogous part, and life continues or is destroyed in accordance to whether this internal heat is preserved or extinguished. More specifically, Aristotle distinguishes two causes for the internal heat destruction through analogy with the two forms of death by which a living being may die: either old age or violence. Transferred to its field of reference (the preservation of internal heat), this analogy means that a living being's connate heat dies out by natural decay or by something that accelerates this natural course and that may be either internal to the body or external. In case of "acceleration," the connate heat may suffer respectively from an excess of internal heat itself or from environmental factors such as an excess of cold in the surrounding environment. Respiration is the process that allows animals to preserve their connate heat by tempering it. The inhaled air cools the body off and hence contributes to maintaining an optimal body temperature for the vital process of nutrition and the ensuing nourishment of the entire body in all its parts.[53] In this way, living beings are equipped with a safety mechanism that consists in their capacity to engage in a continuous thermic exchange with the environment.[54]

Respiration, however, is not indistinguishably shared by all living beings, as some Presocratics had stated or implicitly assumed. Anaxagoras and Diogenes, for instance, had argued that even fish respire introducing the air present in the water, while others, like Empedocles and Democritus, did not bother to assess whether respiration belonged to all living beings.[55] For Aristotle instead only animals that possess lungs and windpipe breathe, that is, only those that have a body structure equipped with nonuniform parts suited to the activity of breathing. Those animals that are deprived of these parts do not breathe, but are cooled off by corresponding mechanisms that allow the introduction in the body of, or, at least, a maximum contact with, a cooling "element," be it water or air. Fish, for instance, introduce water by the gills, while long-lived insects are cooled off by air by means of a thin membrane situated in their stomach's indentation.[56] There is a physiological reason why for Aristotle these animals do not respire like other blooded animals: they possess a lower connate heat and do not need as much cooling as that provided by the inhaling of air. In being less hot than the terrestrial animals, fish do not need to introduce air to cool them off and are suited to inhabit, and thrive in, an aquatic environment. Similarly, even though they lack lungs and windpipe and hence respiration, insects can still afford to live on land.

In line with the theory just now outlined, in *History of Animals* Aristotle does not merely distinguish aquatic animals from the terrestrial ones based on the fact that the first inhabit a liquid environment while the latter live on land. This mode of grouping (along with the identification of birds) was a traditional distinction, which he explicitly endorses in *PA*[57] and which helps him organize broadly the material discussed in *HA*.[58] But when looking at animals' habitats Aristotle ultimately refines this distinction because in extending his analysis from animals' body parts to their activities and lives he considers the connection of animals' body anatomy with their environments in light of the vital processes they engage with. In other words, Aristotle focuses on the cooling-off mechanism, which is enabled at the nonuniform mereological level, and hence comes up with a novel definition of animal groups in relation to their habitats. Accordingly, in the general remarks at the beginning of *HA* he claims that "some are water-animals (*enydra*), others land-animals (*khersaia*),"[59] seemingly excluding from this preliminary division the "winged animals" (*ptēna*). Indeed in his vision, birds are (still) land animals because they breathe. Birds have feet—observes Aristotle—and never merely only fly (living suspended in the air), but they touch the ground and stand, at any rate, on trees and other surfaces.[60] They are equipped with parts that enable them to breathe and maintain an optimal body temperature. In being so structured, birds (along with other flying but nonrespiring creatures, i.e. bees) are for Aristotle winged (*ptēna*) animals of the earth (*khersaia*).[61]

Animals' physiology, however, is not the only meter for determining their habitats. In tune with the ethological development of *HA*, Aristotle further discusses animals on the basis of where, regardless of their "cooling-off" physiological mechanism (i.e. by air or water), they feed, live, and breed. So in relation to water animals he remarks that,

> Some both live and feed in the water, take in water and emit it, and are unable to live if deprived of it: this is the case with many of the fishes. Others feed and live in the water; but what they take in is air, not water; and they breed away from the water. Many of these animals are footed: e.g. the otter, the beaver, the crocodile; many are winged, e.g. the shearwater and the plunger; many are footless, e.g. the water-snake ... Again some water-animals live in the sea, some in rivers, some in lakes, some in marshes (e.g. the frog and the newt).[62]

Water animals (*enydra*) are distinguished into two groups: those that cool off by introducing/emitting water (and live and feed in water) versus those which, while taking in air, live and find their nourishment in water and procreate away from it. In this group Aristotle lists footed (i.e. crocodile), footless (i.e. water-snake), and even winged animals (i.e. the shearwater), reckoning a number of diversified ecosystems.[63] Water animals—he observes— inhabit different aquatic environments: besides the sea, also rivers, lakes, and marshes. Food provision on land too is a crucial factor that qualifies the land animals (*khersaia* or later *peza*), a group including living beings which are lung-equipped and breathe (by introducing and emitting air) and those which cool off without breathing but on account of the surrounding air, whether through indentations (as for the hottest ones among them, like bees and wasps) or not.[64]

While in *HA* Aristotle discusses animals' relation to their environments in terms of physiology and ethology (particularly the feeding habits),[65] we should incidentally note that he does not always pursue the ethological-based definition of water animals. For in the same passage referred to above, Aristotle mentions "land-animals" (*khersaia*) that feed in water without defining them as "water animals," as we would expect from his previous assertion that "water animals" include creatures that take in air but live and feed in water and breed away from it.[66] Perhaps due to the lack of final editing, this failure to conform to a previously established definition also indicates a certain flexibility on Aristotle's part in using the physiological and ethological distinctions and that the ethological-based definition of water animals as encompassing those breathing creatures that feed in water does not completely override the "more fundamental" definition of land animals as creatures that (are warmer and) take in air. [67]

6.2.1 Diet, pleasure, and the fight for survival

The ethologically based distinction is, however, understandably relevant in a treatise which aims at discussing animals in terms of activities and lives in their space of existence. At the beginning of book 7 Aristotle returns to this distinction before undertaking a systematic discussion of animals' nutrition by groups and relative food preferences. Here too, as he did in book 1 with regard to animals' living environments, he appeals to their bodily constitution in order to explain their diet.[68] He writes,

> And their food differ chiefly according to the matter out of which they are constituted. For each one's growth comes naturally out of the same matter (*hylē*). And what is natural is pleasant; and all pursue their natural pleasure.[69]

It is animals' body material constitution that determines their diet. That is why, for instance, some birds feed on berries, others on insects, others again on fish, each group living in the environment that makes it access the suitable source of nourishment. In considering diet compatible with, and dictated by, bodies' material "substratum" Aristotle is both pointing to animals' common nature and likely elaborating on previous Presocratic doctrines. For if each animal requires food that is compatible with its body, then, animals' dietary differences are in fact epiphenomena of a "law" that pertains to them all. In this respect, already Empedocles had understood animals' nourishment in terms of an attraction of "like for like." According to the doxography, he believed that animals (including plants) were nourished "by the remaining (*hypostasis*) in them of that which is proper to their own nature (*oikeion*)."[70] Granted, for Aristotle in *On the Soul* and elsewhere nutrition is a process made possible by what appears to be, from a chronological point of view, both unlike and like: before digestion food is unlike the living being's body, but afterward it is like.[71] But *HA* is unconcerned with nutrition technicalities and pinpoints a "material" compatibility between animals' (elemental) composition and their nourishment: animals' body *krasis* calls for a specific food and a compatible "lifestyle" (i.e. life in aquatic environments making available the suitable food).[72] And while in *HA* Aristotle refrains from going into details and merely points to a qualitative compatibility between animals' (elemental) composition and their nourishment,[73] from *On Respiration* we understand that it ultimately has to do with wet and dry matter. That is, animals with a dry constitution live in dry places, those with a wet in wet ones[74]—all feeding accordingly. So while for Aristotle the mere appeal to animals' elemental constitution is insufficient to understand the more complex relationship they entertain with their environment, it still works broadly to account for animals' localities in relation to their differentiated diet, and becomes the default explanation for the nature of the "water animals" that breathe. On this view, animals with a dry constitution feed on dry food, while those with a water constitution feed on "water food" (i.e. sea-gull on fish)—there being, we may well assume, as many dietary variations as animal constitutions.

Nature dictates pleasure. For in pursuing what is natural to them, that is, what is in accordance with their bodies, animals pursue their natural pleasure (*hēdonē kata physin*). Aristotle expresses this idea in *Nicomachean Ethics* where he endorses Heraclitus' dictum, recalled earlier, that "an ass would prefer chaff to gold, since to asses food gives more enjoyment than gold"[75] and further remarks that horses enjoy different things than dogs or humans. Linked to nutrition, pleasure orients and supports animal life (*bios*), for it is for its sake that each animal looks for food and continues to live. In *HA* pleasure is also said to accompany animals' mating and distinguish them from plants, which reproduce without sensation.[76] But, as was argued in Chapter 5, pleasure plays a more extensive role than promoting these functions of the nutritive soul, i.e. feeding and mating. For a feeling of enjoyment is intrinsic in all animals' activities and coextensive with their life, inasmuch as their activities relate to their nature and fulfill it.

Aristotle reviews animals' diet by large strokes, often describing the body parts involved in the process of nutrition and the practices associated with it, for instance, whether animals have a lair or some kind of dwelling, whether they possess locomotion, and how they relate to their environment or the time of the day,[77] or again whether they have hunting strategies.[78] His analysis moves from the simpler forms of animal sea life, like the stationary testaceans, to more complex forms of moving (*kinetika*) water animals to birds (terrestrial and aquatic), then proceeds to thorny-scale animals, next to quadrupeds and ends with insects.[79] A main, recurrent observation is whether animals feed on plants ("things that grow," *ta phyomena*) or "flesh," (*sarx*),[80] or whether they are omnivorous with the consequent distinction among a vegetarian, carnivorous, and a more diversified diet. If the simpler forms of testaceans feed on drinkable water,[81] the sea-anemones (*akalēphai*) get their nourishment from the small fishes that fall upon them. Strange, elusive creatures, they possess a mouth in the center of their body as the big specimens of this form of life make manifest. On the other hand, while seemingly stationary limpets (*lepades*) detach themselves from the rocks on which they cling to go and get their nourishment. The sea tortoises (*khelonai*) are instead omnivorous (*panphaga*). Endowed with a strong mouth they feed on seashells in the sea but when they come out to the shore they also eat grass (*poa*). Aristotle goes to great lengths to describe the eclectic diet of the crustaceans: it includes stones, wood, sea-weeds, flesh, and, in the case of the rock crustacean, even dung.[82] A member of this group, the crayfishes (*karaboi*) make their lairs (*thalamai*) in rough and stony places from which they hunt small fishes capturing them with their double claws and bringing them to their mouths in the same way as the pincers-equipped crabs do.[83]

As for the birds, Aristotle discusses first the terrestrial ones pointing out the association between specific body features on the one hand and their lifestyle and diet on the other. For instance, all the crook-taloned birds eat flesh, but the others feed on grubs.[84] Another group feeds on thorns, and sleeps in the same place where it feeds, the thorn-bushes.[85] There are then those which feed on grain (*karpophagounta*) and herbage (*poiophagounta*), and of these some are visible all the time (wood pigeons and common pigeons), others at specific seasons because they go into hiding or migrate. The oinas, for instance, appear in the fall when they arrive to Greece with their young. Other birds land instead in the summer, and build their nest and reproduce.[86] As for the aquatic birds (those that find their food in water), they live in a variety of environments, namely seas, lakes, and rivers. Aristotle distinguishes the web-footed (i.e. swans, ducks, etc.), which spend most of their time in the water from those whose toes are separate and live beside the water. The herons stay by lakes and rivers where they feed on the submerged growths with their flat and long beaks. But other aquatic birds are carnivorous, and others, like the sea crows, omnivorous (*panphagon*). Aristotle takes note also of the sea-eagles which spend their time by the sea and hunt for prey by the lakes,[87] hence implying that these birds may fly long distances to acquire their food. As for the horny-scaled animals like lizards and other quadrupeds or snakes, they are omnivorous, their diet including smaller creatures and herbage (*poa*). Among them the snake represents a special case. Aristotle defines it as the most gluttonous (*likhnotatos*) of animals and remarks on its immoderate enjoyment of wine, adding that at his time it was common to capture vipers by leaving pieces of pots containing wine on the dry-stone walls and making them drunk. Another noteworthy fact about snakes is their eating "ceremony," how they normally grab and set straight their source of food (small birds, animals, whole eggs), and then by alternating body stretches and contractions they force it down the gullet.[88]

Finally Aristotle discusses the diet of the viviparous quadrupeds, turning from the wild, saw-toothed (*karkharodonta*)[89] and hence flesh-eating animals, to the domesticated, and mostly herbivorous, ones. We learn about the wolf, which is particularly attracted to eating human flesh (*anthropophagon*) and the only one among the animals to ingest a certain type of earth. Besides, when ill, like dogs also the wolf takes herbs, showing a form of intelligence of which more will be said later in this chapter.[90] The omnivorous bear has the most diverse diet: honey and insects, leguminous crops, flesh, and fruit which he eats climbing up on trees thanks to its supple body. By contrast, the quadrupeds with horns and without saw teeth, whether wild or tame, mainly feed on grain (*karpophaga*) and herbage (*poēphaga*);[91] and so do horses, mules, and asses.[92] An exception, the pig is mostly inclined to feed on roots (*rhizophagon*) because of the conformation of its snout.[93] As for insects, those with teeth are omnivorous, while those equipped with a tongue extract liquids from different sources. Some feed on any juice, others merely on blood, others again on the juices of plants and fruit (like the bee).[94]

Essential to understand animals' subsistence and life as well as their presence in specific habitats (animals tend to live where they feed), nutrition is also crucial to comprehend their interspecific and intraspecific relations. When discussing animals' diet in *HA* 7 Aristotle notes that all fish feed on spawn,[95] and, besides, that the carnivorous ones among them even eat members of their own kind.[96] He returns more forcefully to this claim in book 8 making war a universal condition that pertains to all living beings that share the same place. So he writes,

> Now there is a war (*polemos*) against each other among animals that occupy the same places and get their living (*zōē*) from the same things. For if their food is scarce, even those of the same breed fight against each other, for they say that even the seals living around the same place make war, male against male and female against female, until one kills the other or is driven away; and the pups all do the same. Further, all are at war with the carnivores, and they with the others; for they feed on the animals.[97]

Whether water or land, for Aristotle animals' living environment is irrelevant. For there will be war among animals as long as they inhabit the same places and derive their subsistence from the same food. The fight for life influences animals' dispositions to the point of involving even members of the same breed, who compete with one another to the very end, pups included. Either one is killed or, when overcome, has to leave the area. Moreover an eternal war frames the relation between the carnivores and the other animals. Yet if provisions were available and no fight for survival were necessary, then "the animals that are now frightened (*phoboumena*) and grow wild (*agriainonta*) would probably behave tamely both towards humans and toward each other."[98] Indeed, even the wildest creatures would become tame (*hēmerousthai*) and live with one another in view of the benefits (*ōpheleiai*) they would receive. As a case in point, Aristotle evokes the sacred crocodiles of Egypt, which become docile toward the priests that look after them. On his view, abundance of food tames hostile dispositions. Our philosopher lists a whole series of micro-ecosystems in terms of animals' relations, where species fight one another within kinds and across kinds, but where instrumental alliances also take place. So while the eagle is at war with the dragon-snake,[99] the ichneumon[100] with the spider, and the crow with the owl, the raven and the fox are friends (*philoi*). The raven, we are told, aids the fox when it is being attacked by the predatory merlin. At war are also the birds that live from the sea,[101] while inimical to one another are all the fishes that do not shoal together.[102] War, however, can be an occasional affair. And so is friendship. For if some wild animals are always at war,[103] others are so only at times, like humans.[104] On the other hand, some fish form shoals and are friends only while pregnant, others after giving birth, and still others when the food is abundant.[105]

6.3 Animals' characters and learning

In book 1 of *HA* after giving examples of animal lives and actions,[106] Aristotle proceeds to illustrate their characters (*ēthē*):

> Now here are the sorts of ways in which animals differ from each other in regard to disposition (*kata to ethos*). Some are gentle, and sluggish, and not inclined to be aggressive, e.g. the ox; others are ferocious, aggressive

and stubborn, e.g., the wild boar; some are intelligent and timid, e.g., the deer and the hare; others are mean and scheming, e.g., serpents; others are noble and brave and high bred, e.g., the lion; others are thorough-bred, wild and scheming, e.g., the wolf ... Again, some are mischievous and wicked, e.g., the fox; others are spirited and affectionate and fawning, e.g., the dog; some are gentle and easily tamed, e.g., the elephant; some are bashful and cautious, e.g., the goose; some are jealous and ostentatious, like the peacock. The only animal which is deliberative (*bouleutikon*) is man. Many animals have *in common* (*koinōnei*) the power of memory (*mnēmē*) and can be trained (*didakhē*); but the only one which can recall past events at will is man.[107]

One of the four catalysts of animals' differences (along with body parts, actions, and lives), *ēthē* describes animals' "internal" dispositions, i.e. their inclinations to be in a certain way and hence to act accordingly.[108] In being naturally inclined to be gentle and sluggish the ox behaves and acts differently than the aggressive, ferocious, and tenacious boar; so does the tricky fox in comparison with the affectionate dog. To qualify animals' characters in the above passage Aristotle uses a combination of adjectives (i.e. mischievous and wicked, bashful and cautious, spirited, affectionate, and fawning) indicating that characters are intrinsic and complex, that is, inherent in living beings' nature and made up by a combination of dispositions. They precede action but are manifested by it, contributing to life (*bios*). For instance, an aggressive animal is more prone to fight than a gentle and sluggish one, and arguably more resourceful and mobile in the territory it inhabits. As with the other areas of animals' difference (parts, actions, and lives), also their characters are considered at a level of generality. In other words, all animals belonging to the same species share the same character, a fact that may be explained, as we will see in the next section, on a physiological basis but does not imply a blind behavioral determinism. Indeed in the above passage Aristotle also points out that many animals share the capacity to be trained and that the acquisitive process depends on the possession of memory (*mnēmē*)—in addition to the sense of hearing as he clarifies in *Metaphysics*,[109] quoted below. Now, animals' receptivity to training may influence their natural inclinations and contribute to a plastic behavior. Moreover, it implies the unfolding of an individual's personal experience (admittedly not of interest to Aristotle *qua* scientist intent to pursue a discourse at a general level) denying a rigid, deterministic view of character as if each member of the same species were to be and act the same. If natural dispositions are predetermined by species membership and have a physiological basis, they can still be nuanced and bent or, conversely, accentuated depending on the living beings' intrinsic capacities (memory and hearing), the environmental circumstances, and particular life experiences.[110] In fact, it seems particularly in the area of animal sociality that Aristotle may allow character changes.[111] With abundance of food, we saw, animals may change their natural inclinations and behave differently as normally expected, like the Egyptian crocodile, who adapted to his "confortable" environment and became "gentle,"[112] although one may also wonder to what extent the (original) natural disposition had actually changed or whether it continued to exist but was not acted out.

A passage in *Metaphysics* 1 helps us elucidate the role of memory in animal learning (*mathēsis*) (i.e. the other side of "being trained") in a discussion that appeals to the senses and different degrees of intelligence, and unfolds a comparison with the human beings. Aristotle writes,

> All men naturally desire knowledge. An indication of this is our esteem for the senses; for apart from their use we esteem them for their own sake, and most of all the sense of sight ... The reason of this is that of all the senses sight best helps us to know things, and reveals many distinctions. Now animals are by nature born with the power of sensation (*aisthēsis*), and from this some acquire the faculty of memory (*mnēmē*), whereas some do not. Accordingly the former are more intelligent (*phronimōtera*) and capable of learning (*mathētikōtera*) than those which cannot remember. Such as cannot hear sounds (*mē ... akouein*) (as the bee, and any other similar type of creature) are intelligent (*phronima*), but cannot learn (*aneu to manthanein*); those only are capable of learning (*manthanei*) which possess this sense in addition to the faculty of memory (*mnēmē*). Thus the other animals live by impressions (*phantasiai*) and memories (*mnēmai*), and have about a small share of experience (*empeiria*); but the human race lives also by art and reasoning. It is from memory that men acquire experience, because the numerous memories of the same thing eventually produce the effect of a single experience.[113]

Animals that are capable of learning are more intelligent than those who are not. Key to the learning process is the perception of sounds (*akoē*)—besides memory (*mnēmē*)—and animals, which are deprived of hearing are still intelligent but incapable of learning. Only animals that have hearing *and* memory can learn. As for the comparison between animals and humans, Aristotle insists on the central cognitive role that sensation (*aisthēsis*) plays for both, but, faithful to the stress put elsewhere on animals' perception of sounds, he outlines a contrast between human sight and animal hearing along with their respective cognitive outcomes (*gnōsis* versus *mathēsis*). Sight is for humans a privileged sense that leads to knowledge (*gnōsis*), ultimately conceived of as abstract understanding of reality.[114] Hearing, by contrast, makes animals learn (*mathēsis*) within the physical boundaries of their interpersonal relations and by extension, we may arguably add, through their overall acoustically based experience of the world.[115]

A passage in *HA* 8 complements the above claims from *HA* 1 and *Metaphysics* 1 in that it discusses animal learning (*mathēsis*) in conjunction with teaching (*didaskalia*) and relates them to the sense of hearing as receptive to a spectrum of acoustic sensibles. Aristotle writes,

> The characters of the animals are less obvious to us by perception in the case of the less developed and shorter-lived ones, but more obvious in the longer-lived. For they are seen to have a certain capability in relation to each of the

soul's affections—to intelligence and stupidity, courage and cowardice, to mildness and ferocity, and the other dispositions of this sort. Certain animals at the same time are receptive of some learning (*mathēsis*) and instruction (*didaskalia*), some from each other, some from humans, that is all that partake of some hearing (*akoē*) [not just those that hear sounds (*psophoi*) but also those that distinguish the differences between the signs (*tōn sēmeiōn aisthanesthai tas diaphoras*)].[116]

In this passage Aristotle connects learning and teaching to hearing, subsuming under the same capacity the discrimination of sounds (*psophoi*) and the distinction of the differences between "acoustic signs" (*tōn sēmeiōn aisthanesthai tas diaphoras*).[117] The contrast between sounds and acoustic signs here likely refers to animals' distinct expressions, that is, on the one hand, to their unspellable sounds, which Aristotle elsewhere attributes to wild beasts, and to the articulate language possessed by birds, on the other.[118] And that animal learning stems from each other *and* from humans, let us understand that the "acoustic signs" from which animals learn encompass also human language. Thus learning is for Aristotle an interpersonal and "social" activity, mediated by vocal communication and the relevant sense, and the fact that it happens through acoustic discrimination points, along with *Metaphysics*, to the role of hearing in animals' overall relation to the surrounding world. For animals' capacity to hear and learn from the simple and/ or articulate vocal expressions of other conspecifics (and humans) implies their awareness of the "meaning" of other natural sounds and the ability to recognize, for instance, the movement or cry of a prey or the flowing water of a river (qua incidental sensibles).[119] From this point of view (admittedly, not addressed by Aristotle but still deduced from his discussion), for the more intelligent animals the alertness to soundscape with ensuing *phantasiai* and memories would be key to a comprehension of their environment along with the perception of scents, which, as stated in *On the Soul*, are more acute in nonhuman creatures than in human beings.[120]

6.3.1 Between psychology and ethological physiology

After this interlude on animals' learning we can return to their characters pursuing Aristotle's comparative discussion in *HA* 7.[121] Animals share with human beings gentleness and wildness, courage and cowardness, fears and boldnesses, and likes to a greater or lesser degree. For other traits, however, such as art (*tekhnē*), wisdom (*sophia*), and intelligence (*synesis*), animals possess "another natural capacity similar to this sort" (*hetērē toiautē physikē dynamis*), i.e. by analogy. Now, in relating nonhuman animals' characters to humans' in terms of degree and analogy, Aristotle continues to pursue in respect to character too the articulation of sameness and difference we saw at work in his discussion of animal parts.[122] As for the reason why he distinguishes character traits in terms of degree and analogy it has to do with his psychological doctrine. Animals share with human beings the sensitive soul, hence the difference in those traits that pertain to the sensitive soul and involve body "movements" (like fear or courage) is by degree.[123] But with respect to those traits that are exclusive functions of the rational soul such as art and intelligence, the difference is by analogy. That is, in being deprived of the rational soul, animals can only possess traits analogous to those of humans. So, for instance, also a swallow builds its nest with much care, putting the right materials at each phase of the construction thereby showing the precision of its intelligence (*akribeia dianoias*),[124] but for Aristotle such ability does not derive from the same capacity humans exercise when they plan and construct a house. Animal skill is not (human) art nor is it supported by the type of reasoning characteristic of humans, but it is equivalent in creatures that do not possess *logos* as reason. It may be helpful at this point to recall a passage from *Physics*, discussed in Chapter 1, in which Aristotle reports the difficulties of understanding by what capacities animals "work," how, for instance, spiders, ants,[125] and similar animals can engage in a creative activity (*poiein*) showing, we may add, logic, organization, and focused inventiveness. Do these animals have mind (*nous*)? Some people, Aristotle tells us, were at a loss (and they indeed claimed animals had it).[126] For him, however, nonhuman animals operate in this way without mind and the mental capacities of art, research, and deliberation, which are exclusive to humans. He explains this seeming paradox by appealing to the teleology inherent in nature, and ultimately to the extraordinary status of animals as *composite* natural beings empowered to move, act, and live.[127] So when in *HA* 8 Aristotle appeals to animals' "analogous natural capacities" (*hetērai physikai dynameis*),[128] one should read into the qualification "natural" an ultimate reference to animals' body constitution and understand that it is animals' bodies (as realizations of the sensitive soul via the forms of cognition stemming from sensation, i.e. *phantasia*)[129] that enable them to have the equivalents to human art, wisdom, and understanding—although in *HA* Aristotle points to the phenomenology of these aspects in animals' lives rather than telling us what they are per se.[130]

As one could expect, the physiological basis of animals' characters and intelligence features instead in the etiologically bent *Parts of Animals* where Aristotle discusses blood (or its equivalent) and its elemental composition. In book 2 he writes,

> Now some of these animals also have a more subtle intelligence (*glaphyrōtera dianoia*), not because of the coldness of their blood, but rather because it is thin and pure; for what is earthen (*geōdes*) has neither of these properties. For those animals with finer and purer moisture have quicker perception (*eukinētotera aisthēsis*). Indeed, because of this even some of the bloodless animals have a more intelligent soul (*synetōtera psykhē*) than some of the blooded ones; e.g. the bee and the ant kind and any others there may be of this sort ... The animals that are excessively watery (*hydatōdē*) are more timid (*deilotera*). This is because fear cools; accordingly, those having such a blend (*krasis*) in the heart are predisposed to this affection, since water is solidified by the cold. This is also why the other bloodless animals are, generally speaking, more timid (*deilotera*) than the blooded, and when afraid

become immobile (*akinētizei*), discharge residues, and in some cases change their colors. However, those with excessively fibrous, thick blood are more earthen in nature (*geōdestera physis*), and both spirited in character (*thymodē ethos*) and excitable because of their spirit. For spirit is productive of heat, and solids that have been heated give off more heat than liquids; and the fibres are solid and earthen, so that they become like sparks in the blood, and produce a boiling in the spirit. That is why bulls and boars are spirited and excitable; their blood is most fibrous, and indeed the blood of the bull solidifies more quickly than all ... The nature of the blood is the cause of many features of animals with respect to both character and perception, as is reasonable, since blood is the matter (*hylē*) of the entire body; for nourishment (*trophē*) is matter, and blood is the last stage of nourishment.[131]

In this passage, Aristotle relates sensation and intelligence and makes them depend on the nature of the blood[132] in a discourse that incidentally echoes the Presocratic conception of the continuity of thought with sensation and, more in particular, Empedocles' belief that precision in perception and "mastery" of thinking depend on hematic standards.[133] For Aristotle blood is the matter of living beings' bodies (for the blooded ones)[134] and determines their "psychological" traits and capacities, namely intelligence and character—with its analogue serving the same functions in the bloodless animals. The relevant discrimination pertains to the watery and earthy components of blood.[135] A thinner and clear blood, that is, one containing less fibrous, earthy "ingredients," is both conducive to sensation *and* intelligence. It is consistently with this theory that in *History of Animals* the elephant is said to have both the most perceptive sensation and superior understanding.[136] But characters too are seen in terms of blood's basic compositional formula: animals that are rather (or more) timorous possess a blood that is excessively watery, i.e. with more water in it. Since fear is accompanied by coldness and watery blood is cold, animals with a predominantly watery blood are intrinsically fearful and liable to react with fear and hence, as we subsequently understand, to escape or become motionless. Excessive moisture of the body is appealed to also for the bloodless animals, including insects[137] whose blood-equivalent part makes them inherently timid and react accordingly.[138] By contrast, those animals with an earthy constitution, that is, with more fibers in their blood,[139] are prone to passion and, we may add, to anger and action.[140] For in being more fibrous and thick (i.e. earthy) their blood is particularly conducive to be heated up triggering (aggressive) movement rather than stillness or retreat. Bulls and boars have such a constitution and hence a fiery and passionate character, and so does any other animal similarly constituted. Two interesting observations on the relation between animal physiology and their natural dispositions (emotional or more strictly cognitive) should be added at this point. On the one hand, to account for animals' dispositions Aristotle uses comparatives and whether expressing intensity or degree these comparatives allow a certain flexibility of attribution. That is, to say that animal intelligence is "rather subtle" or "more subtle" and that (some) animals are "more timorous" or "rather timorous" seems to leave it open (or at least does not firmly deny) that other animals might also be somewhat intelligent or timorous.[141] Aristotle selects "charged" dispositions (in the examples just discussed, intelligence and fear) to describe the character of specific animal groups, but this does not mean that other animals may not present those types of traits.[142] They are simply not as prevalent and attitudinal in their range of psychological affections. If hunted by a number of men, a lion, we are told in *HA*, shows cautiousness and circumspection (and does not attack).[143] On the other hand, Aristotle connects his blood-centered theory of animal characters, including intelligence, to the quality of their nourishment. For he reminds us that while blood is the material composing animal bodies it is in fact nourishment that has underwent its final stage of concoction.[144] Hence it seems to be implied that food may contribute to dispositions (certainly abundance of food does!), although, as we saw earlier in this chapter, animals' specific nourishment is in turn suited to, and required by, their specific constitutions.[145]

In an *ad hoc* study Fortenbaugh has argued that for Aristotle animals' emotions are passive, noncognitive body responses to pleasant and painful situations, adopted under a Platonic influence, anticipating a later Peripatetic trend, and betraying a humanization of animals.[146] For this author, in not possessing the intellect, animals cannot have the emotions and emotional dispositions of human beings and, in the zoological treatises, Aristotle's attribution of emotions to animals is due to a different (biological) framework that associates emotions with sensation rather than with the intellect. In sum, animals' emotions cannot be cognitive.[147] This denial is problematic, I believe, because it ultimately relies on a logocentric perspective that takes as standard for emotion Aristotle's conception of *human* emotions along with their association with human intellect, judgment, and morality.[148] True, in the zoological treatises Aristotle does not address emotions as resulting from the intellect, which we saw is irrelevant to understand animals' lives nor does he intend emotions as moral as Fortenbaugh initially claims,[149] but such disconnection from the intellect and morality does not mean that for Aristotle animals' emotions are not cognitive. Sensation in its aspects and "architectures," including incidental sensibles[150] and "positive" *phantasia*, are indeed forms of understanding. As we have seen in Chapter 5, Aristotle pairs sensation and thought as discriminative operations while he calls *phantasia* a form of *noēsis*.[151] Acknowledging the cognitive value of sensation and its derivatives (completely ignored in Fortenbaugh's study) does support a cognitive value of animals' emotions, *in separation* from morality and *logos*, which do not apply to them. Moreover, in a compelling study directed to explore the intersection of Aristotle's ethics and biology, Lennox has underscored how for Aristotle animals and children alike share character traits such as courage or temperance (i.e. natural virtues), which in adult humans would develop into virtuous states and full-fledged moral virtues by resorting to the intellect.[152] Admittedly, animal humanization may be present in that the emotions attributed to animals are called in the same way as those felt by humans. Animals are said to feel fear or anger on the basis of their behavior, not because one among them has revealed what it actually feels. Yet, because humans and animals share the same physiology and emotions have a physiological basis, there is no theoretical reason that prevents us from attributing to animals the same emotions as humans given that living beings other than humans manifest them (i.e., fear, anger, or courage) with a revealing, "meaningful" behavior. A deer that withdraws manifests fear, and a lion that advances courage—and they do so based on their natural dispositions in combination with their perception of reality. Indeed, several descriptions of animal behavior in *HA* indicate that animals' emotions are connected to situational and self-awareness. For instance, whether it is true or not that crayfish die out of fear (*dia ton phobon*) when trapped in a net along with octopuses,[153] the fact that in *History of Animals* Aristotle refers so indicates that for him the crayfish in question senses (*aisthanesthai*) the octopus (as a destructive enemy) among the other unfortunate creatures and that such a "realization" in whatever way it occurs (by smell or sight or a combination of the relevant senses, "accompanied" by *phantasia* or memory) triggers a negative emotion that makes them die. So arguably when crayfish change their usual way of locomotion and move backward with the horns raised forward this behavior depends on a feeling of fear that is not merely a passive, automatic reaction, but an emotion caused by the awareness of an impending danger.[154] Likewise, that the deer feels confident (*tharrein*) to leave its hiding place because its horns have started solidifying and can be finally used as means of defense indicates for Aristotle that the animal is aware of his own body and how it can interact with the environment, when equipped with horns or not. To solidify the newly sprouted horns, we learn, the deer has previously sunbathed and tested them against trees, acting again in a way that reveals understanding of the beneficial effects of the heat and "purposeful" intelligent action.[155] In light of these examples, then, rather than accompanying passively sensations, emotions (such as fear and courage) seem to rise through the awareness provided by sensation and trigger a change in behavior, which is ultimately relevant for the individual's self-preservation. Crayfish move backward and, after the due tests of its defensive equipment (i.e. the horns), the deer undertakes more adventurous paths. By contrast, trained by the shepherds, the herds react to sudden noises with their acquired behavior even inside the stable[156] and in this automatic response to acoustic stimuli they show a lack of understanding of the actual situations and their ultimate stupidity—a revealing example that not only indicates Aristotle's species-based differentiation of animals' emotional dispositions but also their different (i.e. inferior) cognitive grasps. Be that as it may, in *History of Animals* Aristotle adopts two ways to discuss animals' emotions. If, on the one hand, in his general discussions of book 1, 7, and 8 he encodes emotions in animals' character dispositions (in positive or comparative degrees), on the other, he is also attentive to their reactions to specific events, in which case emotions appear to be triggered as *ad hoc* responses to external situations rather than being merely intrinsic in animals' constitutional make up—granted that such a constitution (i.e. the blood and its equivalent) is conducive to such emotions. In sum, emotions are conceived as part of animals' species-related nature, which may be fearful or courageous, mild or ferocious, etc. (and may actually combine a variety of traits) or are pursued in their emergence revealing animals' different degrees (or even lack, i.e. in the case of sheep) of understanding.[157]

It is in this framework, which relates animals' characters, including emotions and intelligence, to the (sensitive) soul and anchors them to a physiological basis (admittedly, only discussed in *Parts of Animals*)[158] that in *History of Animals* animals are defined *phronimoi* ("practically" intelligent). Lacarrière remarks that in the zoological treatises *phronesis* is used interchangeably with *synesis* (intelligence),[159] and such an overlapping indicates that animal *phronesis* pertains to those capacities (i.e. *synesis*, *sophia*, and *tekhnē*) that in *HA* are analogous to the ones possessed by humans. Attributed to animals, *phronesis* and cognate adjectives (and semantically related words)[160] capture their practical intelligence whose corresponding capacity in humans stems from the rational soul and has to do with action.[161] A passage from *Nicomachean Ethics* is key to understanding how *phronesis* plays out in the animal sphere. Aristotle writes,

> Hence even some of the animals are said to be *phronimoi* namely those which display a capacity for forethought (*dynamis pronoētikē*) as regards their own lives (*peri ton autōn bion*).[162]

Phronesis enables animals to act in view of their own lives and it too, along with the rise of emotions like fear and courage discussed earlier, reveals animals' situational and self-awareness. *HA* abounds in examples of this capacity, whose concrete application differs from species to species but always points to practical success for the sake of one's own life and/or to that of the offspring.[163] Called *phronimos* and opening Aristotle's discussion of animal intelligence, the deer is in this respect emblematic. The female shows its intelligence by giving birth alongside the roads as it is aware that wild beasts do not approach human habitats. After birth she eats seseli, a medicinal herb. We may also recall the behavior of the male in relation to the growth and shedding of its horns, discussed above in relation to fear and confidence.[164] But many other quadrupeds too act intelligently (*phronimōs*) helping themselves (*pros boētheia hautois*). Aristotle mentions the wild goats on Crete who, struck by arrows, eat dittany or the tortoise who after swallowing a viper takes some oregano. We learn that dittany works as an expeller, while oregano is a poison antidote. *Phronesis* is also seen in the hunting technique of panthers and ichneumons and in the symbiotic relations of crocodiles, which again show a more gentle side. More specifically, panthers are aware of their alluring scent and hide to attract their preys. On the other hand, the Egyptian ichneumon attacks the asp only after having summoned other conspecifics and heals from the reptile's bites by plastering itself with mud "created" for the occurrence.[165] As for the crocodile, it allows the trochilos to clean its teeth without harming it, aware that the bird is doing something beneficial.[166] *Phronesis* is species-related and blood-dependent,[167] and in some examples of animals' practices and lives it is also manifestedly acquired through learning and memory (*mnēmē*)[168] inasmuch as it relies on the transmission of knowledge from mother to offspring.[169] So, we read, the deer accustoms her young to find refuge in a lair, located on impervious rocks with a single access.[170] Mindful of this information, the fawns learn not only the spot where to seek safety while they are young and vulnerable, but also, arguably, in which places to lead their own offspring in turn (i.e. places that are similarly inaccessible).[171] Thus their mother teach them by habituation to comprehend their environment and to use it to their own advantage and life preservation. Similarly, about 12 days after having given birth on land the seals take their young to the sea to accustom them to the aquatic environment from where the seal pups will eventually be able to secure their food and where they will spend the great part of their life.[172] Born on earth, the baby seals learn to live in water.

6.4 The nonhuman paradox: Being political in Aristotle's zoology

An important factor that differentiates animals' lives and activities (and is connected to their intelligence)[173] is their sociability. In *History of Animals* 1 Aristotle writes,

> Here are some further differences with respect to animals' manner of life, *bioi*, and activities, *praxeis*. Some are gregarious, *agelaia*, some solitary, *monadika*: this applies to footed animals, winged ones, and swimmers alike. Some of the gregarious animals are "political," *politika*, whereas others are more dispersed. Examples of the gregarious animals are: birds—the pigeon class, the crane, the swan (no crook-taloned bird is gregarious); swimmers— many groups of fishes, *e.g*, those called migrants (*dromadas*), the tunnies, the pelamy, and the bonito. And man dualizes (*emphaterizein*). The political animals (*politika*) are those which have some common activity (*koinon ergon*); and this is not true of all gregarious animals. Examples of "political" animals are man, bees, wasps, ants, cranes. Some of them live under a ruler (*hup' ēgemona*), some have no ruler (*anarkha*); examples, cranes and bees live under a ruler, ants and innumerable others live not (*anarkha*).[174]

Aristotle distinguishes between gregarious and solitary animals, and acknowledges the existence of a subgroup of gregarious animals that are *politika*, "political."[175] These are living beings that, in addition to the physical proximity with the members of their group, undertake a common activity (*koinon ergon*). Such distinctions in animal sociability cut across animal kinds and habitats: birds, fishes, and terrestrial creatures all include species that live in these different ways, i.e. alone, in groups, or "politically." Among the *politika zōa*, Aristotle lists human beings, bees, ants, wasps, and cranes. He complements in this way his reference to political animals in *Politics* in which the human being is presented as "more political than any other gregarious animals like bees" on account of his possession of *logos* intended as speech and means to communicate right and wrong.[176] Unlike in *Politics*, however, in *History of Animals* Aristotle is not interested to pursue the higher degree of humans' political nature. He lists the human being as one of the political animals and points out in turn an internal division between the political animals that live under a ruler (*hup' hēgemona*) and those without (*anarkha*) like the ants. And while in the *HA* passage quoted above Aristotle does not refer to intelligence, he reverts to this notion in commenting on cranes' political behavior or describing animals' relations that are "more or rather political." For instance, he discusses cranes' migrations and "societal power structure," calling them *phronima*,[177] and looks at animal reproduction and the rearing of the young noting that animals that are more intelligent (*synetōtera*) and equipped with a better memory (*mnēmē epi pleon*) continue their association with their offspring after having accomplished the task of nourishing them. In doing so, he tells us, they entertain a more "political" relation (*khrasthai politikōteron*).[178] On the other hand, foremost in the list of political animals, bees are said to be intelligent (*phronima*) without being able to learn.[179] Thus, on his view, there appears to be a strong connection between animal intelligence and developed organization and sociability.[180]

As the above examples show, Aristotle attaches to the adjective *politikos* a wider semantic range than one would expect considering its etymology from the word *polis*. For even though they do not form the political association peculiar to the human being, i.e. the community of citizens called *polis*,[181] nonhuman animals still constitute, and live in, "political" partnership, from the prolonged interaction of parents and offspring among certain species to the communal work performed by a group like migrations, on which and other activities more will be said in the next pages. Here suffice it to point out that however strange this label (*politikos*) for the continuing association of family members may appear, it shows that in his study of animals Aristotle regards as "political" any interpersonal relation that goes beyond the individual's mere subsistence and fulfillment of nutrition and reproduction. Some animals for him are capable of a relationally richer life.[182] Moreover, the use of the comparative *politikōteron*, "rather politically" or "more politically" to describe the continuing association of "family" members (parents versus offspring)[183] suggests that an *incipit* of political life (*politikōs*) is already inherent in the basic (and enduring) association of male and female. Such a use of *politikos* is compatible with the conception of the *polis'* natural development presented in *Politics* 1, from the basic partnership of male and female to the association of households and then villages to the political community of citizens.[184] In his account of animals' activities and lives Aristotle singles out behaviors that express recognition of family membership and the fulfillment of parental or filial roles. He remarks, for instance, that male and female pigeons form lifelong partnerships;[185] wild birds build their nests in view of the offspring;[186] swans are good parents (*euteknoi*);[187] and that both male swallows and pigeons take care of the nutrition of their nestlings.[188] And while young storks feed their parents[189] and sheep and goats lie in family groups,[190] bears put in front their cubs and carry them when fleeing a danger.[191]

As for the type of common activity (*ergon*) to which nonhuman animals attend to, it may vary, but it is directed to the welfare of the community,[192] each member playing a role that contributes to the success of the common enterprise. Among the *politika zōa* mentioned in the passage from *History of Animals* quoted above, bees form the most complex community. Its members' activities center around the production and preservation of food, and the foundation of colonies, and require different forms of interaction, from the construction and protection of their dwelling place to the many complementary tasks that lead to the production of honey. Aristotle observes that bees are the only animals that produce their food, for they feed on honey.[193] Ants cooperate to collect food and store it in a place underground prepared to that effect, while wasps construct their nest together.[194] Cranes, on the other hand, cooperate in the migration they undertake every year to find a more livable habitat, from the Scythian plains in the far North to the marshes above Egypt in the South.[195] They fly very high in order to command a wide view, and if they see clouds and stormy weather they fly down and rest.[196] And while Aristotle does not say so, it is ultimately cooperation and leadership that enables them to succeed in their journey. Among the birds cranes' migration is the longest one as it extends from *eskhata* to *eskhata*, furthest points to farthest points.[197] And it may be perhaps on account of this superior command of distant journeys that in his list of political animals Aristotle mentions only the cranes among the migratory species. For in moving between extreme areas of the world cranes are the most representative of a successful, yet highly challenging, joint enterprise. *History of Animals* presents other migratory animals: not only birds like pelicans, quails, wood pigeons, and turtle-doves,[198] but also fish. For instance, the coly-mackerel spends the summer in the Propontis and winter in the Aegean, while the majority of shoaling and gregarious fish migrate for the summer into the Pontus because it is richer in food, inhabited by fewer predators and hence more favorable to reproduction, and the fresh, sweet water suitable to nourishing the embryos.[199]

Linked to nutrition and reproduction, migration too is a crucial area of animals' activities. Aristotle interprets it in light of their sentient nature noting that,

> All animals (*panta*) have an innate perception of change in respect of hot and cold and just as among humans some move indoors during the winter while others who command extensive territory spend the summer in the cold parts and the winter in the warm sunny parts, so it is with those animals that are able to change their location.[200]

All animals are sensitive to the change of temperature. Those who are physically equipped to travel long distances escape from the impending cold winter and the hot spell of summer and move to regions with a milder climate. So do the human beings who have the economic possibility: they spend their summers in cool areas, and the winter in sunny, warm ones. Others with limited means simply protect themselves from the "inhospitality" of winter by moving inside their houses (and so, we may add, do those animals who are not equipped for a long journey). In sum, migration is due to the sense of touch and constitutes a phenomenon of political nature when it engages animals' communities travelling long distances but is a "private" one in the case of humans.

6.4.1 *The plasticity of the political animals*

Apart from their ability to cooperate, nonhuman political animals share also a sense of space as a community. They identify spatially. In the case of bees, wasps, and ants their common activity centers around a settlement, which the community builds, uses, and inhabits. Political animals are able to carve out in nature an animal-made environment. Thus, as humans live and cooperate in the polis, so bees do in the hive (or honeycombs), ants in underground colonies, and wasps in combs. Aristotle does not give many details about ants. He labels them with the bees as the most industrious (*ergatikōtata*) of insected animals,[201] so much so that they even work at night, if the moon is full.[202] And he mentions their action of unearthing the soil.[203] As for wasps, at the onset of summer they "choose a place with a good look-out and start fashioning the combs" which will keep growing in number as the procreation by the mother-wasps takes place inside them.[204] Feeding and food provision also happen there. The bees, on the other hand, construct the wax combs of their hive as if weaving a web. Starting from the top and working downward they build adjacent areas for different dwellers, working bees, kings, and drones.[205] To prevent the infiltration of other creatures they fortify the floor with the gummy substance they draw from flowers and especially trees, and they narrow the entrances.[206] Their life centers on the hive, which is a place for dwelling, generation, feeding, honey production, and storage and which, as humans do with their *polis*, bees eagerly defend. Already Homer, in book 12 of the *Iliad*, presents a simile with bees fighting by "their homes" against hunters[207] and Aristotle adds that, while outside they do not attack other animals, in proximity of the hive bees become very fierce and kill anyone they would overcome.[208]

If bees along with wasps and ants share a spatial identity, cranes do not. Their inclusion in the list of political animals, Mulgan remarks, makes clear that in order to be "political" nonhuman animals do not need "permanent territorial unity"[209] since these birds, one of the four species cited in *History of Animals*, migrate between North and South and do not live in a compact settlement. Cranes are "political" on account of their organized navigation across continents as a group. It is tempting then to consider them and other migratory animals as the ultimate embodiment in the realm of zoology of that precept of Athenian ideology[210]—later endorsed by Aristotle in *Politics* 3[211]—that as long as they existed as a community of men the Athenians existed as a polis, no matter how far they were from the space of their city or whether it be destroyed.[212] For the Athenians, however, this deracinated mode of political life could only be temporary, and lived out in a time of extreme danger, while for migratory animals like cranes it is a permanent condition of their political life.[213]

Cooperation with an eye to the community welfare, a sense of spatial identity, or, by contrast, the ability to survive and even live well independently of it, however, are not the only elements that characterize the lives of nonhuman political animals. There is one more fundamental element, which Aristotle provides but which does not seem to have received adequate attention in subsequent studies. Political animals are considered also in terms of power (*arkhē*). For some of them are under a ruler (*hup' ēgemonou*), others "anarchical" (*anarkha*). Bees, wasps, man, and cranes, along with many gregarious migratory fish, belong to the first subgroup, while the "anarchical" one comprises ants and "innumerable other animals." In power-structured communities leaders are in charge of activities on behalf of all the other members. For instance, in the case of wasps it is the leaders who decide where to establish the nest. A leader directs the cranes in their migrations and, when they settle, keeps guard while the other fellows are sleeping. Subordinate to the leaders, but still playing a special role in the community of cranes, are the "signalers" which whistle among the last birds of the flock so that everybody may hear their cry and not get lost.[214] Fish too migrate under leaders.[215] Bees offer an interesting scenario of leadership, and class division in general. Aristotle reports that different *genē* of bees coexist in the same community: working bees, drones, leaders, and thieves,[216] each playing a definite role. It is the leaders, at times called kings (*basileis*), who lead out the swarms. Without leaders swarms do not fly, and, in the event a leader dies, sooner or later the swarm, which was under him, perishes too.[217] Thus, for political animals living in a power structure, leadership cannot be easily replaced and the absence of a leader results in the destruction of the community. When not fulfilling their tasks, the leaders remain in their combs letting the regular bees attend to their differentiated work in the chain of honey production.[218] And yet, despite their position the leaders are still subjected to the crude control of the community. If there are too many of them or if they lead the swarm astray, the other bees kill them. Superfluity of roles and mistakes are fatal while food accessibility is regulated and monitored. Indeed, death is the penalty also for the thieves caught in the act of stealing the honey, or when trapped in a comb and unable to leave because they ate in excess. If there is a shortage of honey, bees will not only expel, or presumably kill, the drones, but will also destroy their combs.[219] When food is abundant, however, drones are taken care of and fed because their parasitic presence—later observes Aristotle—increases the bees' productivity and the health of the hive.[220]

As in the case of emotions,[221] also the attribution of a political life to animals has been seen as an interference of the human in Aristotle's zoology. Drawing on his previous studies and underscoring an "anthropic perspective," Lloyd has recently approached Aristotle's discussion of animals' "natural sociability, skills, and intelligence" to claim that,

> The influence of the use of humans as the yardstick for other animals extends beyond the physical and the physiological and permeates Aristotle's discussion of animal psychology and intelligence through and through.[222]

There are reasons to disagree with this position. True, as also noted earlier in relation to animal emotions, in the case of animal sociability too Aristotle borrows from a human relevant (let alone made) vocabulary: to consider cranes, bees, ants, or wasps *politika* seems to gauge them by human standards as it is humans that, strictly speaking, live in *poleis*. In *Politics* the emphasis on *logos* as humans' exclusive capacity to indicate the advantageous and the harmful, right and wrong does indeed stress and pursue the higher degree of human political life as opposed to nonhumans'. In line with this perspective, a number of scholars have inquired into the positive difference that qualifies humans *vis-à-vis* their nonhuman fellows[223] and, in proceeding so, they too have implicitly approached nonhuman animals "with the yardstick of humans"—an appropriate step in the context of *Politics*. But in *History of Animals* the framework is different than in Aristotle's human-centered treatises building on man's exclusivity. For here, faithful to a zoological project that unfolds as an inquiry into animals' common nature,[224] anchoring it in their sensitive soul and aiming at the animal as such, Aristotle pursues animals' differences from a wider perspective that transcends a myopic human/animal parallel with the rational animal taken as the measure. At the beginning of the treatise Aristotle presents animals' facts in terms of sameness articulated through differences by degree and analogy.[225] And the point of view from which he assesses the differences is variable. As noted earlier, animals may be compared to humans and humans to animals,[226] and animals among themselves, regardless of the human. Remember the "more political" behavior of those "more intelligent" creatures that continue their association with the offspring once they have carried out their nourishment. And if humans along with their vocabulary and assets such as art, wisdom, and intelligence or the "political" seem at times to inform Aristotle's account it is a matter of "currency:" the human animal is the most familiar to him.[227] The reason is "naturally" pragmatic, not anthropocentric. Regarding animals' political life, in *HA* Aristotle does not measure their lesser attainment of the political with regard to man's exemplary paradigm. In fact in his description of political animals' lives and practices he even glosses over differences by analogy and degree. Rather he pursues animals' success in working together for a common goal that grants them to survive and live in their habitats (and we may add) according to their body constitutions and fulfilling the most recondite, natural, and universal of all desires. Taken as such, the political defines a form of cohesion that equally exists among (some) nonhuman animals and their fellow humans and that can be already identified in the association between male and female and ensuing offspring along with their continued association. In fact what emerges as surprising in Aristotle's discussion of the political animals and equally emblematic of his wider perspective is his distinction between political animals under a leader and those without. For not only do animals cooperate, and live successfully, in a power structure (as human beings do) but they are successfully political and undertake a common work for the good of the community even *without* leaders. And it is this *without* which contributes in his zoological discourse to a "positive difference" that distinguishes the realization of the political among ants and similars from that of other political animals, including the human being. Ultimately, in *History of Animals* nonhuman animals are not merely less political than the human being as claimed in *Politics* but realize the political in conditions in which the human being does not, i.e. by working successfully together without a ruler, by keeping cohesion flying across continents, or by building underground complex "settlements"—thus living a political life that ultimately stems from the realization of the functions and capacities inherent in their marvelous complex bodies.

Notes

1 More specifically, this set of books is devoted to the body parts involved in the process of sexual reproduction, but extends to animals' mating habits and temperaments, and includes the processes of spontaneous generation pertaining to creatures of both land and sea.

2 For a discussion of the relation between *bios* and activities, see section 6.2 below.

3 Book 9 considers human generation. For the order of books 7–9 I follow Balme's 1991 edition, which reflects the manuscripts prior to the rearrangement of Theodore of Gaza's Latin translation of the treatise (see Balme, 1991, 56, n.a.). In fact, *HA* includes also a tenth book on "the failure to generate," which, according to ancient catalogues dating back to the 3rd cent. BC Peripatos, was a separate work (Moraux, 1951, 107).

4 See, for instance, the discussion of pigeons (*HA* 5 562b3–25).

5 These books have been considered compilations by other peripathetic authors such as Theophrastus or Eudemus. For a discussion of the scholarly critique, targeting internal contradictions, variations of vocabulary, and disorderliness, see Balme's introduction to the Loeb edition (1991, 3–13).

6 For a review of the different interpretations and a discussion of Balme's revisionist and pioneering role, see Gotthelf, 1988.

7 See Balme, 1987b, 13–17; 1991, 13–20.

8 On this point and the connection of *History of Animals* with *Posterior Analytics*, see Lennox, 1987, 90–109; Gotthelf, 1988, 104–1.

9 In fact, while converging in interpreting *HA* in light of Aristotle's philosophy of science there is no absolute agreement between Balme, on the one side, and Gotthelf and Lennox, on the other. The question of the relation of *HA* to the other zoological treatises is intertwined with its relative chronology and remains a vexed one. In terms of content, Lennox argues the priority of the "*hoti*" (i.e. facts) over the "*dioti*" (i.e. causes) taking *HA* to precede the other etiologically bent treatises. See *HA* 1 491a7–17 and Lennox' article in the *Stanford Encyclopedia of Ancient Philosophy*, which appeals to Aristotle's principles of scientific explanations laid out in *Posterior Analytics* (*APo* 1, 10). The language of *HA* seems to confirm this interpretation as it often presents "classificatory terms" such as "identical" or "difference" (*tauton, diaphora*), but almost never explicatory terms such as "substance," "cause," or "end" (*ousia, aitia, telos*) (Natali, 2014, 142–3). As for the time of composition, the name-places *HA* refers to indicate that the treatise dates back to the period after Plato's death (347 bC) when Aristotle left the Academy and went to Asia Minor, first to Assos and then Lesbos (Lee, 1948, 85). Balme however attributes *HA* to a later phase of Aristotle's career, after *PA* and *GA*, pointing out that *HA* often presents brief references to the *phainomena* which *PA* and *GA* describe more diffusely and that it offers information that is discordant or not included in them (1991, 21–6). For a recent comprehensive discussion of the different positions on the question of *HA*'s thematic and chronological relation to the other treatises, including the reception and redaction of Aristotle's work, see Feola, 2017, 35–57.

10 Admittedly, in *HA* Aristotle does not say so, but such case can be made because his descriptions of animals' practices and lives have greatly to do with their nutrition and reproduction (in addition to other actions aiming at self-preservation, see 6.3.1 below), which in *On the Soul* are attributed to the nutritive soul (combined in animals with the sensitive soul). In Aristotle's systematic view, nutrition is for the sake of reproduction, which expresses animals' (and plants') desire to live forever (see Chapter 3).

11 See Chapter 3.

12 See 6.2 below.

13 Significantly, the order of discussion (first reproduction, then nutrition) reflects Aristotle's teleological understanding of life processes (see above note 10).

14 I mean body-oriented psychology on account of two related features: first, because of Aristotle's general definition of the soul as "the first actuality of a living (hence organic) body" (*DA* 2, 1) and second, because animals' psychological traits derive from their specific body constitution; see 5.3.1 and 6.3.1.

15 *Pol.* 1 1253a7–9. In other words, the political nature of some animal species has to do with their mode of interaction with other conspecifics along with their impact on, and approach to, the environments in which they live.

16 *PA* 2 650b14–651a19; see 6.3.1. Body physiology is also relevant in *DA* to account for a vast range of psychological affections (1 403a4–403b15).

17 As Natali remarks, in moving from animals' parts to their "lives" *HA*'s thematic development has an intrinsic finality (Natali, 2014).

18 See n. 10 above and Section 2.3.

19 Indeed not only do the strictly ethological books (7 and 8) deal with these issues, but also books 1, 5, and 6.

20 Aristotle's comment in book 6 is illuminating of this way of proceeding. For he claims that in describing the reproduction of fish he has aimed at presenting what happens "for the most part" (*to epi poly gignomenou*) (*HA* 6 571a26–7).

21 Aristot. *PA* 1 644a24–644b10; cf. *GA* 1 715a1–3 where, reviewing the subject of *PA*, Aristotle underscores the method adopted, based on the discussion of the parts in common (*koinēi*) to all animals and those that belong to distinct groups.

22 It is noteworthy, however, that while focusing on differences that are reducible to various kinds of identity in *History of Animals* Aristotle discusses some psychological capacities like memory and learning as "partaken of" by animals in common (*koinonein*) (see *HA* 1 488b12–27, quoted at the beginning of section 6.3 below). A synonym for *metekhein* (to partake) (see Bonitz, 1955, 400), *koinonein* is related to *koinonos*, "the one who takes part in with or shares" (see Montanari, *sv koinonein, koinonos*). This etymological relation makes it compelling to interpret *koinonein* as indicating a possession shared in common rather than a mere possession as this verb is usually translated.

23 *HA* 1 487a10–11. In this list, animals' characters or dispositions lead to their actions, contribute to their modes of life (see Peck, 1965, xci), and are dependent on their physiology and, ultimately, "body parts" (v–vii); on the thematic organization of *History of Animals*, see Lennox, 2001.

24 Balme remarks that Aristotle's focus on difference is an application to the study of animals of an important principle of his philosophy of science according to which the identification of differences is instrumental to definitions, and definitions in turn to scientific knowledge. But, as the same author continues, in *HA* the notion of difference (*diaphora*), along with the complementary notion of attributes (*symbebēkota*), does not have the same technical value as in more theoretical treatises such as *Analytics*, *Metaphysics*, and *PA* (1, 2–4) in which Aristotle discusses it in the context of logical division (*diaeresis*). *History of Animals* remains untouched by logical concerns (see Balme's introduction to books 7–9 in the Loeb edition, 1991, 3–6; 1987a, 69–89; cf. Gotthelf, 1988, 103–6).

25 *HA* 1 486a15–23, transl. by A.L. Peck (italics by the author) and further 486b17–22; cf. *PA* 1 644b11–16 along with its discussion in Chapter 1. In the *PA* passage Aristotle shows to conceive animals' body parts as the same in terms of morphological identity (*eidos*), of degree (by the more and the less) and by analogy (the fish spine is the same as a terrestrial animal's bone). It is especially in this last case (i.e. in absence of morphological similarity) that the notion of function becomes relevant; cf. *HA* 2 497b6–14 where Aristotle articulates sameness and difference along the same lines and further stresses that some of the parts are common (*koinon*) to all, while others are found in certain classes only.

26 Aristot. *Mete.* 4 387b4–6/DK 31 B 82 "hairs, leaves, the dense feathers of birds *are the same* (*tauta*),/And the scales (*lepides*) on sturdy limbs" (transl. by Laks and Most, italics by the author).

27 For a discussion of DK 31 B 82 and its implicit analogies in relation to Empedocles' understanding of the living world, see Zatta, 2019, 52–3, 103–4; on Empedocles' use of metaphors and Aristotle's critique, Zatta, 2018a, especially 67–9.

28 *Logos*, we saw, indicates the rationality immanent in animals' body constitutions, and is manifested and operative in all aspects of their existence that, anchored in their bodies, express finality within a discourse that equates the proceeding of nature with that of art.

29 Gotthelf notes that Aristotle discusses characters alone while he typically associates lives (*bioi*) and activities (*praxeis*) (1988, 107). This association is due to the fact that activities contribute to and support animals' lifestyles, i.e. whether they are land or water animals.

30 In his account of animals' lives, Aristotle seems to maintain a neutral position in this respect. He uses comparatives to describe animals' characters (see 6.3.1, 6.4 below) but not to evaluate them in terms of worth.

31 At the same time, though, Aristotle is highly selective and does not account for everything that could be observed in his own time about animals. As Gotthelf insists, to have a fully working hypothesis about the role of *HA* one should understand why Aristotle mentions some facts and leaves out others (1988, 113).

32 For a systematic review of Aristotle's sources of information on animals' bodies, activities, and behaviors, see Lanza 1971, 16–23; cf. Tricot, 1957, 10–12.

33 Indeed special attention is devoted to the marine life of the lagoon of Lesbos, where Aristotle had lived for a couple of years after he left Plato's Academy (344–2BC) (see n. 9 above).

34 These aspects are all amenable to the four main categories of difference, namely, body parts, characters, activities, and lives, which Aristotle presents in book 1; see above.

35 *HA* 1 492a25–7.

36 *HA* 4 533a–533b14.

37 See *HA* 1 487b10–2, 5 548b11–15, 549a4–12.

38 *HA* 4 534b21–3. Also the octopuses showed a similar sensitivity to scents. For they would leave the rock which they were holding down to if some fleabane was put near them (*HA* 4 534b25–9).

39 By habitats here I mean animals' space of existence, namely the environments in which they live and accomplish their different activities, first and foremost those related to nutrition. There is a surprising silence in scholarship about Aristotle's conception of habitat, due to his teleological outlook on animals' bodies and nature's proceeding as well as the parallel denial, in his doctrine, of animals' successful adjustment to their living environment in contrast with evolutionary theories, ancient and modern (such a denial is implicit in Aristotle's teleological model, but also explicit in his critique of Empedocles' zoogony in *Phys.* 2.8). Besides, scholarship's silence, I believe, is also due to the pursuit of Aristotle's philosophy of biology as stemming from his philosophy of science at the expenses of his empirical orientation, which is especially present in *History of Animals*, but in a way that, as was remarked at the beginning of this chapter, resonates with his overall methodological approach to animals as "unique" natural bodies *vis-à-vis* elements and plants. Now in *History of Animals* the notion of habitat is intrinsic to Aristotle's definition of animals as water and land animals (water animals live in liquid environments, land animals on "earthy" ones), discussed below, and apparent in the numberless remarks on specific environments, like rivers, marshes, the sea, or coast, hills, mountains, forests, and fields or cities where distinct species live; see also his general observations supported by examples at *HA* 7 605b23–607a35 where he discusses animals' differences according to localities (*topoi*) and notes that in some places certain animals do not occur at all, while in others they do occur but are smaller and shorter-lived (for the habitats of fish, see also *HA* 7 602a16–22, for those of birds, including their dwelling, *HA* 8 614b32–615b16). For a recent discussion of Aristotle's lack of engage-

ment with "questions of fit" in light of animals' "essential capacities," see Gelber, who argues that "habitat is partially constitutive of the vital capacities that comprise a kind's essence" (2015, 269 and 278–97). Rather than taking a metaphysical approach as this author does, in the following pages I address Aristotle's conception of habitat in terms of animals' complex bodies and compositional layers and the "movement/s" such bodies enable them to have (i.e. in terms of animals' *logos* as discussed in Chapter 1). Approached from the perspective of the study of animals as creatures of nature, the diversity of habitats (at least in relation to the major distinction between land and water environments but arguably also to micro-ecosystems) has to do with the difference in animals' physiology (the body's intrinsic heat), body parts (nonuniform, uniform, and "elements"), and the constitution-required nutrition. Indeed, to anticipate the foregoing discussion (at 6.2 below), even those animals which "dualize" (namely, the animals that breathe, but feed and spend time in water) and are considered water animals from an ethological point of view (rather than a strictly physiological, i.e. breathing) behave in the way they do (looking for food in aquatic environments) because of their elemental composition (which represents the first level of body *synthesis*) in addition to their conformation as resulting of specific nonuniform parts (the third level of body *synthesis*); in this respect, see Chapter 1 and the discussion of animals' diet below.

40 DK 12 A 11.
41 DK 12 A 30 ("The first animals were born in moisture, surrounded by thorny bark, but as they increased in age they moved to where it was drier, and when the bark burst open they changed their way of life in a short time," transl. by Laks and Most).
42 For the different positions on this subject, see Gregory, 2016, and the bibliography there cited. For a discussion of Anaximander's testimonies on the origin of animal life in the context of Presocratic doctrines and a cosmological framework, see Zatta, 2019, 18–9, and n. 26.
43 DK 12 A 10 ("He also says that at the beginning human beings were born from animals of different species, because of the fact that the other animals nourish themselves quickly by themselves, while only human beings are in need of a long period of nursing; that is why, being of this sort, they could not have survived at the beginning," transl. by Laks and Most); cf. DK 12 A 30.
44 True, Anaximander does not speak of elements in relation to animal bodies. But in the same testimony addressing the birth of human beings from a different species, he is attributed a view, which points to an elemental basis for the process of formation of the cosmos and is therefore relevant to understand the formation of animal bodies as they took shape in the unfolding of cosmological dynamics. On this view, the birth of the world is due to the separation from eternity of the "seed of warm and cold" (*gonimon thermou kai te psykhrou*) and subsequently "a certain sphere of fire grew around the air surrounding the earth like the bark around a tree;" cf. the first organisms' bark breaking out once they lived in a drier environment (see n. 40 above).
45 DK 31 A 72, transl. by Laks and Most.
46 Diod. Sic. 1, 7, 5/DK 68 B 5; for the attribution to Democritus of the doctrine in Diodorus' account of the origin of life and biological "development," see Vlastos, 1946, 51–9 and Luria 2007, 1221–7.
47 Admittedly, Empedocles and Democritus' allotment of animals to different habitats implied some organic considerations, which tended however to be generally valid for all animals without differentiations, as for instance, for Aristotle the *exclusive* relation terrestrial animals entertained with their earthy environment or fishes with their aquatic ones. Respiration was attributed to all living beings (for Empedocles, see DK 31 B 100; for Democritus, DK 68 A 106) and likely explained why for Empedocles the fiery fish (see below) migrated from land to water (see Longrigg, 1965, 314–5).
48 *PA* 1 640b16–23.

49 See, for instance, the changing interplay of heat and earthy components in the discussion of small animals and the number or lack of feet at *PA* 4 686b27–30.

50 See below 6.2.1.

51 *PA* 2 646a13–21; see Chapter 1.

52 Aristot. *de Resp.* 14 477a32/DK 31 A 73; in this respect, Guthrie remarks that this version relating the emergence of fishes to a first instance of animal migration may well belong to a zoogony prior to the last one under the rule of Strife (1965, 206). With regard to respiration, Aristotle criticizes Empedocles on multiple fronts, namely, for the impossibility that "hot" fishes escaped into an aquatic environment (and hence for both the evolutionary adjustments and elemental incompatibility this belief entailed) and for the lack of an adequate discussion of the respiratory mechanism (see below p. 182 and n. 54). But Aristotle may be simplifying and/or misunderstanding Empedocles for whom respiration actually involved the entire body and happened through the pores in the skin (*rhis*, see DK 31 B 100) while the intake of air through the nose was a circumscribed phenomenon, parallel to the more diffused respiratory process involving the surface of the body.

53 *de Iuv.* 469b7–20.

54 Aristotle observes that the air exhaled is warm in an implicit contrast with the air inhaled (*de Resp.* 472b34–5).

55 See, respectively, *de Resp.* 470b31–471b6; 470b6–11, 28–31 on Democritus; 473a15–8 on Empedocles (who also seems to have asserted that all living beings breathe) (see DK 31 B 102 and note 52 above). On the ignorance of animals' internal anatomy (i.e. body parts) as the cause for his predecessors' wrong accounts on respiration, see *de Resp.* 471a24–9. For a systematic analysis of Aristotle's critique toward the Presocratics' approach to living beings, see Chapter 1.

56 *de Resp.* 474b25–475a6. In this passage, Aristotle mentions also the short-lived insects, which in being less hot than the long-lived ones cool off only by means of their surroundings (*to periekhon*), and not by a finely covered cleft at their waist as it happens to bees and wasps, among others.

57 See *PA* 1 642b10–15 where Aristotle remarks that birds and fishes are regular names for animals with distinct characteristics and that it is a mistake to split animals' groups up and consider, for instance, certain birds (who live in aquatic environments) as water creatures (*enydra*). As for the traditional nature of the distinction of land, water, and "air" creatures (respectively, *khersaia, thalassia, peteina*), see Herodotus' account of metempsychosis (2, 123), likely related to Pythagoras, and Plato's *Timaeus* on the emergence of living beings other than man (41B–42D, 92A–C).

58 For instance, when in *HA* 7 Aristotle discusses animals' diet, he focuses first on fishes and then on birds.

59 *HA* 1 487a15–6.

60 See *HA* 1 487b24–31 and, importantly, 487b22 (*ptēnon de monon ouden estin*).

61 *HA* 1 487b19–20.

62 *HA* 1 487a16–28, transl. by A.L. Peck. Aristotle returns to these two ways of identifying water and land animals at the beginning of book 7, which is dedicated to their food-acquisition strategies and other activities (see 7 489a10–32).

63 Aristotle qualifies animals that breathe but live in food-providing aquatic environments (i.e. otters and shearwaters) as "dualizing," (*HA* 7 589a10–32), adopting the verb (*epanphoterizein*). In Peck's eloquent words, Aristotle uses this verb for a living being that "plays a double game [...], manages to be in some respects on both sides of whatever fence is under consideration" (see Peck, 1965, lxxiii–lxxv). Such a "doubleness" can pertain to animals' body anatomy, to their diet or way of life. For instance, besides applying this notion to animals that partake of traits of both land and water inhabitants like the seal, Aristotle mobilizes it for "pairs of anatomical alternatives, as when … deciding whether the sea anemone is an animal or a plant (*PA* 681b1), whether the pig as a class is solid-hoofed or cloven-hoofed (*HA* 499b12, b22) … Apes, monkeys

and baboons dualize between man and quadruped (*PA* 689b32), the ostrich dualizes between bird and quadruped (*HA* 697b14)." Besides, only creatures that breathe (and hence take in air) can feed and live in water but no water animal (i.e. that takes in water) feeds on land. Thus environmental versatility pertains more to terrestrial (breathing) creatures than to water creatures. In discussing animals' habitats in relation to their body constitutions ("elemental" and mereological) and feeding habits Aristotle singles out the cetaceans. This group includes dolphins and whales and all its members have the unique characteristic of taking in both air and water. While provided with snout and lungs through which they breathe, the cetaceans also have a blow-hole through which they expel the water they have previously introduced in the act of swallowing their aquatic preys. Yet for Aristotle these animals do not contradictorily embody two cooling mechanisms (by air and water) and are to be considered within the larger group of water animals. For they introduce water as a by-product of the activity of nutrition rather than as a process parallel to breathing. Aristotle further adduces many observations to support the idea that their respiration is the true vital process. When caught into nets, for instance, dolphins suffocate for the lack of air while at night they have been seen sleeping with their snout above water in order to breathe (*HA* 7 589a33–589b23).

64 *HA* 1 487a27–35. Besides these groups, Aristotle's ethological bent leads him to account for another group of animals that change habitat and spend the first part of their life in water but upon developing end up being land creatures. He mentions the bloodworm, which as a larva lives in the rivers, but when it becomes a gnat, it moves to dry land (*HA* 1 487b3–8, and Peck's commentary on this passage).

65 See discussion below.

66 Among the footed, he had mentioned otters, crocodiles, and beavers, and among the winged, shearwaters and plungers.

67 In this respect, cf. Aristotle's division of living things/beings in *GA* where he distinguishes between water and land animals in addition to plants, correlating them to the environments with which they "physiologically" engage (namely, water, air, and earth): "plants (*phyta*) belong to the earth, aquatic creatures (*enydra*) to the water, and land animals (*peza*) to the air" (3 761a14–6, transl. by A.L. Peck).

68 See above.

69 *HA* 7 589a6–10.

70 See DK 31 B 90 and DK 31 A 77/Aët. *Plac.* 5, 27; cf. also Hicks who attributes to Democritus a similar view (1965, 346).

71 *DA* 2 416a19–b11; cf. *Phys.* 8 260a29 and Hick's commentary (1965, 345).

72 See *HA* 7 589b23–4 where Aristotle discusses "water animals" that breathe in terms of their *sōmatos krasis* (bodily blend) and life (in water); cf. also 7 590a13–7, where he sums up the three criteria by which animals are called water and land animals and lists 1) physiology, 2) bodily blend, and 3) feeding. Ultimately, as Balme remarks, bodily blend and feeding overlap.

73 *hai de trophai diapherousi malista kata tēn hylēn ex hoias synestēkasin* (*HA* 7 589a7–9).

74 *de Resp.* 477b25.

75 *NE* 10 1176a7–8/DK 22 B 9, all passages from *Nicomachean Ethics* are translated by H. Rackham; on animals' pleasure, see Chapter 5.

76 *HA* 7 588b28–30. In being devoid of a sensitive soul, for Aristotle plants nourish themselves and reproduce without any sort of sensation.

77 For instance, river fishes forage mostly during the night, while during the day they retreat into the deep water (*HA* 7 592a24–7).

78 In this respect, see the crayfish mentioned below.

79 *HA* 7 590a19–596b18.

80 For instance, among aquatic animals endowed with locomotion the purpuras, crustacea, cephalopods, and other fishes are flesh-eating (*sarkophaga*) (see, respectively, *HA* 7 590b1–3, 590b14, 590b20–1, and 591a10–13 where Aristotle mentions a num-

ber of exclusively flesh-eating fish like selachians, congers, and tunny). Grey mullets do not eat flesh at all, but sea-weeds (*phykia*) and sand (*HA* 7 591a19–25); for other "plant-eating" fishes, see also *HA* 7 591b11–14. Among birds, the crook-talons (i.e. eagles, kites, and hawks) are flesh-eating (*HA* 7 592a29–592b1–15), others eat grubs (*HA* 7 592b18–29), while others feed on thorns (*HA* 7 592b30–593a3).

81 On drinkable water as eels' staple food, see *HA* 7 591b31–592a6.
82 *HA* 7 590b10–14.
83 *HA* 7 590b21–26.
84 See, respectively, *HA* 7 592a29–592b1–15, 592b18–29.
85 *HA* 7 592b30–593a1–4.
86 *HA* 7 593a15–24.
87 *HA* 7 593a25–593b24.
88 *HA* 7 593b29–594a25.
89 At *HA* 2 501a18 Aristotle describes the conformation of saw teeth and attributes them to quadrupeds like lions and dogs, to all the fishes, and to the seal (inasmuch as like fishes it acquires its food from an aquatic environment).
90 *HA* 7 594a25–32 (*ta agria kai karkharodonta panta sarkophaga*). Another flesh-eating animal, called glanos or hyena, has a liking for human flesh to the point of even digging corpses up from graves. As to its hunting practices, the glanos hunts dogs by producing a sound as if it were vomiting and hence attracting his prey (*HA* 7 594a30–594b5).
91 *HA* 7 595a13–16.
92 *HA* 7 595b23–4.
93 *HA* 7 595a16–18.
94 *HA* 7 596b10–16.
95 *HA* 7 591a7–8.
96 *HA* 7 591a18 (*allēlophagein*).
97 *HA* 8 608a19–28, all translations of *History of Animals* 7 and 8 are by D.M. Balme; for the influence of Aristotle's conception of animals on Hobbes' philosophy of man, see Zatta, 2018b.
98 *HA* 8 608b30–33.
99 This animal is of difficult identification inasmuch as *drakaōn* is a generic term for land and sea creatures that have the form of a snake (see Li Causi, Pomelli, 2015, 429, n. 80).
100 Ichneumon here possibly stands for a type of wasp rather than for the quadruped described in *HA* 8 612a16 (see Li Causi, Pomelli, 2015, 433, n. 10).
101 *HA* 8 609a23–24.
102 *HA* 8 610b1–19.
103 This statement does not contradict the previous observation that if food were abundant animals would be kind toward one another and at peace. For animals' state of war is a natural condition that emerges from their desire to live as they cope with the availability of resources.
104 Among the nonhuman animals Aristotle discusses the ass and the akanthis, a bird that lives (and feeds) on thorn-bushes. The two animals fight only during the limited time in which the ass feeds on the tender thorns (*HA* 8 610a3–6).
105 *HA* 8 610b1–19.
106 See discussion at 6.1, 6.2.
107 *HA* 1 488b12–27, transl. by A.L. Peck (italics by the author).
108 On characters' impact on activities and lives, see *HA* 7 588a17–9.
109 In the *Met.* 1 980a21–980b21 Aristotle uses the qualifier *mathētikos*.
110 It is noteworthy that by using the passive voice of *didaskō* Aristotle has in mind a specific interpersonal setting for animals' training (parents/offspring and shepherds/herds, etc.). Yet, animals' receptivity to "being taught" arguably transcends such circumstances and applies to their life as unfolding "in touch" with the environment (see upcoming discussion).

111 Besides the crocodile's example, see Aristotle's observation that fishes become friends and shoal together when there is abundance of food. Animals' sociality pertains to their modes of life (*bioi*) and is distinguished in "solitary" and "gregarious," with the "political" being a subcategory of the gregarious (see 6.4 below), and arguably presupposes a consistent set of character traits. For instance, in the *HA* passage above Aristotle qualifies the ox as gentle and not inclined to be aggressive, a character trait that implicitly supports the gregarious nature of cattle (*HA* 6 575b19).

112 See above.

113 Aristot. *Met.* 1 980a21–980b21, translation by Hugh Tredennick.

114 In respect to the emphasis on human sight in this passage, Whiting notices that in rational animals "sight is informed by the intellect and sees things as triangle, eclipses, and other objects of knowledge" (2002, 188); on human knowledge and the emergence of reason, see also 193.

115 By "experience" here I do not intend Aristotle's technical notion in *Metaphysics* 1 (i.e. a collection of memories crystallized into a single relevant body of "knowledge") but more generally animals' conscious reception of reality via *phantasia* and memory.

116 *HA* 8 608a11–21.

117 The fact that learning depends equally on hearing sounds and the capacity to differentiate between acoustic signs indicates that Aristotle considers the sense of hearing central to learning but also that there are different levels of animal learning.

118 In the background of this distinction between sounds and "acoustic signs" there is the discussion of *psophos*, *phonē*, and *dialektos* in *HA* 4 535a28–9. See also *de Int.* 16a29 (*agrammatoi psophoi*) and *PA* 2 660a29–660b1 where Aristotle attributes to beasts' inarticulate sounds meaning (*dēlousi ti*) and points out that in vocalizing more than other animals some birds seem to convey learning (*mathēsis*); cf. Zirin, 1980, especially 337–43. As this author remarks, bird song is learned. For Aristotle writes that when some of the small birds are reared away from their parents they sing like other birds they have heard, and adds that a nightingale has been seen to teach its chicks (*HA* 4 536b, 14–20). See also *Pol.* 1 (1253a8–12) where Aristotle states that animals' voices indicate (*sēmainein*) pain and pleasure in contrast to the human capacity to convey right and wrong.

119 But even when hearing, not all animals possess the same awareness. Aristotle remarks that "shepherds teach the flocks to run together after a sudden noise; for if one is caught in a thunderstorm and does not run with the others, it miscarries if it happens to be pregnant. This is why if there is a sudden noise in the house, they run together out of habit" (*HA* 8 610b34–611a2). This report shows that animals learn from the shepherd to react with an acquired behavior (i.e. to stay together) to sudden sounds. Sheep, however, fail to discriminate between sudden sounds in respect to whether they happen outside or inside the shelter and behave out of habit in the same way as if the noises represented the same danger. Hence Aristotle defines flocks' character as stupid (*euēthes*) and simple-minded (*anoēton*).

120 See Aristotle's *DA* 2 421a10–2.

121 *HA* 7 588a16–31. Here is the passage in full, "For even the other animals mostly possess traces of the characteristics to do with the soul, such as present differences more obviously in the case of humans. For tameness and wildness, gentleness and roughness, courage and cowardice, fears and boldnesses, temper and michievousness are present in many of them together with resemblances of intelligent understanding, like the resemblances that we spoke of in the case of the bodily parts. For some characters differ by the more-and-less compared with man, as does man compared with a majority of the animals (for certain characters of this kind are present to a greater degree in man, certain others to a greater degree in other animals), while others differ by analogy: for corresponding to art, wisdom, and intelligence in man, certain animals possess another natural capacity," with a slight modification.

122 See 6.1.1.

123 On this point, see also Fortenbaugh, 2006, 185.
124 *HA* 8 612b21–8.
125 Ants are the most laborious of animals and build complex underground chambers where they live, store their food, and perform other activities; see 6.4 and 6.4.1.
126 With this general statement Aristotle is likely referring to the Presocratics. See, for instance, DK 28 A 45 ("Parmenides, Empedocles and Democritus say that the mind and the soul are the same thing; in their view no animal would be altogether lacking in reason," transl. by Laks and Most).
127 See Chapter 1.
128 *HA* 7 588a30–1.
129 For the involvement of *phantasia* in animals' lives with regard to *phronesis*-related activities, see Labarrière, 1990, 407.
130 On this point, see Lennox, 1999a, 18, 22–3.
131 *PA* 2 650b19–651a15, transl. by J. Lennox.
132 For the nature of blood in light of the "biochemistry" of *Meteorologics* 4 (and hence animals' subjection to "physical laws" and inscription in the realm of nature), see Lennox, 2001, 201–2.
133 On the continuity for the Presocratics between thought and sensation see Chapter 5, and Zatta, 2019, Chapter 3, along with the fragments and testimonies there discussed; for Empedocles, DK 31 B 105 and DK 31 A 86.
134 Blood is matter (*hylē*) inasmuch as it nourishes the body (cf. Lennox, 2001, 203), enabling the growth and preservation of its parts.
135 For the introduction of these parameters, see *PA* 2 650b14–8.
136 *HA* 8 630b20 (*euaisthēton kai tēi synesei tēi allēi hyperballon*).
137 For the connection between coldness and the symptoms of fear, see also *MA* 701b28–32.
138 See, for instance, *PA* 4 679a13 (on the octopus), 682b25 (on the dung-beetles), and Lennox, 2001, 202.
139 On the watery and earthy components (respectively, sierum and fibers) of the blood, see *PA* 4 651a17 and 4 650b18, 650b36–651a1, 651a7; Mingucci, 2015, 93. Further, Lennox underscores the relevance of Aristotle's discussion to modern scientific accounts which also distinguish a component in plasma called fibrines, its various levels in different species and predominance in the ones mentioned by Aristotle (2001, 201).
140 See the physiological account of anger as a surging of the blood in *On the Soul* (1 403a25–403b3).
141 Significantly, both in *PA* 1 and *HA* 7, Aristotle's use of comparatives points to animals' membership to the same *genos* hence articulating their "sameness;" see 6.1.1. In the *HA* 7 passage (588a25–27) the "direction" of the comparison is double: from animals to man and from man to animals, cf. Balme, 1991, 59, n.d, Lennox, 1999a, 17.
142 In this respect, see the beginnings of books 7 and 8 in which Aristotle attributes generally to animals a number of psychological affections without restrictions as to which animals feel what (7 588a22–24, 608a15–8). He then proceeds to give examples of specific animals' behaviors showing their dispositions, i.e. speaking of the stupidity of the herds or the parental affections of the mares (8 610b23–611a15).
143 *HA* 8 629b10–17. In fact, with regard to the lion's behavior Aristotle indicates that even in the same wild animals there can be differences in gentleness and wildness (in connection with nutrition). For when the lion is feeding it is very dangerous but when it is not hungry and has been fed, very gentle.
144 See the end of the passage from *PA* quoted above (2 651a12–15).
145 See above.
146 On Aristotle's adoption in zoology of an "anthropic perspective," see more recently also Lloyd (2012, 292–3). With regard to *HA*, the methodological artic-

ulation of sameness via differences discussed at 6.1.1 denies that humans are taken as the standard by which nonhuman animals are judged and so does the two-way-comparison (i.e. of animals to humans and of humans to animals) by which Aristotle discusses the difference in character at the beginning of *HA* 7 (588a25–28); see n. 121. Aristotle appears to adopt the animal-to-human comparison in book 8 (612b18–20) when he claims that nonhuman animals of small size show many imitations (*mimēmata*) of human life and illustrates this claim by referring to the swallow's architectonic skills. Yet not literally referring to actual imitations, this term likely expresses the relation by analogy between humans' art and its animals' equivalent. In lacking a proper name for animal skills, Aristotle defines them in terms of *mimēmata*. In addition, this word choice may convey also a critical allusion to Democritus who did claim that human beings learned the most important activities from animals by imitation (*mimēsis*) (DK 68 B 154; cf. Aristot. *Phys.* 2, 8, 199a20–9, and Chapter 1).

147 2006, 185–6, and 160 (and more generally, section III).
148 For this author, in *Nicomachean Ethics* emotions are cognitive and involve moral virtue (2006, 159, and, more extensively, 107–30). When a human being is angry or fearful s/he has become so on the basis of a judgment, *thinking* that s/he has been spited or that a danger is imminent and hence using the faculties of the rational soul. And since moral virtue has partly to do with the disposition to feel emotions like anger and fear, it is involved in the judgments leading to the experience of such emotions.
149 See 2006, 183.
150 Note that the incidental sensibles enable animals to recognize reality in its elements on the basis of having had previous contacts with it; see Chapter 4.
151 *DA* 3 433a10–14; on *phantasia* as a cognitive faculty enabling animals to have an integrated view of reality and envisioning prospects, see also Chapter 5.
152 Lennox, 1999a, 12–6. A key passage in Lennox' interpretation is *NE* 6 1141b1–16 in which Aristotle asserts that the states of character are present in all human beings *from birth* and even in children and beasts.
153 *HA* 7 590b11–7. Fear features prominently in Aristotle's description of animal emotional dispositions and emotions; see, for instance, *HA* 8 608b31, 609a34, b17, 622b14, 627a18–19, 629b21, 630b12; cf. Fortenbaugh, 2006. 283.
154 *HA* 7 590b21–9.
155 *HA* 8 611b15–19. Aristotle's reference to the deer's confidence is inserted in a longer passage that discusses the deer's practical intelligence (*phronesis*); see below.
156 See note 119.
157 So, for instance, Aristotle claims that the male octopus is courageous because he tries to bring help to the female when trapped by the net. Whether or not the octopus understands that he puts itself in danger, from Aristotle's account it is clear that he understands that its female is in danger (*HA* 8 608b16–9).
158 For *phronesis* as depending on light, hot, and pure blood, see *PA* 2 648a10–1; in addition, *PA* 2 648a4–5 presents light and cold blood as conducive to intelligence (*noerōteron*) (and sensation), while for 650b19–20 watery blood confers to animals a "specially subtle intelligence" (*dianoia*).
159 Labarrière, 1990, 406.
160 For the fluid range of words expressing animals' intelligence, see Labarrière, 1990, 405–6.
161 Yet Lennox remarks that in Aristotle's ethics the intellectual virtues of *phronesis* and *synesis* are not the same. For only the first has to do with action while the second involves a judgment disconnected from action (1999a, 19 along with *NE* 1143a5–7). Aristotle's inclination to blend the two and refer to *phronesis* in the zoological treatises squares with his interest in animals' actions and lives.
162 Aristot. *NE* 6 1141a26–28. Fortenbaugh downplays this statement by remarking that with "*phasin*" (they say) Aristotle is reporting a traditional opinion, not necessarily

endorsing it (2006, 167–8). Yet the many examples relating to animal *phronesis* in *History of Animals* show that Aristotle took this opinion seriously and that he considered *phronesis* a capacity that gives rise to animals' different self-preserving strategies and practices.

163 Labarrière remarks that *phronesis* enables animals to live well, and not only to survive. He supports this claim with *Sens.* 1 436b12–437a17 in which Aristotle identifies two functions of the distant senses for the animals endowed with locomotion according to whether they have *phronesis* or not. Distant perception empowers the first ones to live well (*eu heneka*) by means of the discrimination of differences and knowledge of "objects of thought and things to do," while it enables the seconds to preserve themselves (*sōteria*) through searching for what is useful and beneficial and escaping from what is destructive (1990, 416–7).

164 See above.

165 Aristotle explains that they soak themselves in water and roll in the ground.

166 For these and other examples of animals' intelligent behavior, see *HA* 8 611a15–612b9; cf. Labarrière, 1990, 420–5 and Lloyd, 2012, 282–4, who discuss the discrete areas of animals' lives in which *phronesis* is at work. Because *phronesis* is in fact applied to a variety of areas in animals' lives Labarrière argues for a plurality of intelligences rather than a single capacity.

167 It is important to point out, however, that Aristotle acknowledges differences among individuals (not only animal groups). In the same passage in which he recognizes the difference in the nature of blood of a single individual (in the upper part versus the lower parts) he also claims that "the blood of an animal differs from that of another," a claim that applies both across species and within species with consequent differences in their cognitive capacities and characters (*PA* 2 647b34–648a2).

168 Aristotle calls the animals that do not learn (*aneu to manthanein*) *phronima* and deems more intelligent (*phronimōtera*) those that possess memory (*mnēmē*) and are therefore also more apt to learning (*mathētikōtera*) (*Met.* 1 980b21–25); on animal learning see 6.3.

169 In this respect, Labarrière argues for animals' possession of a "sous-universal, préprédicatif" which allows animals to act with cognition without leaving their life to chance (1990, 418).

170 *HA* 8 611a20–24.

171 On animal *phronesis* as relating to experience and habit, see Labarrière, 1990, 410.

172 *HA* 6 567a5–7.

173 For the presence of this association also in Aristotle's ethics, see *NE* 7 1141b23–4 (in respect to *HA*, Lennox speaks of a "pattern of relationships among differences in activities, ways of life and characters," 1999a, 19).

174 Aristot. *HA* 1 487b33–488a11, transl. by A.L. Peck with slight modifications.

175 While it is a scholarly convention to call Aristotle's *politika zōa* social, in this study I remain faithful to his qualification and render *politika* as political. I proceed so because, as Mulgan remarks, "political" refers to the quintessential feature of which all political animals, humans included, partake, namely the engaging as a community in a common activity (1974, 439).

176 See *Pol.* 1 1253a7–19. "And why man is a political animal (*politikon zōon*) in a greater measure than any bee or any gregarious animal is clear. For nature, as we declare, does nothing without purpose; and man alone of the animals possesses speech (*logos*). The mere voice, it is true, can indicate (*esti sēmeion*) pain and pleasure, and therefore is possessed by the other animals as well (for their nature has been developed so far as to have sensations of what is painful and pleasant and to signify (*sēmainein*) those sensations to one another), but speech is designated to indicate the advantageous and the harmful, and therefore also the right and the wrong; for it is the special property of man in distinction from the other animals that he alone has perception (*aisthēsis*) of good and bad and right and wrong and the other moral qualities,

and it is partnership in these things that makes a household and a polis," transl. by H. Rackham with slight modifications.

177 *HA* 8 614b18–27; see below.
178 *HA* 7 589a1–3; cf. also 3 *GA* 753a7–14, where he qualifies those animals that care for their youngs' nutrition until their development as *phronimotera* while those that continue the association (*koinonein*) with their fully developed offspring showing intimacy (*synētheia*) and friendship (*philia*) are said to possess the greatest intelligence (so are humans and some quadrupeds).
179 See 6.3 above.
180 But this connection does not mean that animals which are solitary or gregarious are not intelligent. In the *Metaphysics* passage quoted above Aristotle remarks that animals endowed with the sense of hearing are more intelligent (*phronimotera*) in respect to the intelligent (*phronima*) bees which are without it.
181 For Aristotle's definition of the *polis*, see *Pol.* 3 1274b31–1275a1.
182 It should be emphasized, however, that Aristotle uses the comparative adverb *politikōteron* to qualify the relations parents entertain with their offspring, not the animals themselves. In other words, *politikos* as a trait qualifies those animals which undertake a common activity (i.e. bees, ants, wasps, and cranes, among others) for the sake of the community.
183 *HA* 7 589a1–3.
184 *Pol.* 1 1252a24–1252b31.
185 See *HA* 8 612b33–34, even though *politikos* does not appear in his discussion of lifelong partnership, but rather qualifies in a comparative form the parents/offspring relationships.
186 *HA* 8 614b32.
187 *HA* 8 615b25.
188 See, respectively, *HA* 8 612b27–8, 613a3–6.
189 *HA* 8 615b25.
190 *HA* 8 611a15.
191 *HA* 8 611b33–5.
192 *Contra* Cooper for whom nonhuman animals as well as human ones "live together in cooperative communities in which each benefits from the work of the others as well as from his own," but only humans work together for the good of the community (2005, 79). That animal communities themselves are benefited is clear if we think of the work organization of ants and bees. For in the events some members of these societies fail to perform their tasks, the entire community would suffer. See discussion below.
193 *HA* 623b18–20.
194 See, respectively, *HA* 623b13–4 and 628a13–5. In the case of wasps, it is a specific category—that of "leaders"—within the larger group which attends to the construction of the "wasperies." For a discussion of bees, wasps, and ants, their lives and varieties, in the ancient sources at large, see Beavis, 1988, 187–217.
195 *HA* 7 597a5–7.
196 *HA* 8 614b19–22.
197 *HA* 7 597a30–2; cf. Her. 2.2; Op. *H.* 1.621. On cranes' migration, see also Arnott, 2007, 52–3.
198 *HA* 7 597a9–597b30.
199 *HA* 7 598a24–598b6.
200 *HA* 7 596b20–8.
201 *HA* 8 622b19–20
202 *HA* 8 622b27–8.
203 *HA* 8 629a7–9.
204 *HA* 8 628a12–9.
205 Apparently, in speaking about the "kings," the beekeepers contemporary to Aristotle ignored the existence of the queen bees.

206 *HA* 8 623b30–3; bees show high architectonic skills and hence possess a natural capacity analogous to human *tekhnē*. In this way they are like the swallows who build their nests paying attention to the building materials in relation to the structure of their dwelling.

207 *Il.* 12.167–82.

208 *HA* 8 626a15–8.

209 Mulgan, 1974, 439.

210 On this claim voiced by Nikias during the expedition in Sicily (Thucydides 7.77.7) and its endorsement by different Athenians politicians, see Zatta, 2011, 341–4.

211 *Pol.* 3 1274a39–1275a1; cf. 1275b20–22.

212 This precept proved to be true when, in 480 BCE, facing the Persian invasion, the Athenians abandoned their polis to fight the enemy from the fleet (Hdt. 7.141–3; 8.41; 8.61).

213 Cranes' ability to feel and act like a community, independently of a sense of place, transpires well from the myth, already present in Homer, narrating their fight against the pigmies upon their arrival onto the Southern lands of Egypt (*Il.* 3.3–7).

214 *HA* 8 614b22–7.

215 *HA* 7 598a29–30.

216 *HA* 8 624b22–7.

217 *HA* 8 624a27–34.

218 *HA* 8 624a26–7.

219 *HA* 8 625a15–25.

220 *HA* 8 627b9–10.

221 See Fortenbaugh, 2006, and discussion above.

222 Lloyd, 2012, 292–3, see also n. 146 above; Lloyd, 1983, 26–43, and especially 33 where the human being is taken as the "model" for Aristotle's discussion of animal physiology.

223 For the "more" that qualifies humans' political nature, it has been differently pin-pointed in "consciously pursued gain," the "latitude and indeterminacy" of their behavior, the "combination of reason and passions," and the role of spiritedness (*thumos*) and, again, in the presence of civic friendship or the "reasonable perception" of interest; see respectively Kullmann, 2005, 99–102; Lord, 1991, 57–60; Cooper, 2005, 74–80; and Salkever, 2005, 40.

224 See section 6.1.1 above.

225 See 6.1.1 where Aristotle discusses sameness by differences by degree and analogy. The difference *kat' eidos* is instead introduced in *Parts of Animals*; see Chapter 1.

226 *HA* 7 588a16–31, quoted in n. 121 above.

227 See *HA* 1, 6 491a20–23 and discussion in Chapter 1.

Conclusion

> For since the living animal is a body possessing soul, and every body is tangible, and tangible means perceptible by touch, it follows that the body of the animal must have the faculty of touch if the animal is to survive. For the other senses, such as smell, vision and hearing, perceive through the medium of something else; but the animal when it touches, if it has no sensation, will not be able to avoid some things and seize others. In that case it will be impossible for the animal to survive (*sōzesthai*). This is why taste is a kind of touch; for it relates to food and food is a tangible body … The other senses are for the means of well-being (*tou eu heneka*) … for instance, the animal has sight in order that it may see, because it lives in air or water, or generally in a transparent medium, and it has taste because of what is sweet and bitter, in order that it may perceive these qualities in food, and may feel desire and be set in motion; and hearing that it may have significant sounds made to it, and a tongue that it may make significant sounds to another animal.
>
> (Aristotle, *On the Soul* 3.12–13)

Any exegesis with "if" is doomed to lead astray from the facts and issues at stake. But one can still ask how Aristotle's thought on animals and their lives would have been taken and what conceptual paths it would have supported had not it been received through the lenses of the influential tradition that came after him. This is not the place to reconstruct Aristotle's reception but only for a last set of reflections, within the scope of this book. Whether explicitly or implicitly, assumed or unassumed, also Aristotle's successors, i.e. the thinkers and writers that came after him in the Western tradition, had to face a choice. It might not have been the same one that Aristotle confronted when inquiring about animals within the framework of his study of nature (*physis*) and vis-à-vis the inanimate beings (and plants). Still they had to choose from which perspective to look at animals and their lives, where to put the value and what to appreciate, and such a choice was often ripened and/or asserted as they read and engaged with Aristotle often transmitting, in turn, his thought to future generations. In *On the Intelligence of Animals*, for example, Plutarch claimed that animals possess reason and understanding on the basis of many examples of animal behaviors in *History of Animals*, from the hunting strategies of the cuttlefish to the cooperative nature of

DOI: 10.4324/9780367816001-8

the purpuras which in spring gather together in one place to build their underwater colony as bees do with their honeycomb on earth.[1] Plotinus, on the other hand, recognized that "no doubt all lives are thoughts—but qualified as thought vegetative, thought sensitive and thought psychic."[2] An intellectual heir of Plato, in this claim Plotinus seems to have rather appealed to Aristotle and his partition of the soul along with the fundamental functions of life it delineated—nutrition, sensation, and reason. But Plotinus also transcended this position, for he redeemed plants from utter alterity and further unfolded the idea that life in any of its forms (vegetal, animal, or human) implies a form of rationality.

Aristotle, we saw in the introduction, praised Anaxagoras who detected *nous* in all animals, a category that for the Presocratic included also plants. But instead of *nous*, it was argued in this book, Aristotle spoke of *logos*, which he considered inscribed in animals' complex bodies and tacitly informed the lives such bodies could afford them to live (and in a way that distinguished them from plants). Beyond the hierarchy due to animals' different degrees of internal heat and the status of their offspring at birth (whether complete or not), beyond the idiosyncratic, godlike posture of man and beyond the difference between the complete animals endowed with the five senses and *phantasia*, on the one hand, and those possessing merely touch and *phantasia* indeterminately, on the other, for Aristotle, student of nature, each animal was in fact perfect and worthy of study. Even in Heraclitus' humble kitchen there were gods—he reminded his reader, hence illuminating with an anecdote the form of egalitarianism that lay at the core of his zoology. For each living body revealed an artful, functional structure that allowed it to move and live in its space of existence, to sense its food and to recognize its mate, to eat, preserve itself, and to reproduce; to take care of its offspring and to communicate, to take action at the changing of seasons and migrate, or to choose the optimal place for a lair and much more—for those that did so. By carrying out these activities, each living being was empowered to live, striving to live forever, according to its nature, whether in water or land, leading a life directed to perpetuate the movements that constituted it, even if only in another living being like oneself. Plants too strove to live but in their silent, effortless, and bare existence.

For Aristotle, the range of animals' activities and thus their lives relied on their power of sensation, which he denied to both the simple bodies of the elements and the basic bodies of plants. In addition to desire, pain, and pleasure, sensation implied knowledge of the qualities of the world and its entities, in a structured way that depended on the body (as realization of the sensitive soul) and reflected the individual's sensorial capacities and experience. But sensation also implied the awareness of being the subject of sensation, and of the pain and pleasure associated with it. With regard to the special sensibles, the tangibles and tastes, the colors, scents, and sounds sensation was infallible inasmuch as the senses were apt to grasp the *logoi* intrinsic in the embodied qualities. And it is likely because of the sophistication inherent in the mechanism of sensation, its receptive precision, that in *On the Soul* Aristotle expressed a doubt about the sensitive soul, namely that "one could not assign it easily to the rational (*logon ekhon*) or irrational part (*alogon*)."³ A passing and often-neglected comment, this claim reveals that if Aristotle marginalized from his study of animals human reason conveyed by speech and dealing with universals (and right and wrong) he was still ready to question whether animals possessed a form of reason that was intrinsic in sensation and its products. He did after all refer to sensation-stemming *phantasia* as a kind of *noēsis*. Still Aristotle did not pursue this interpretation (i.e. the relation of sensation to *logos* as reason) because it would have undermined the clarity and order of his systematic view on life, based on nutrition, sensation (with or without locomotion), and reason as complementary living faculties among which reason stood, in Plato's style, as the capacity exclusive to the human being. Aristotle pursued instead the equivalence of sensation to thought. Like thought, sensation also provided animals with the capacity to discriminate (*krinein*) and live intelligently, in different degrees. And in affording discrimination, sensation realized one of the two irreducible functions of the living being's soul, the other being locomotion.⁴ In the end, for Aristotle animals did not need human *logos* to live nor, for those endowed with the distant senses and locomotion, to live well (*eu zēn*). Sensation and its apparatus abundantly sufficed because they were essential to life, absolutely more so than reason which needed them in order to work, at least at first to secure the intelligibles, and still, more generally, to subsist.⁵ And it is not that the realm of sensation was expanded because animals were denied reason, as some scholars have argued. Rather sensation and its apparatus were given the due crucial importance because as creatures of nature for Aristotle animals, humans included, were first of all living bodies exposed to, and moving in, a physical reality the awareness of which rose through sensation.⁶ Life immanence was central to his approach to animals because it alone secured life continuity and fulfillment.

Hence, to resume the initial thread, it is not surprising that in *On Abstinence from Killing Animals*, Porphyry, another "interpreter" of Aristotle, claimed that "Aristotle and Plato, Empedocles and Pythagoras and Democritus, and all who have sought to grasp the truth about animals, have recognized that they share in *logos*."⁷ This comment emerges in a long section in which Porphyry sets out to demonstrate that "every soul is rational in that it shares in perception and memory."⁸ To this end Porphyry fights a battle over *logos* against the Stoics and to claim that animals possess both "expressive" and "internal *logos*"⁹ he resorts to Aristotle,¹⁰ as did Plutarch before for a similar cause. Porphyry also deploys Aristotle's methodological principle in *Parts of Animals* that animals should not be looked at as carved in stone¹¹ but living. It was their lives that manifested *logos*. So by advocating Aristotle's claims and method Porphyry forces him to recognize that animals "share in *logos*."¹² Hence, he aligns Aristotle with Empedocles and Democritus who did in fact claim that no living being was without *logos*.¹³

This Aristotle did not claim but his relation to the Presocratics has been an important one for this book to pursue. For embedded in a philosophical tradition that considered animals within the realm of nature Aristotle asserted and defined the methodological boundaries of his own study against the Presocratics', locating the *arkhē* of animals' movement in their own bodies rather than in an original beginning; affirming the role of *telos* versus *tykhē*; and insisting on the composite, organic nature of animals' bodies beyond the elemental composition—to cite a few. And as for his critique of Democritus that shape and color were insufficient criteria to look at animals, it went deeper than asserting that one had to study them in relation to their soul and its relevant parts and comprehend the animal as such (*toiouto to zōon*), for this critique also pointed to the monolithic conception of the soul Democritus (and other Presocratics) held and to the lack of differentiation between sensation and thought such a view of the soul led to. Animals did not have the thought provided by the rational soul, disconnected from the body and movement but for Aristotle they still possessed the form of thought granted them by *phantasia*, which stemming from sensation maintained a strong connection with the body and the living environment, and remained anchored in the particularity of each living being's individual existence. In this claiming, a body-rooted form of thought resides, I believe, a striking point of convergence between Aristotle and his predecessors, regardless of the specific differences both within the Presocratic doctrines and between these and Aristotle's.

***

A critical turn in the interpretation of Aristotle came with Thomas Aquinas and his Christian, Platonizing approach. Aquinas directed the compass of Aristotle's conception of animals to the human/animal alterity rather than to the sentient/lifeless alterity, which informed it and which this book pursued. He put the accent on *logos* (as reason) and mind rather than life. The emphasis rested on what the human being alone possessed in respect to what animals did not, hence crystallizing a point of view that has surfaced over and over in successive readings of Aristotle and one that reflects his ethical discourse concerned with human life. Aquinas' commentary on *On the Soul*, which constitutes a foundational text for Aristotle's philosophy of life and particularly animal life, is emblematic of his switched perspective—already from the outset. Aquinas approaches Aristotle's conception of the soul in terms of his apology for the study of animals in *Parts of Animals* 1, according to which, we saw in Chapter 1, mortal animals are better known than the celestial, immortal ones and zoology acquires legitimacy because human beings live among animals and share with them a more akin nature (*oikeiotera physis*). Let us recall that for Aristotle animals' knowability and shared nature balanced out the higher value inherent in the immortal, celestial, rational beings. But Aquinas breaks Aristotle's arguments down and circumscribes knowability and sharedness to the human being alone severing its tie with animals in a discourse that ultimately aims at underscoring the nobility of the intellect. He writes,

> It [the science of the soul] is certain for it is experienced in the human being itself, namely that s/he has a soul and that the soul vivifies. It is also more noble, because the soul among inferior creatures is more noble.[14]

For Aquinas Aristotle's study of the soul is an accurate science because *human beings* themselves experience the soul and it is noble because its object, the soul, is "more noble" among the lower beings that are the animals. Strikingly, the nobler soul, Aquinas proleptically means here, is the rational, separable-from-the-body soul, which only human beings among the animals possess and which they share with the higher, immortal beings. Hence Aquinas, interpreter of *On the Soul*, subverts Aristotle's framework radically separating the human being from the other animals, electing him to be the noble one among the worthless, and providing enduring food for thought for the generations of Aristotle's interpreters and others to come. In this interpretation, the life of the mind rises above life in the world injuring it, the human being projected toward higher spheres and eternity, its animality suppressed. Anthopocentrism tramps over that zoocentrism which was at the core of Aristotle's study of living beings and life, obliterating his appreciation of sensation as the cement of all animals' existence, and how it actually enabled living beings to live through their ensouled bodies and ensuing knowledge, partaking of eternity in this world. And true, these two perspectives (i.e. zoocentrism and anthropocentrism) coexisted in Aristotle's corpus itself and ultimately overlapped his division between theoretical and practical sciences, with physics on the one side of the fence, ethics on the other.[15] Accordingly, living beings' existence was conceived to be for the sake of their very existence[16] or for the existence of the rational and moral being that is man. But even in the ethics Aristotle's zoocentrism breaks in asserting the epistemological validity of the theoretical—the grand look on animal life is appealed to contextualize within the domain of nature the focused treatment of the life that is proper to humans. For even in a treatise devoted to unveil the roots of human happiness, all living beings, humans included, are claimed to find pleasure in the different activities they pursue because in the end they all indifferently fulfill their recondite and intrinsic desire to live.

Notes

1 See respectively Plut. *de Sol. An.* 978b and Aristot. *HA* 8 621 b33; *de Sol. An.* 980d and Aristot. *HA* 5 546b18–22.
2 Plot. *Enn.* 3.8.
3 *DA* 3 432a30–a32b1.
4 *DA* 3 427a17, cf. Pellegrin, 1996, 471. In this passage Aristotle ignores the nutritive function which animals share with plants. He is referring instead to sensation and the sensitive soul, which defines animals and which, according to his geometrical view of the soul, presupposes the faculty of nutrition.
5 Let us recall that nutrition is fundamental for the preservation of the living being and all its faculties, that for all animals it happens through the contact senses (i.e. touch and taste), and for those endowed also with locomotion with the aid of the distant senses too.
6 It bears noting that while Aristotle points out his different take on sensation with respect to the Presocratics (for him sensation is neither simply an encounter between likes nor between unlikes, but of unlikes that become likes), he ultimately pins down animal life

in an encounter with the physical world, sharing in this way the Presocratics' approach; cf. Chapter 4.

7 Porph. *De Abst.* 3.6.7, transl. by G. Clark.

8 3.1.4.

9 These two forms of *logos* correspond to a differentiation between language as expressing thought and internal thought (3 188).

10 A discussion of Porphyry and the Stoics is out of the scope of this conclusion, the reference to him aiming at showing how the crisis about the human/animal divide supposedly initiated by Aristotle lies more with the interpretations of Aristotle than his own work (see introduction). For the sake of precision let us remark that Porphyry's argument against the Stoics unfolds through stages; first he claims that animals possess "expressive logos," then "silent logos," and last the rational soul. Porphyry supports each of these claims with evidence from Aristotle appealing (among other points), respectively, to his observations on animal learning (3.6.5), perceptual and physical nature (3.8), and actions in view of self-preservation (3.9.5).

11 This principle is used to object to the ignorant man who instead of researching about animals among different categories of people that deal with them (i.e. hunters, herdsmen, mahouts, trainers of wild animals and birds) chooses "to carve them up as if they were stone," thereby misrepresenting them (*De Abst.* 3.6.6, cf. *PA* 1 641a19–21; see Chapter 3).

12 It bears noting that Porphyry is aware of forcing Aristotle to assert what he never did but was for him implicit in his ethological observations. Accordingly, Porphyry states that "anyone who says that animals have these qualities by nature fails to realize that he is saying they are rational by nature, or else that *logos* does not exist in us by nature" (3.10.1). As to what *logos* Aristotle is made to have attributed to animals, whether expressive or internal, it is clear from Porphyry's argumentative strategy that it retroactively covered both types: animals used language to communicate and acted intelligently for their self-preservation.

13 DK 28 A 45.

14 Aquinas, *Commentary of Aristotle's On the Soul*, ad 402a1–5, 5–6 (Kocourek 3).

15 See *Met.* 6 1025b.

16 Admittedly, this claim comes from *On the Soul* and not from the strictly zoological treatises. Still, as observed in the course of this book, in *On the Soul* the soul is presented as the principle of animal life hence bearing relevance for the comprehension of animals as beings that are by nature (see n. 13 of the introduction and Chapter 3).

Bibliography

Editions, translations, commentaries

Adam, J., 1902, *The Republic of Plato*, vol. II, Cambridge: Cambridge University Press.

Amigue, S., 2006, *Theophrastus, Recherches sur le plantes*, vol. 1. Paris: Les belles lettres.

Aquinas, T., 1946, *The Commentary of St. Thomas on Aristotle's Treatise on the Soul*, transl. by R.A. Kocourek, St. Paul: College of St. Thomas.

Balme, D.M., (prepared for publication by A. Gotthelf), 1991, *Aristotle History of Animals Books VII-X*, Cambridge, MA/London: Harvard University Press.

Balme, D.M., 1992, *Aristotle's De Partibus Animalium I and De Generatione Animalium I*, Oxford: Clarendon Press (originally published in 1972).

Bloch, D., 2007, *Aristotle on Memory and Recollection. Text, Translation, Interpretation, and Reception in Western Scholasticism*, Leiden/Boston, MA: Brill.

Bury, R.G., 1926, *Plato Laws Volume II: Books 7–12*, Cambridge, MA: Cambridge University Press.

Bury, R.G., 1929, *Plato Timaeus, Critias, Cleitophon Menexenus Epistles*, Cambridge, MA: Harvard University Press.

Carbone, A.L., 2002, *Aristotele. Le Parti degli Animali*, Milan: Rizzoli.

Charlton, W., 1970, *Aristotle's Physics I and II*, Oxford: Clarendon Press.

Cole, A., 2004, "On Generation and Corruption 1.5," in F. De Haas, J. Mansfeld (eds.), *Aristotle: On Generation and Corruption, Book 1. Symposium Aristotelicum*, Oxford: Clarendon Press, 171–94.

Cross, R.C., Woozley, A.D., 1964, *Plato's Republic*, New York: St. Martin Press Inc.

Diels, H., Kranz, W., 1951–1952, *Die Fragmente der Vorsokratiker*, vol. 3, 6th ed., Berlin, Grunewald: Weidmannsche Buchhandlung.

Drossaart Lulofs, H.J., Poortman, E.L.J. (eds. and trans.), 1989, *Nicolaus Damascenus, De Plantis: Five Translations*, Amsterdam: North Holland Publishing Company.

Düring, I., 1961, *Aristotle's Protrepticus*, Göteborg: Acta Universitatis Gothoburgensis.

Emlyn-Jones, C., Preddy, W., 2013a, *Plato Republic Volume I: Books 1–5*, Cambridge, MA: Harvard University Press.

Emlyn-Jones, C., Preddy, W., 2013b, *Plato Republic Volume II: Books 6–10*, Cambridge, MA: Harvard University Press.

Forster, E.S., Furley, D. J., 1989, *Aristotle. On Sophistical Refutations, On Coming-to-Be and Passing Away, On the Cosmos*, Cambridge, MA: Harvard University Press.

Fowler, N.H., Lamb, W.R.M., 1925, *Plato Statesman, Philebus Ion*, Cambridge, MA/London: Harvard University Press.

Freese, J.H., Striker, G., 2020, *Aristotle. Art of Rhetoric*, Cambridge, MA/London: Harvard University Press.

Giardina, G.R., 2008, *La Chimica Fisica di Aristotele*, Rome: Aracne editrice.

Giardina, G.R., 2009, *Aristotele. Sull'anima II. La fisica dell'anima e delle sue facoltà Sensoriali*, Roma: Aracne.

Gain, F., 2011, *Aristote. Les parties des animaux*, Paris: Librairie Générale Française.

Gauthier, R.A., Jolif, J.Y., 1959, *L'Étique à Nicomaque. Introduction, traduction et commentaire*, vols. 1–3, Louvain/Paris: Peteers Publishers.

Gulik, C.B., 1989 (revis.), *Athenaeus: The Deipnosophists*, Cambridge, MA: Harvard University Press.

Guthrie, W.K.C., 1939, *Aristotle On the Heaven*, London: Heinemann.

Halliwell, S., 1988, *Plato Republic 10*, Warminster: Aris & Phillips.

Hamlyn, D.W., 2002, *Aristotle's De Anima*, Oxford: Clarendon Press.

Heinze, R. (ed.), 1899, "THEMISTIUS," in *Libros Aristotelis De Anima Paraphrasis*, Berlin: G. Reimeri.

Hett, W.S., 1936, *Aristotle. On the Soul Parva naturalia On Breath*, Cambridge, MA: Harvard University Press, (2000 repr).

Hicks, R.D., 1965, *Aristotle De Anima*, Amsterdam: A Hakkert.

Joachim H. 1922, *Aristotle On Coming-to-be and Passing-away*, Oxford: Clarendon Press.

Laks, A., Most, G., 2016, *Early Greek Philosophy*, vol. 9, Cambridge MA: Harvard University Press.

Lamb, W.R., 1927 (revised), *Plato. Charmides, Alcibiades, Lovers, Theages, Minos, Epinomis*, Cambridge, MA: Harvard University Press.

Lamb, W.R., 1925, *Plato. Lysis, Symposium, Gorgias*, Cambridge, MA: Harvard University Press.

Lang, P., 1964 repr., *De Speusippi Academici Scriptis. Accedunt Fragmenta*, Frankfurt: G. Olms (first ed. Bonn 1911).

Lanza, D., Vegetti, M. (eds.), 1971, *Opere biologiche di Aristotele*, Torino: UTET.

Le Blond, J.-M., 1945, *Aristote, philosophe de la vie. Le livre premier du Traité sur les Parties des Animaux*, Paris: Éditions Montaigne.

Lee, H.D.P., 1952, *Aristotle: Meteorologica*, Cambridge, MA: Harvard University Press.

Lennox, J., 2001, *Aristotle: On the Parts of Animals*, Oxford: Oxford University Press.

Lonie, I.M., 1981, *The Hippocratic Treatises, "On Generation," "On the Nature of the Child," "Diseases IV"*, Leiden/Boston, MA/Köln: De Gruyter.

Louis, R., 1956, *Aristote. Les parties des animaux*, Paris: Les Belles Lettres.

Luria, S., Krivushlna, A., Fusaro, D., 2007, *Democrito. Raccolta dei frammenti, interpretazione e commentario*. Milan: Bompiani.

Movia G., 1979, *Aristotele. L'Anima*, Naples: Loffredo Editore.

Movia, G., 2001, *Aristotele. L'Anima*, Milan: Bompiani.

Newman, W.L., 1887, *The Politics of Aristotle*, vols. 1–4, Oxford: Clarendon Press.

Nussbaum, M., 1978, *Aristotle's De Motu Animalium*, Princeton, NJ: Princeton University Press.

Ogle, W., 1882, *Aristotle. On the Parts of Animals*, London: Kegan Paul.

Patillon, M., Segonds, A.P., 1995, *Porphyre, De l'abstinence, Tome III, Livre IV*, Paris: Les Belles Lettres.

Peck, A.L., 1942, *Aristotle. Generation of Animals*, Cambridge, MA: Harvard University Press.

Peck, A.L., 1965, *Aristotle. Historia Animalium*, vol. I, Cambridge, MA/London: Harvard University Press.

Peck, A.L., 2006, *Aristotle: Parts of Animals, Movement of Animals, Progression of Animals*, Cambridge, MA/London: Harvard University Press (originally published 1937).

Pellegrin, P., 1995, *Aristote. Les Parties des Animaux,* transl. and notes by J.-M. Le Blond, Paris: Flammarion.

Polansky, R., 2010, *Aristotle's De Anima,* 1st reprint ed., Cambridge: Cambridge University Press.

Rackham, H., 1956, *Aristotle. The Nicomachean Ethics,* Cambridge, MA: Harvard University Press.

Rackham, H., 1959, *Aristotle. Politics,* Cambridge, MA/London: Harvard University Press.

Rashed, M., 2005, *Aristote. De la génération et corruption,* Paris Les Belles Lettres.

Rose, V., 1886, *Qui Ferebantur Librorum Fragment,* Lipsiae: Teubner.

Ross, W.D., 1936, *Aristotie's Physics. A Revised Text with Introduction and Commentary,* Oxford: Clarendon Press.

Ross, W.D., 1955, *Aristotelis Fragmenta Selecta,* Oxford: Clarendon Press.

Rowe, C., Broadie, S., 2002, *Aristotle Nicomachean Ethics,* Oxford: Oxford University Press.

Saunders, T.J., 1995, *Aristotle Politics Books I and II,* Oxford: Oxford University Press.

Shields, C., 2016, *Aristotle De Anima,* Oxford: Oxford University Press.

Simpson, P.L.P., 1998, *A Philosophical Commentary on the Politics of Aristotle,* Chapel Hill/London: University of North Carolina Press.

Sorabji, R., 1972, *Aristotle On Memory,* London: Bloomsbury Academic.

Sorabji, R., 2006 (revised), *Aristotle On Memory,* Chicago, IL: Chicago University Press.

Steward, J.A., 1892, *Notes on the Nicomachean Ethics of Aristotle,* vol. 2, Oxford: Clarendon Press.

Stratton, G.M., 1917, *Theophrastus and the Greek Physiological Psychology before Aristotle,* London: Macmillan.

Tarán, L., 1981, *Speusippus of Athens: A Critical Study with a Collection of the Related Texts and Commentary,* Leiden: Brill.

Taylor, C.C.W., S.Y Luria, *Demokrit,* English Translation by C.C.W. Taylor in https://www.academia.edu/25014428/S.Y_Luria_Democrit_English_Translation_by_C.C.W._Taylor

Todd, R.B., 1996, *Themistius. On Aristotle's On the Soul,* Ithaca, NY: Cornell University Press.

Torraca, L., 1961, *Aristotele. Le parti degli animali,* Padova: CEDAM.

Torstrick, A., 1862, *De Anima libri III,* Hildesheim: G. Olms.

Tredennick, H., 1933, *Aristotle Metaphysics. Books 1–9,* Cambridge, MA/London: Harvard University Press.

Tricot, J., 1957, *Aristote. Histoire des Animaux,* Paris: Librarie Philosophique J. Vrin.

Vegetti, M., 1976, *Opere di Ippocrate,* Torino: UTET.

Vegetti, M. (ed.), 1998, *Platone. La Repubblica,* vols. I–VII, Napoli: Bibliopolis.

Wallace, E., 1882, *Aristotle's Psychology,* Cambridge: Cambridge University Press.

Walzer, R., 1934, *Aristotelis dialogorum fragmenta,* Florence: G. C. Sansoni.

References

Algra, K., 1995, *Concepts of Space in Greek Thought,* Leiden: Brill.

Ambler, W.H., 1985, "Aristotle's Understanding of the Naturalness of the City" *The Review of Politics* 47.2: 163–85.

Arnott, G.W., 2007, *Birds in the Ancient World from A to Z,* London/New York: Routledge.

Balme, D.M., 1975, "Aristotle's Use of Differentiae in Zoology," in J. Barnes, M. Schofield, R. Sorabji (eds.), *Articles on Aristotle. 1. Science*. London: Duckworth, 183–93.

Balme, D.M., 1987a, "Aristotle's Use of division and differentiae," in A. Gotthelf, J.G. Lennox (eds.), *Philosophical Issues in Aristotle's Biology*, Cambridge: Cambridge University Press, 69–89.

Balme, D.M., 1987b, "The Place of Biology in Aristotle's Philosophy," in A. Gotthelf, J.G. Lennox (eds.), *Philosophical Issues in Aristotle's Biology*, Cambridge: Cambridge University Press, 9–20.

Balme, D.M., 1990, "Anthrōpos Anthrōpon Genna: Human is Generated by Human," in G.R. Dunstan (ed.), *The Human Embryo: Aristotle and the Arabic and European Tradition*, Exeter: University of Exeter Press, 20–31.

Baluška, F., Mancuso, S., Volkmann, D. (eds.), 2006, *Communication in Plants. Neuronal Aspects of Plants' Life*, Berlin: Springer.

Baluška, F., Mancuso, S. (eds.), 2009, *Signaling in Plants*, Berlin: Springer.

Bénatouïl, T., 2019, "Mouvements et vie chez Aristote : quelques remarques "autour" des plantes"*Anais de filosofia clássica* 13.25: 1–20.

Baltussen, H., 2000, *Theophrastus against the Presocratics and Plato: Peripatetic Dialectic in the Sensibus*, Leiden/ Boston, MA/Köln: De Gruyter.

Beare, J., 1906, *Greek Theories of Elementary Cognition*, London/Edimburgh/New York/ Toronto: Clarendon.

Beavis, I.C., 1988, *Insects and Other Invertebrates in Classical Antiquity*, Exeter: University of Exeter.

Bernardete, S., 1975, "Aristotle *de anima* III. 3–5" *Review of Metaphysics* 28: 611–22.

Blundell, S., 1986, *The Origins of Civilization in Greek and Roman Thought*, London: Croom Helm.

Bodson, L., 1990, *Aristote. De Partibus animalium. Index verborum. Liste de fréquence*, Liège: Université de Liège. Centre informatique de Philosophie et Lettres.

Bolton, R., 1987, "Definition and Scientific Method in Aristotle's *Posterior Analytics* and *Generation of Animals*," in A. Gotthelf J.G. Lennox (eds.), *Philosophical Issue in Aristotle's Biology*, Cambridge: Cambridge University Press, 120–166.

Bolton, R., 2005, "Perception naturalized in Aristotle's De Anima," in J. Salles (ed.), *Metaphysics, Soul, and Ethics in Ancient Thought*, Oxford: Clarendon Press, 209–224.

Bolton, R., 2009, "Two Standards for inquiry in Aristotle's *De Caelo*" in A.C. Bowen, C. Wildberg (eds.), *New Perspectives in Aristotle's De Caelo*, Leiden: Brill, 51–82.

Bonitz, H., 1955, *Index Aristotelicus*, Berlin: De Gruyter.

Bradshaw, D., 1997, "Aristotle on Perception: The Dual-Logos Theory," *Apeiron* 30: 143–161.

Brink, C.O., 1955, "Οἰκείωσις and Οἰκηιότης: Theophrastus and Zeno on Nature in Moral Theory," *Phronesis* 1.2: 123–145.

Broadie, S., 1993, "Aristotle's Perceptual Realism," *Southern Journal of Philosophy* 31: 137–159.

Burneyat, M.F., 1976, "Plato on the Grammar of Perceiving," *Classical Quarterly* 26: 29–51.

Burneyat, M.F., 1992a, "Is an Aristotelian Philosophy of the Mind Still Credible? A Draft," in M. Nussbaum, A. Rorty (eds.), *Essays on Aristotle's De Anima*, Oxford: Oxford University Press, 15–26.

Burneyat, M.F., 1992b, "How Much Happens When Aristotle Sees Red and Hears Middle C? Remarks on *De Anima* 2.7–8," in M. Nussbaum, A. Rorty (eds.), *Essays on Aristotle's De Anima*, Oxford: Oxford University Press, 421–34.

Burneyat, M.F., 2002, "De Anima II 5," *Phronesis* 47: 28–90.

Burneyat, M.F., 2001, "Aquinas on Spiritual Change," in D. Perler (ed.), *Ancient and Medieval Theories of Intentionality*, Leiden/Boston, MA/Cologne: Brill, 129–53.

Burneyat, M.F., 2004, "Aristotle on the Foundations of Sublunar Physics," in F. De Haas, J. Mansfeld (eds.), *Aristotle: On Generation and Corruption*, Oxford: Clarendon Press, 7–24.

Cashdollar, S., 1973, "Aristotle's Account of Incidental Perception," *Phronesis* 18: 156–175.

Carbone, A.L., 2011, *Aristote illustré. Représentations du corps et schématisation dans la biologie aristotélicienne*, Paris: Classiques Garnier.

Caston, V., 1996, "Why Aristotle Needs Imagination," *Phronesis* 41.1: 20–55.

Caston, V., 2005, "The Spirit and the Letter: Aristotle on Perception," in R. Salles (ed.), *Metaphysics, Soul, and Ethics in Ancient Thought: Themes from the Work of Richard Sorabji*, Oxford: Clarendon Press, 245–320.

Charles, D., 1990, "Aristotle on Meaning, Natural Kinds and Natural History" in D. Devereux, P. Pellegrin (eds.), *Biologie, Logique et Métaphysique chez Aristote*, Paris: CNRS Éditions, 145–167.

Cimatti, F., 2015, "Ten theses on Animality," *Lo Sguardo-Rivista di Filosofia* 18: 41–59.

Claus, D., 1981, *Toward the Soul*, New Haven/London: Yale University Press.

Code, A., 1997, "The Priority of Final Over Efficient Cause in Aristotle's *Parts of Animals*," W. Kullmann, S. Föllinger (eds.), *Aristotelische Biologie*, Stuttgart: Franz Steiner Verlag, 127–43.

Code, A., 2004, "On Generation and Corruption I.5," in F. de Haas J. Mansfeld (eds.), *Aristotle: On Generation and Corruption, Symposium Aristotelicum*, Oxford: Clarendon Press, 171–93.

Collobert, C., 2002, "Aristotle's Review of the Presocratics: Is Aristotle Finally a Historian of Philosophy?" *Journal of the History of Philosophy* 40: 281–95.

Cole, E.B., 1992, "Theophrastus and Aristotle on Animal Intelligence, in W.W. Fortenbaugh, D. Gutas (eds.), *Theophrastus: His Psychological, Doxographical, and Scientific Writings*, New Brunswick, NJ: Transactions Publishers, 44–62.

Connell S.M., 2016, *Aristotle on Female Animals*, Cambridge: Cambridge University Press.

Cooper, J., 1975, *Reason and Human Good in Aristotle*, Cambridge, MA/London: Harvard University Press.

Cooper, J., 2005, "Political Animals and Civic Friendship," in R. Kraut, S. Skultety (eds.), *Aristotle's Politics: Critical Essays*, Lanham, MD: Rowman and Littlefield, 65–89.

Corcilius, K., Gregoric, P., 2010, "Separability *vs.* Difference: Parts and Capacities of the Soul in Aristotle," *Oxford Studies in Ancient Philosophy* 39: 81–120.

Corcilius, K., 2011, "Aristotle's Definition of Non-Rational Pleasure and Pain and Desire," in J. Miller (ed.), *Aristotle's Nicomachean Ethics. A Critical Guide*, Cambridge: Cambridge University Press, 117–40.

Corcilius, K., Perler, D. (eds.), 2014, *Partitioning the Soul. Debates from Plato to Leibniz*, Berlin/Boston, MA: De Gruyter.

Deltel, W., 1999, "Aristotle on Zoological Explanation," *Philosophical Topics* 27.1: 43–68.

Ebert, T., 1983, "Aristotle on What Is Done in Perceiving," *Zeitschrift für philosophische Forschung* 37: 181–198.

Everson, S., 1995, "Proper Sensibles and *kath'auta* causes," *Phronesis* 40: 265–92.

Everson, S., 1997, *Aristotle on Perception*, Oxford: Oxford University Press.

Falcon, A., 2001, *Corpi e Movimenti. Il De caelo di Aristotele e la sua fortuna nel mondo antico*, Napoli: Bibliopolis.

Falcon, A., 2005, *Aristotle and the Science of Nature: Unity Without Uniformity*, Cambridge: Cambridge University Press.

Falcon, A., 2015, "Aristotle and the Study of Animals and Plants," in B. Holmes, K.-D. Fisher (eds.), *The Frontiers of Ancient Science. Essays in Honor of Heinrich Von Staden*, Berlin/Münich/Boston, MA: De Gruyter, 75–91.

Feola, G., 2017, "Alcune considerazioni sull'ordinamento del *corpus biologico di Aristotele*, in M.M. Sassi, E. Coda, G. Feola (eds.), *La zoologia di Aristotele e la sua ricezione dall'età ellenistica e romane alle culture medievali*, Pisa: Pisa University Press, 35–57.

Festugière, A.J., 1936, *Aristote. Le plaisir (Ethique à Nicomaque VII 11–14, X 1-5), introduction, traduction et notes*, Paris: repr. 1960.

Fortenbaugh, W.W., 1970, "On the Antecedents of Aristotle's Bipartite Psychology," *Greek, Roman, and Byzantine Studies* 11: 233–50.

Fortenbaugh, W.W., 2006, *Aristotle's Practical Side: On his Psychology, Ethics, Politics and Rhetoric*, Leiden: Brill.

Frede, D., 1992, "Disintegration and Restoration in Plato's *Philebus*," in R. Kraut (ed.), *The Cambridge Companion to Plato*, Cambridge: Cambridge University Press, 311–37.

Frede, D., 1992, "The Cognitive Role of *Phantasia*," in M.C. Nussbaum, A.O. Rorty (eds.), *Essays in Aristotle's De Anima*, Oxford: Oxford University Press, 279–95.

Frede, D., 2004, "Aristotle's Account of the Origin of Philosophy," *Rhizai: A Journal for Ancient Philosophy and Science* 1: 9–44.

Frede, D., 2009, "Nicomachean Ethics VIII.11-12: Pleasure," in C. Natali (ed.), *Aristotle's Nicomachean Ethics, Book Vii Symposium Aristotelicum*, Oxford University Press, 183–208.

Frede, M., 1987, "Observations on Perception in Plato's Later Dialogues," in M. Frede, *Essays in Ancient Philosophy*, Minneapolis, MN: University of Minnesota Press, 3–8.

Frede, M., 1996a, "Introduction," in M. Frede, G. Striker (eds.), *Rationality in Greek Thought*, Oxford: Oxford University Press, 1–29.

Frede, M., 1996b, "Aristotle's Rationalism," in M. Frede, G. Striker (eds.), *Rationality in Greek Thought*, Oxford: Oxford University Press, 157–73.

Frey, C., 2007, "Organic Unity and the Matter of Man," *Oxford Studies in Ancient Philosophy* 32: 167–204.

Frey, C., 2018, "Aristotle on the Intellect and the Limits of Natural Science," in J.E. Sisko (ed.), *Philosophy of Mind in Antiquity*, London/New York: Routledge, 160–74.

Furley, D.J., 1956, "The Early History of the Concept of the Soul," *BICS* 3: 1–18.

Furth, M., 1988, *Substance, Form and Psyche: An Aristotelian Metaphysics*, Cambridge: Cambridge University Press.

Gastaldi, S., 1998, "La *mimesis* e l'anima," in M. Vegetti (ed.), *Platone La Repubblica*, vol. VII, Napoli: Bibliopolis, 93–150.

Gauthier, R.A., 1970, *L'Éthique à Nicomaque*, Louvain: Publications Universitaires/Paris: Beatrice-Nauwelaerts.

Gelber, J., 2015, "Aristotle on Essence and Habitat," *Oxford Studies in Ancient Philosophy* 48: 267–293.

Gill, M.L., 2009, "The Theory of the Elements in *De Caelo* 3 and 4," in A. Bowen, C. Wildberg (eds.), *New Perspectives in Aristotle's De Caelo*, Leiden: Brill, 139–61.

Gosling, J.C.B., Taylor, C.C.W., 1982, *The Greeks on Pleasure*, Oxford: Oxford University Press.

Gotthelf, A., Lennox, J.G. (eds.), 1987, *Philosophical Issues in Aristotle's Biology*, Cambridge: Cambridge University Press.

Gotthelf, A., 1987, "First Principles in Aristotle's *Parts of Animals*" in A. Gotthelf, J.G. Lennox (eds.), *Philosophical Issues in Aristotle's Biology*, Cambridge: Cambridge University Press, 167–98.

Gotthelf, A., 1988, "Historiae I: Plantarum et Animalium," in Fortenbaugh W.W., Sharples, R.W. (eds.), *Theophrastean Studies: On Natural Science, Physics, and Metaphysics, Ethics, Religion and Rhetoric*, New Brunswick, NJ: Transaction Publishers, 100–35.

Granger, H., 1981, "The differentia and the per se accident in Aristotle" *Archiv für Geschichte der Philosophie* 63: 118–29.

Granger, H., 1985, "The Scala Naturae and the Continuity of Kinds" *Phronesis* 30.2: 181–200.

Gregory, A., 2016, *Anaximander: A Reassessment*, Bungay, Suffolk: Bloomsbury.

Gregoric, P., 2001, "The Heraclitus Anedocte: *De Partibus Animalium* i 5.645a17-23," *Ancient Philosophy* 21: 73–86.

Gregoric, P., Grgic, F, 2006, "Aristotle's Notion of Experience," *Archiv für Geschichte der Philosophie* 88: 1–30.

Gregoric, P., 2007, *Aristotle on the Common Sense*, Oxford: Clarendon Press.

Grosso, R., Zanatta, M., 2005, *La forma del corpo vivente. Studio sul De Anima di Aristotele*, Milano: Unicopli.

Guthrie, W.K.C., 1965, *A History of Greek Philosophy. The Presocratic Tradition from Parmenides to Democritus*, vol. II, Cambridge: Cambridge University Press.

Hankinson, R.J., 1991, *Galen, On the Therapeutic Method*, Oxford: Oxford University Press.

Hutchinson, D.S., Johnson, M.R., 2005, "Authenticating Aristotle's *Protrepticus*," *Oxford Studies in Ancient Philosophy* 29: 193–294.

Irwin, T., 1977, *Plato's Moral Theory: The Early and Middle Dialogues*, Oxford: Clarendon Press.

Johansen, T.K., 1998, *Aristotle on the Sense Organs*, Cambridge: Cambridge University Press.

Johansen, T.K., 1999, "Myth and Logos in Aristotle," in R. Buxton (ed.), *From Myth to Reason? Studies in the Development of Greek Thought*, Oxford: Oxford University Press, 279–91.

Johansen, T., 2000, "Body, Soul and Tripartition in Plato's *Timaeus*," *Oxford Studies in Ancient Philosophy* 19: 87–111.

Johansen, T., 2005, "In Defense of Inner Sense: Aristotle on perceiving that one perceives," *Proceedings of the Boston Area Colloquium in Ancient Philosophy* 21: 235–276.

Johansen, T., 2009, "From Plato's *Timaeus* to Aristotle's De Caelo," in A.C. Bowen, C. Wildberg (eds.), *New Perspectives on Aristotle's De Caelo*, Leiden/Boston, MA: Brill, 1–9.

Johansen, T., 2012, *The Powers of Aristotle's Soul*, Oxford: Oxford University Press.

Johansen, T., 2014, "Parts in Aristotle's Definition of the Soul: *De Anima* Books I and II," in K. Corcilius, D. Perler (eds.), *Partitioning of the Soul from Ancient to Early Modern Philosophy*, Berlin: De Gruyter, 39–61.

Johnson, M.R., 2008 (2005), *Aristotle on Teleology*, Oxford: Oxford University Press.

Johnston, M.A., 2012, "Aristotle on Odour and Smell," *Oxford Studies in Ancient Philosophy* 43: 143–83.

Kahn, C.H., 1960, *Anaximander and the Origin of Greek Cosmology*, New York: Columbia University Press.

Kahn, C.H., 1974/1979, "Sensation and Consciousness in Aristotle's Psychology, in J. Barnes, M. Schofield, R. Sorabji (eds.), *Articles on Aristotle, vol. 4: Psychology and Aesthetics*, London: Bloomsbury, 1–31.

Karbowski, J., 2014, "Empirical *Eulogos* Argumentation in *GA* III, 10," *British Journal of the History of Philosophy* 22: 25–38.

Kraut, R., 1991, *Aristotle on the Human Good*, Princeton, NJ: Princeton University Press.

Kraut, R., 2012, "Aristotle on Becoming Good: Habituation, Reflection, and Perception," in C. Shield (ed.), *The Oxford Handbook of Aristotle*, Oxford: Oxford University Press, 529–57.

Kraut, R., 2001, "Aristotle's Ethics," https://plato.stanford.edu/entries/aristotle-ethics/#Plea

Kullmann, W., 1974, *Wissenschaft und Methode*, Berlin: De Gruyter.

Kullmann, W., 2005, "Man as a Political Animal in Aristotle," in D. Keyt, F. Miller (eds.), *A Companion to Aristotle's Politics*, Oxford: Oxford University Press, 94–117.

Kullmann, W., 2007, *Aristoteles: Über die Teile der Lebewesen*, Berlin: De Gruyter.

Kupreva, I., 2005, "Aristotle on Growth: A Study of the *Argument of* On Generation and Corruption I 5," *Apeiron* 24(2): 104–159.

Labarrière, J.-L., 1984, "Imagination humaine et imagination animale chez Aristote," *Phronesis* 29.1: 17–49.

Labarrière, J.-L., 1990, "De la Phronesis Animal," in D. Devereux, P. Pellegrin (eds.), *Biologie, Logique et Métaphysique chez Aristote*, Paris: CNRS Éditions, 405–28.

Labarrière, J.-L., 2005, *La condition animale. Études sur Aristote et les Stoiciens*, Louvain-La-Neuve: Éditions Peteers.

Le Blond, J.-M., 1938, *Eulogos et l'argument de convenance*, Paris: Taffin-Lefort.

Lee, H.D.P., 1948, "Place-Names and the Date of Aristotle's Biological Work," *Classical Quarterly* 42: 61–7.

Lee, H.D.P., 1985, "On the Fishes of Lesbos, Again," in A. Gotthelf (ed.), *Aristotle on Nature and Living Things*, Pittsburg, PA: Mathesis Publications, 3–8.

Lennox, J.C., 1985, "Demarcating Ancient Science. A discussion of G.E.R. Lloyd, Science, Folklore, and Ideology: the Life Sciences in Ancient Greece," *Oxford Studies in Ancient Philosophy* 3: 307–24.

Lennox, J., 1987, "Divide and Explain: The Theory of the *Analytics* in Practice," in A. Gotthelf, J.C. Lennox (eds.), *Philosophical Issues in Aristotle's Biology*, Cambridge: Cambridge University Press, 90–119.

Lennox, J.O., 1991, "Between Data and Demonstration; The *Analytics* and the *Historia Animalium*," in A.C. Bowen (ed.), *Science and Philosophy in Classical Greece*, New York/London: Garland, 1–37.

Lennox, J.C., 1999a, "Aristotle on the Biological Roots of Virtue: The Natural History of Natural Virtue," in J. Maienschein, M. Ruse (eds.), *Biology and the Foundation of Ethics*, Cambridge: Cambridge University Press, 10–31.

Lennox, J.C., 1999b, "The Place of Mankind in Aristotle's Zoology," *Philosophical Topics* 27: 1–16.

Lennox, J.C., 2011, "The Unity and Purpose of *On the Parts of Animals I*," in R. Bolton, J. Lennox (eds.), *Being, Nature, and Life in Aristotle, Essays in Honor of Alan Gotthelf*, Cambridge: Cambridge University Press, 56–77.

Leunissen, M., 2010, *Explanation and Teleology in Aristotle's Science of Nature*, Cambridge: Cambridge University Press.

Leunissen, M., 2011, "'Crafting Natures': Aristotle on Animal Design," *Philosophic Exchange* 41.1: 28–49.

Leunissen, M., 2018, "Order and Method in Aristotle's *Generation of Animals*," in A. Falcon, D. Lefevbre (eds.), *Aristotle's Generation of Animals. A Critical Guide*, Cambridge: Cambridge University Press, 56–74.

Li Causi, P., Pomelli, R., 2015, *L'anima degli animali. Aristotele, frammenti stoici, Plutarco, Porfirio*. Torino: Einaudi.

Lloyd, G.E.R., 1961, "The Development of Aristotle's Theory of the Classification of Animals," *Phronesis* 6: 59–81.

Lloyd, G.E.R., 1966, *Polarity and Analogy*, Cambridge: Cambridge University Press.

Lloyd, G.E.R., 1981, "Necessity and Essence in the Posterior Analytics," in E. Berti (ed.), *Aristotle on Science*, Padova: Editrice Antenore, 157–71.

Lloyd, G.E.R., 1983, *Science, Folklore and Ideology*, Cambridge: Cambridge University Press.

Lloyd, G.E.R., 1992, "Aspects of the Relationship Between Aristotle's Psychology and his Zoology," in M.C. Nussbaum, A.O. Rorty (eds.), *Essays on Aristotle's De Anima*, Oxford: Oxford University Press, 147–67.

Lloyd, G.E.R., 1996, *Aristotelian Explorations*, Cambridge: Cambridge University Press.

Lloyd, G.E.R., 2012, *Being, Humanity and Understanding*, Oxford: Oxford University Press.

Long, C.P., 2011, *Aristotle on the Nature of Truth*, Cambridge: Cambridge University Press.

Longrigg, J., 1965, "Empedocles' Fiery Fish," *Journal of the Warburg and Courtauld Institutes* 28: 314–5.

Longrigg, J., 1993, *Greek Rational Medicine: Philosophy and Medicine from Alcmaeon to the Alexandrians*, London/New York: Routledge.

Lord, C., 1991, "Aristotle's Anthropology," in C. Lord, D. O'Connor (eds.), *Essays on the Foundations of Aristotelian Political Science*, Berkeley, CA/Los Angeles, CA/Oxford: University of California Press, 49–73.

Lorenz, H., 2006, *The Brute Within: Appetitive Desire in Plato and Aristotle*, Oxford: Oxford University Press.

Louis, P., 1952, "Le traité d'Aristote sur la nutrition," *Revue de Philologie* 26: 29–35.

Lovejoy, A.O., 1936, *The Great Chain of Being: A Study of the History of an Idea*, Cambridge, MA: Harvard University Press.

Machamer, P., 1978, "Aristotle on Natural Place and Natural Motion," *Isis* 69: 377–87.

Manuli, P., 1977, *Cuore, sangue, cervello: biologia e antropologia nel pensiero antico*, Milano: Episteme editrice.

Marmodoro, A., 2014, *Aristotle on Perceiving Objects*, Oxford: Oxford University Press.

Matthen, M., 2009, "Why Does the Earth Move to the Center? An Examination of Some Explanatory Strategies in Aristotle's Cosmology" in A. Bowen, C. Wildberg (eds.), *New Perspectives in Aristotle's De Caelo*, Leiden: Brill, 119–38.

Menn, S., 2002, "Aristotle's Definition of Soul and the Programme of the *De Anima*," *Oxford Studies in Ancient Philosophy* 22: 81–139.

Mingucci, G., 2015, *La fisiologia del pensiero in Aristotele*, Bologna: Il Mulino.

Modrak, D.K.W., 1987, *Aristotle: The Power of Perception*, Chicago, IL: Chicago University Press.

Moraux, P., 1951, *Les listes anciennes des ouvrages d'Aristote*, Louvain: Éditions Universitaires.

Moreau, J., 1959, "L'éloge de la biologie chez Aristote," *Revue des études anciennes* 61: 57–64.

Morison, B., 2002, *On Location: Aristotle's Concept of Place*, Oxford: Oxford University Press.

Moss, J., 2005, "Shame, Pleasure and the Divided Soul," *Oxford Studies in Ancient Philosophy* 29: 137–170.

Moss, J., 2008, "Appearances and Calculations: Plato's Division of the Soul," *Oxford Studies in Ancient Philosophy* 34: 35–68.

Moss, J., 2012, *Aristotle on the Apparent Good*, Oxford: Oxford University Press.

Mulgan, R.G., 1974, "Aristotle's Doctrine that Man is a Political Animal," *Hermes* 102, 438–45.

Murphy, D., 2005, "Aristotle on Why Plants Cannot Perceive," *OSAP* 29:295–339.

Naddaf, G., 2005, *The Greek Concept of Nature*, Albany, NY: SUNY Press.

Naddaf, R., 2002, *Exiling the Poets: The Production of Censorship in Plato's "Republic,"* Chicago, IL: Chicago University Press.

Natali, C., 2014, *Aristotele*, Roma: Carocci.

Nussbaum, M., 1978, "The Role of *Phantasia* in Aristotle's Explanation of Action," in M. Nussbaum, *Aristotle's De Motu Animalium*, Princeton, NJ: Princeton University Press.

Nussbaum, M., 1983, "The Common Explanation of Animal Motion," in P. Moreaux, J. Wiesner (eds.), *Zweifelhaftes im Corpus Aristotelicum*, Berlin: De Gruyter.

Osborne, C., 2007, *Dumb Beasts and Dead Philosophers: Humanity and the Humane in Ancient Philosophy and Literature*, Oxford: Oxford University Press.

Papachristou, C., 2013, "Three Kinds or Grades of *Phantasia* in Aristotle's *De Anima*" *Journal of Ancient Philosophy* 7.1: 19–48.

Pellegrin, P., 1986, *Aristotle's Classification of Animals: Biology and the Conceptual Unity of the Aristotelian Corpus* (transl. by A. Preus), Berkeley, CA: University of California Press.

Pellegrin, P., 1996, "Le De Anima et la vie animale: trois remarques," in C. Viano, G. Romeyer Dherby (eds.), *Corps et âme. Études sur le De Anima*, Paris: J. Vrin, 464–92.

Pellegrin, P., 2019, "Le plaisir animal selon Aristote," *Chora* 17: 145–62.

Penner, T., 1971, "Thought and Desire in Plato," in G. Vlastos (ed.), *Plato 2: Ethics, Politics, Religion and the Soul*, Garden City, NY: Doubleday Anchor Books, 96–118.

Preus, A., 1990, "Animal and Human Soul in the Peripathetic School," *Skepsis* 1: 67–99.

Preus, A., 1990, "Man and Cosmos in Aristotle: Metaphysics Lambda and the Biological Works," in D. Devereux, P. Pellegrin (eds.), *Biologie, Logique, et Métaphysique*, Paris: C.N.R.S. Éditions, 467–87.

Price, A.W., 2017, "Variety of Pleasures in Plato and Aristotle," *Oxford Studies in Ancient Philosophy* 52: 177–208.

Rabinoff, E., 2018, *Perception in Aristotle's Ethics*, Evanston, IL: Northwestern University Press.

Randall, J.H., 1960, *Aristotle*, New York: Columbia University Press.

Rapp, C. 2009, "Nicomachean Ethics VIII.11-12: Pleasure and *eudaimonia*," in C. Natali (ed.), *Aristotle's Nicomachean Ethics, Book VII Symposium Aristotelicum*, Oxford University Press, 209–35.

Rashed, M., 2011, "Aristote à Rome au IIe siècle: Galien *De Indolentia* && 15–18," *Elenchos* 32: 55–77.

Rees, D.A., 1957, "Bipartition of the Soul in the Early Academy" *JHS* 77:112–8.

Repici, L., 1985, "Il paradigma animale nella botanica di Teofrasto," *Rivista di filosofia* 76: 367–98.

Rorty, A.O., 2003, repr. "*De Anima*: Its Agenda and Its Recent Interpreters," in M. Nussbaum, A. Rorty (eds.), *Essays on Aristotle's De Anima*, Oxford: Oxford University Press (1st ed. 1992).

Salkever, S.G., 2005, "Aristotle's Social Science," in R. Kraut, S. Skultety (eds.), *Aristotle's Politics: Critical Essays*, Lanham, MD: Rowman and Littlefield, 27–64.

Saunders, T.J., 1962, "The Structure of the Soul and the State in Plato's *Laws*," *Eranos* 60: 37–55.

Saunders, T.J., 1995, *Aristotle: Politics I & II*, Oxford: Clarendon Press.

Scaltzas, T., 1994, 2010, *Substances and Universals in Aristotle's Metaphysics*, 2nd ed., Ithaca, NY: Cornell University Press.

Scheiter, K.M., 2012, "Images, Appearances, and Phantasia in Aristotle," *Phronesis* 57.3 251–78.

Schofield, M., 1992, "Aristotle on The Imagination," in M. Craven Nussbaum, A. Rorty (eds.), *Essays in Aristotle's De Anima*, Oxford: Oxford University Press, 249–77.

Shields, C., 2010, "Plato's Divided Soul," in M. McPherran (ed.), *The Cambridge Companion to Plato's Republic*, Cambridge: Cambridge University Press, 147–70; repr. In K. Corcilius, D. Perler (eds.), 2014, *Partitioning the Soul: Debates from Plato to Leibniz*, Berlin: de Gruyter, 15–38.

Shields, C., 2013a, "The Science of Soul in Aristotle's *Ethics*," in K.M. Nielsen, D. Henry (eds.), *Bridging the Gap between Aristotle's Science and Ethics*, Cambridge: Cambridge University Press, 232–53.

Shields, C., 2013b, "The Grounds of *Logos*: the Interweaving of Forms," in F. Miller, G. Anagnostopoulos (eds.), *Reason and Analysis in Ancient Greek Philosophy*, Heidelberg/ New York/London: Springer, 211–30.

Shields, C., 2014, *Aristotle*, 2nd ed., London and New York: Routledge.

Schirren, T., 1998, *Aisthesis vor Platon. Eine semantisch-systematische Untersuchung zum Problem der Wahrnehmung*, Stuttgart/Leipzig: De Gruyter.

Sedley, D., 1991, "Is Aristotle's Teleology Anthropocentric?" *Phronesis* 36:179–96.

Solmsen, F., 1955, "Antecedents of Aristotle's Psychology and Scale of Being," *Classical Philology* 2: 148–64.

Sorabji, R., 1971, "Aristotle on Demarcating the Five Senses," *Philosophical Review* 80.1: 55–79.

Sorabji, R., 1972a, "Aristotle, Mathematics and Colour," *Classical Quarterly* 22.2: 293–308.

Sorabji, R., 1972b, *Aristotle on Memory*, Providence, RI: Brown University Press.

Sorabji, R., 1974, "Body and Soul in Aristotle, *Philosophy* 49: 63–89.

Sorabji, R., 1988, *Matter, Space and Motion: Theories in Antiquity and Their Sequel*, Ithaca, NY/London: Cornell University Press.

Sorabji, R., 1992, "Intentionality and Physiological Processes: Aristotle's Theory of Sense-Perception," in M. Nussbaum, A. Rorty (eds.), *Essays on Aristotle's De Anima*, Oxford: Oxford University Press, 195–226.

Sorabji, R., 1993, *Animal Minds and Human Morals*, Ithaca, NY: Cornell University Press.

Sorabji, R., 2001, 'Aristotle on Sensory Perception and Intentionality: A Reply to Myles Burneyat," in D. Perler (ed.), *Ancient and Medieval Theories of Intentionality*, Leiden/ Boston, MA/Cologne: Brill, 49–61.

Spargue, R.K., 1989, "Aristotle and Divided Insects," *Méthexis* 2: 29–40.

Stalley, R.F., 1975, "Plato's Argument for the Division of the Reasoning and Appetitive Elements within the Soul," *Phronesis* 20: 110–128.

Stenzel, J., 1929, "Speusippos," in *Pauly-Wissowa, Real-Encyclopädie der classischen Altertumswissenschaft*, 2nd ser., vol. III, Stuttgart: Metzler, 1636–69.

Stigen, A., 1961, "On the Alleged Primacy of Sight—With Some Remarks on *Theoria* and *Praxis*—In Aristotle", *Symbolae Osloenses* 37: 15–44.

Strauss, L., 1959, "The Liberalism of Classical Political Philosophy, " *Review of Metaphysics* 12.3: 390–439.

Strevell, E., 2016, *Memory, Phantasia, and the Perception of Time: A Commentary on Aristotle's De Memoria Et Reminiscentia*. Doctoral dissertation, Duquesne University. Retrieved from https://dsc.duc.edu/edt/31

Taylor, C.C.W., 2003, "Pleasure: Aristotle's Response to Plato," in R. Heinaman (ed.), *Plato and Aristotle's Ethics*, Aldershot: Ashgate, 1–20.

Tipton, J.A., 2014, *Philosophical Biology in Aristotle's Parts of Animals*, Cham: Springer.

Vander Waerdt, P.A., 1985, "The Peripatetic Interpretation of Plato's Tripartite Psychology," *Greek, Roman, and Byzantine Studies* 26: 283–302.

Vander Waerdt, P.A., 1987, "Aristotle's Criticism of Soul-Division," *American Journal of Philology* 108: 627–43.

Vlastos, G., 1946, "Ethics and Physics in Democritus, II" *Philosophical Review* 55: 53–64.

Vlastos, G., 1952, "Theology and Philosophy in Ancient Greek Thought," *The Philosophical Quarterly* 2: 97–123.

Ward, J., 1988, "Perception and Logos in *De anima* II.12," *Ancient Philosophy* 8: 217–33.

Warren, J., 2007, "Anaxagoras on Perception, Pleasure, and Pain," *OSAP* 33: 19–54.

Wedin, M., 1988, *Mind and Imagination in Aristotle*, New Haven/London: Yale University Press.

White, K., 1985, "The Meaning of *Phantasia* in Aristotle's *De Anima*, III, 3–8," *Dialogue* XXIV: 483–505.

Whiting, J.E., 2002, "Locomotive Soul: The Parts of the Soul in Aristotle's Scientific Works," *Oxford Studies in Ancient Philosophy* 22: 141–200.

Wicksteed, P., Cornford, M.F., 1957, *Aristotle. The Physics*, Cambridge, MA: Harvard University Press.

Witt, C., 2012, "Aristotle on Deformed Animal Kinds," *Oxford Studies in Ancient Philosophy* 43: 83–106.

Wolfsdort, D., 2013, *Pleasure in Ancient Greek Philosophy*, Cambridge, UK: Cambridge University Press.

Wyckoff, J., 2015, "Analysing Animality: A Critical Approach," *The Philosophical Quarterly* 65.260: 529–546.

Yack, B., 1993, *The Problems of a Political Animal*, Berkeley, CA/Los Angeles, CA/London: de Gruyter.

Zatta, C. 2011, "Conflict and City-Space: Some Exempla from Thucydides' History," *ClAnt* 30.2: 318–350.

Zatta, C., 2018a, "Aristotle on the Sweat of the Earth," *Philosophia* 48: 55–70.

Zatta, C., 2018b, "Aristotle's Animals in Hobbes' Philosophy of Man," in C. Doherty, B. King, *Thinking the Greeks. Festschrift in Honor of James Redfield*, New York/London: Routledge, 216–28.

Zatta, C., 2019, *Interconnectedness. The Living World of the Early Greek Philosophers*, Baden Baden: Academia Verlag (second revised edition).

Zatta, C., 2022, "Memory and Imagination: From Aristotle's Silent Speech to Euripides' Tragic Utterances," *Medicina nei Secoli. Journal of Medicine and Medical Humanities*, 34.1.

Zatta, C., (forthcoming), "Plants' Physiology and Life from the Presocratics to Aristotle," *Archives Internationales de l'Histoire des Sciences*.

Zirin, R., 1980, "Aristotle's Biology of Language," *TAPA* 110, 325–47.

Index

air 24, 27, 52–53, 80, 83, 85, 117–21, 145–46, 180–84, 215

aisthētikon zōon (sentient animal) 109–14, 125

alteration (*alloiōsis*) 3, 16, 25, 34, 51, 80–82, 85, 110–14, 121, 160; *see also* sensation

Anaxagoras 2, 5, 24, 34, 36, 182, 216

Anaximander 179–80

animality/animalness 4, 7, 78–80, 140, 142, 153, 161, 219

animals: activities in the external world (*praxeis*) 5, 25, 150–51, 158, 177, 187–88, 199, 216; *aloga zōa* 32–33, 149; analogous natural capacities 191; analogous to human 190, 195; anarchical 8, 199; aquatic 182; *see also* birds; environment; articulation of differences and sameness 8, 36, 177–79; blooded 15, 17–18, 22, 87, 91, 179, 181–82, 191–92; bloodless 15–16, 36, 118, 124, 179, 181, 191–92; body parts 5–6, 14, 18–22, 24–26, 28, 32–33, 36, 78, 80–81, 83, 85–86, 88, 90–91, 95, 117–18, 120, 149–51, 175–78, 181–83, 185, 188, 190; celestial 12–14, 36, 218; character/s (*ēthos/ēthē*) 8, 13, 27, 59, 110, 125, 175–77, 187–95; common activity (*koinon ergon*) 196–97, 199; communication 8, 119, 150, 153, 190; *see also phantasia*; dualizers 195; dwarf-like (*nanōdes*) 2, 17; elements as the matter of the body parts 30, 32, 35, 117–18; emotions 2, 8, 151, 176; equality/egalitarianism 4, 34–36, 216; experience (*empeiria*) 4, 153, 161, 189; feeding habits 33, 183; fight for survival 184, 187; finality 7, 25, 28, 30, 32–33, 35, 83, 94–96, 175; *see also* teleology; intelligent (*phronimoi*)

195; under a leader 201; learning (*mathēsis*) 8, 189–90; less intelligent (*aphronestera*) 17–19; life/mode of (*bios*) 33, 179, 185, 188; migration 8, 151, 175–76, 197–98, 200; nonuniform parts 15, 26, 30, 32, 85, 179, 181–82; *phronesis* (intelligence) 2, 8–9, 195; *physis* 27–28 (*geodestera* (more earthen) 192; *koinē* (common) 14, 20–21, 36, 51, 80, 91, 116, 139, 177, 184, 201; *see also* nature; *theic* (divine) 17; *thnētē* (mortal) 94–95); *politika zōa* 8, 34, 176, 196–97; power of forethought (*dynamis pronoētikē*) 161, 195; practical life 8, 151; *see also* activities (*praxeis*); life as *polymerēs praxis*; purposeful action 161; quadrupeds 17, 90, 185–86, 195; and self-awareness 193–95, 200; spatial identity 199; stationary 8, 19, 159, 185; *synesis* (comprehension) 161; teaching (*didaskalia*) 189–90; terrestrial 12, 109, 146, 180–82, 196; uniform parts 15, 30, 32, 85–86, 118, 177, 181; (function 32–33, 36); voice (*phonē*) 152–54; work 32–33, 191, 197, 199–201; work/activity (*ergon*) 5, 15, 32–33; *see also* Theophrastus; worthier (*timiōteron*)/more worthless (*atimōteron*) 34; *see also* blood; diet

ants 2, 8, 33, 191, 196, 198–201

Aquinas, T. 1, 218–19

arkhē: as intrinsic origin of movement and rest 6, 24, 26, 34, 89, 217; as material origin (*hylikē arkhē*) 27, 53; as origin of the cosmos 23–24; as power 61, 199; *see also logos, telos/arkhē*; soul

art (*tekhnē*) 4, 12, 28, 30, 32–33, 35, 145–46, 161, 189–91, 201

automata 89

Balme, D. 2, 78–79, 175
Beare, J. I. 153
bears 197
bees 2, 8, 125, 183, 196–200, 216
birds 17, 20–22, 29, 181–86, 196, 198;
    aquatic 185–87; care for the young 197;
    crook-taloned (rapacious) 175, 185, 196;
    eubiotos 147; language 190; migratory
    8, 186, 198–200; terrestrial 185–86;
    web-footed 186
blood 15–16, 36, 89–90, 155, 181, 186,
    192, 194; and character 191–92; and
    intelligence 191–92; and sensation
    191–92; *see also* animals, blooded;
    animals, bloodless

Caston, V. 115
cause/causes (*aitia, aitiai*): coincidence of
    30, 82–84; efficient 27, 81; final 6, 27,
    81; formal 6, 27, 81; material 27, 81;
    see also arkhē, hylikē arkhē; conditional
    necessity
chronology, ontological *vs.* linear 31
Code, A. 86
color/s 4, 15, 31, 114–16, 118–19, 121–26,
    131, 141, 192, 216, 218
common sense (*koinē aisthēsis*) 7, 114,
    124–26, 156–58
complex bodies 7, 51, 80–87, 145, 201;
    bone/s 20–21, 85–87, 90, 117–18, 122,
    178, 181; simple bodies (elements) 28,
    80, 83, 85–86, 117, 120, 216; *see also*
    animals, body parts
conditional necessity 16, 82
contemplation (*theōria*) 34, 67
Corcilius, K 62
cranes 2, 8, 179, 196–98, 205–6
crayfishes 191
crocodile 183, 187–88, 195

death 20, 182, 200
deer 148, 188, 194–95
Democritus 6, 15, 24, 31, 33, 78, 83, 87,
    179–82, 217–18
desire/s 5, 7–9, 14–15, 35–36, 56–62, 79,
    82, 91–93, 95, 110, 119, 127, 137–40,
    142–44, 149–52, 154, 159, 176, 201,
    215–16, 219
diet 146, 179–81, 184–86; meat-based 175;
    omnivorous 186; "vegetarian" 186; *see
    also* nutrition
Diotima 93–95
dreams 154–56

earth 12, 14, 22, 24, 80, 86, 89–90,
    117–18, 120, 145, 160, 181, 183, 186,
    196, 216
elements 3, 7, 13–14, 24–25, 28, 30, 32,
    36, 80–82, 85–87, 110, 114, 116–18,
    120–22, 145, 150, 180–81, 216
elephant/s 33, 179, 188, 192
embryo 7, 15, 58–59, 89–91, 198; *see also*
    fetus/fetation
embryogenesis 35
Empedocles 5, 21, 24–26, 31–32, 53, 83,
    160, 178–82, 184, 217
environment 3, 5, 7, 33, 79–80, 83, 90,
    110, 113, 121–23, 126–27, 138, 146,
    158, 160–61, 176–88, 190, 194, 196,
    199, 218
ephēxes ("next in succession") 64–65
Eudoxus of Cnidus 148

fetus/fetation 89–91
finality 7, 25, 28, 30, 32–33, 35, 83, 94–96,
    175; *see also* teleology
fire 24, 52–53, 80, 87–88, 111, 117–18,
    126, 144–45, 155, 160, 181
food 16, 19, 25, 27, 63, 83, 85, 87–93, 96,
    119, 122, 137–38, 142–43, 146, 151,
    155–56, 175–77, 180, 183–88, 193,
    196–200, 215–16, 219; production 176,
    197–98
form (*morphē/eidos*) 6, 21–22, 26–28,
    30–31, 34, 51, 53, 81, 83–86, 88–89,
    92–93, 95, 113–14, 118; *oikeia morphē*
    90; *see also* sensation
Fortenbaugh, W.W. 193

generation/formation (*genesis*) 26–27, 31
good life (*eu zēn*) 18, 66–67, 119, 147; *see
    also* birds, *eubiotos*
Gotthelf, A. 175
Gregoric, P. 61–62, 124–25
growth (*auxēsis*) 3, 14, 16–18, 20, 25–27,
    34, 51, 53, 58–59, 63, 80–82, 84–90,
    92–93, 113, 157, 159, 176, 184, 195

hand/s 15, 19, 78, 85, 89, 178
heart 15–16, 36, 81, 88, 90, 124, 155, 176,
    181, 191
heat 16, 89, 117, 150, 155–56, 192, 194;
    emphyton thermon 17, 35, 87, 176,
    181–82, 216
Heraclitus 12, 87, 147, 149, 185, 216
History of Animals 5, 8, 13, 18, 20–21, 29,
    33, 36, 94, 96, 123, 125–26, 138, 143,

146–47, 153, 161, 175–79, 181–82, 192, 194, 197–99, 201, 215
horses 146–47, 185–86
human 90; of the embryo 59; plants 29, 85–88; quadrupeds 90
human being/s 2–3, 6–7, 17–18, 21–22, 31–31, 35, 56, 59–60, 62–63, 66–67, 78, 90–93, 96–97, 109–12, 139–42, 145, 147–49, 151–53, 156, 160–61, 176, 180, 189–90, 193, 196–98, 201, 217–19
human intellectual faculties (*to phronein, to noein*) 19; *see also logismos* (calculation); *logos* as reason
hylomorphism 85–86

immortality 4, 12, 35, 83, 91, 93–96; *see also* desire/s

Kahn, C. 23–24
knowability 13, 218
knowledge (*gnōsis*) 189
koinon (*to*) 3, 6, 20, 22–23, 52, 63; by analogy (*kat' analogian*) 22, 177–78; by degree (*kata genos*) 20, 23, 177; *ta koinēi symbebēkota vs.* animals' *ousia* 20–22; *see also* animals, common activity (*koinon ergon*); animals, *koinē physis*

Lennox, J. 13, 18–19, 94, 175, 193
Leunissen, M. 35, 65
life as *polymerēs praxis* (complex action) 31–32, 149
lion 13, 20, 125–26, 141, 152, 188, 193–94
Lloyd, G.E.R. 200
locomotion/movement across space 6, 8, 16–17, 19, 25–26, 51, 63, 66, 79–81, 109, 138–39, 146, 150–52, 154, 159–60, 176–77, 181–85, 194, 217
logismos (calculation/reason/reasoning) 17, 79, 94, 156, 161
logos: as account/discourse 57; decentralization/banishment of 17, 35; as definition 26, 29, 31, 63; expressive (*prophorikos*) and internal (*endiathetos*) 217; as form/artful structure/immanent finality and rationality, embodied *telos* 3, 6–9, 28, 30–32, 34, 66, 81, 87, 89, 110, 118, 158, 178–79, 218; *logoi* (rational grounds) 28, 32; *logoi* of sensible qualities 115; and *ousia* 31; and *peras* 7, 87–88; *see also* plants; sensation; as a polyvalent word 29; as

proportion/ratio 7, 86–88, 114–16, 122; as reason and soul bipartition 55, 57, 62; as reason and soul tripartition 56–57, 62; as reason and speech 1–3, 7, 17, 29, 32, 34, 55, 58, 63–64, 66, 79, 138, 150, 153, 157, 161, 191, 196, 200, 217–18; as reason *vs.* life 62–63; as *telos/arkhē* 28, 30; *vs.* matter and *tykhē* 6–7, 88

Magna Moralia 55
matter (*hylē*) 3, 6–7, 23–25, 27, 30–31, 36, 53, 79–80, 82–83, 85–88, 93–95, 114–16, 120, 184, 192
memory (*mnēmē*) 4, 8, 79, 110, 123, 127, 154–58, 161, 188–89, 194–95, 197
Metaphysics 1–2, 4, 34, 113, 156, 188–90
Meteorologics 3, 81
methodological principles 11–36, 119
mind (*nous*) 1–2, 7, 18, 24, 34, 63, 79–80, 109–11, 151, 155, 159–61, 191, 218–19; objects of (*ta noēta*) 79, 153, 159–61
moral choice (*proairēsis*) 25
movement/s (*kinēsis*) 3–4, 6–8, 16–19, 110, 118, 122–23, 126, 139, 142–43, 151, 155–57, 159, 175–76, 184, 191, 216–18; *see also* growth; locomotion/movement across space; *phantastikai kineseis* (imaginative movements); sensation

nature (*physis*) 2–3, 23, 26, 95–97, 139, 215; *oikeiotera* ("more familiar nature") 11, 13–14, 22, 218; *see also* animals, *physis*
Nicomachean Ethics 5–6, 8, 55, 58–60, 62, 64–65, 110, 139–40, 152, 161, 185
nutrition 5–8, 14–15, 22, 29, 34, 36, 53, 58–59, 63, 66–67, 79, 83, 86–93, 96, 119, 137–40, 142–47, 159, 175–78, 180, 182, 184–86, 197–98, 216–17; *see also* plants

On Generation and Corruption 5, 7, 23, 34, 81–82, 84–89
On Respiration 179, 181, 184
On Sleep and Wake 124–25
On the Generation of Animals 1–2, 7, 15, 27, 35, 79, 89, 91–92, 176
On the Heavens 12, 23
On the Movement of Animals 5, 8, 159
On the Sense 115, 125, 138, 154
On the Soul 1, 4–8, 14, 20, 51–55, 57, 59–64, 66, 78–83, 86, 88–89, 91–93,

95–96, 109, 111, 115, 117–22, 124–26, 137

pain/s 5, 8, 15, 66, 79, 88, 90–91, 110, 119, 127, 137–40, 143–48, 150, 152–55, 177, 216
Parts of Animals 1, 3–7, 11–12, 14–15, 17, 19–23, 25–28, 30, 32–33, 36, 51, 63, 66, 78–80, 82, 89–90, 93, 118–20, 141, 149, 176–77, 181, 191, 195, 217–18
Parva Naturalia 5, 8, 20, 154, 161
Peck, A.L. 29, 31
Pellegrin, P. 21, 23, 25, 138
Phaedrus 55
phantasia (imagination) 4–5, 8, 36, 66, 79, 82, 109–10, 114, 127, 138, 141, 150–61, 191, 193–94, 216–18; *aisthētikē vs. bouleutikē* 151; *aoristōs* (in an indeterminate way) 152; etymology of 152; as a response to the Presocratics 159–61; as synthetizing capacity 153
phantasia as *noēsis* 165, 167, 199
phantasiai (imaginations) 151–52, 154, 159, 161, 189–90
phantastikai kineseis (imaginative movements) 155–56; role in animals' communication 153
philosophy of mind 110
Physics 18, 23–25, 27, 32–33, 65, 80–85, 191
physiologoi (students of nature) 24, 87, 118, 140
plants 3–7, 11, 13–14, 16–17, 19–20, 24–25, 27–29, 33, 35–36, 66, 80–81, 88–89, 94–96, 110, 114, 145, 179, 181, 185–86, 215–16; inability to sense 116–17, 120–21; lack of *mesotēs* 120; *logos* 28–29; mode of living 29, 81, 216; nutrition 87–88, 137, 184; nutritive soul 54, 64, 66, 88–89, 92; uniform parts 28, 120; *see also* growth
pleasure 5, 8, 15, 35, 79, 88, 90–91, 110, 119, 137–50, 152–54, 177, 180, 184–85, 216; of the body (*sōmatikai*) 56, 92, 140–43; common pleasure (*koinē hēdonē*) 142, 144; communication of 66, 154; definition/s of 8, 142–43, 150; denial of pleasures 138, 142; desire for 8, 110, 138; and the desire to live 35, 149, 219; naturalness of 8, 142–43; natural pleasure (*kata physin*) 184–85; and pain 144; provided by the knowledge of animals' causes 11; of the soul (*psykhikai*) 140, 143; special

pleasure (*oikeia hēdonē*) 146–47; universal pursuit of 146
Polansky, R. 94, 111, 125
Politics 1, 5, 7–8, 29, 55, 60, 64–65, 90, 95–96, 119, 153, 176, 196–97, 199–201
politikōs 197
Porphyry 217
presentness 127, 151
Presocratic philosophers/early Greek philosophers 52, 83, 181; *see also physiologoi*
Protrepticus 26, 55, 58, 60–61
Pythagoreans 53

ravens 153–54
reproduction 6–7, 14–15, 35–36, 88–96, 139, 148, 175–79, 197–98; as *physikotaton tōn ergōn* 92
Republic 6, 36, 54–55, 57, 60
respiration 20, 182

scala naturae 64
sea anemones 152
seals 187, 196
sea tortoises 185
semen 89–93
sensation (*aisthēsis*) 1–8, 14–18, 22–23, 34–36, 52–53, 59, 62–64, 66–67, 79, 81–82, 88, 109–14, 116–27, 137–41, 150–61, 176, 185, 189, 191–94, 215–19; awareness of sensing 7, 109, 124, 127; as inborn power 113; medium of 117, 119, 121, 215; objects of (*aisthēta*) 79, 115, 153 (common (*koina*) 7, 121–23, 125–26, 152, 157; incidental (*kata sym*) 7–8, 114, 121–23, 125–26, 145, 152, 190, 193; proper (*idia*) 3, 7, 115, 121, 123, 125, 152; perception of time (*aisthēsis chronou*) 157–59; perceptual discrimination 123–25, 127; sensation between likes 121–22, 160; between unlikes 121
sensation and *logos* 114–16, 122
sensation and pain 140
sensation as actuality 111–14
sensation as potentiality (*dynamis*): developed 111–14; undeveloped 111–13
sense as *mesotēs* (midpoint) 116–17, 139; contact senses 66, 119, 121, 124, 140, 142, 152; distant senses 91, 119, 121, 125, 140–42, 146–48, 152, 217
sensible form/s (*aisthēta eidos/eidē*) 115–17, 120–21

sensorial experience/s 66, 123, 152–53; *see also* alteration (*alloiōsis*); common sense; pleasure

shape (*skhēma*) 15–31, 86

sight 57, 66, 113–14, 117–19, 121, 124, 141, 145, 152, 155, 189, 194, 215

sleep 20, 58, 111, 124, 151, 154–56

snakes 186

soul: as *arkhē* of the *zōon aisthētikon* 109; as *arkhē zōōn* 51; bipartition 6, 54–55, 57–62, 91–92; division, *aplōs* (simpliciter) 61, 88; geometrical model of 65–66, 88, 93, 125; homogeneity 16, 51–52; human 4, 52, 54–57, 59, 61; locomotive 6, 16, 19–20, 29, 51, 63–66, 78, 83, 125, 137–38; *logōi* (in account) 61, 88, 90; "most common definition" of (*koinotatos logos*) 63; nutritive (*threptikon*) 6–7, 14–16, 19–20, 29, 51, 54, 58–59, 62–66, 78, 81–82, 86, 88–94, 109, 120, 125, 137, 139, 148; partition 4, 6, 51–52, 54, 57, 60, 62, 65–66, 154, 159–60, 216; part/s 3, 6, 16, 18–19, 25, 29, 51–66, 78, 92, 110, 125, 137, 140, 160, 218; political model of 63, 65; *prōtē kai physikotatē* 177, 185; rational (*noētikon, logon ekhon*) 4, 6–7, 16–18, 22, 30, 33, 36, 51, 54–66, 78–80, 95, 109–10, 137, 140, 143, 160–61, 191, 195, 216–19; sensitive (*aisthētikon*) 4, 6, 14, 16, 19–20, 29, 51, 54, 62–66, 78, 82, 88–91, 94, 109, 113, 121, 123–26, 137, 154, 156–57, 159–60, 177, 190–91, 195, 201, 216; sensitive (*aisthētikon*) qua imaginative (*phantastikon*) 156; tripartition 6, 36, 54–55, 57, 61–62, 91; *see also* plants

space of existence 5, 8, 32, 67, 80, 161, 174, 177, 184, 216; *see also* environment; spider/s 32–33, 146, 187, 191

spirit (*thymos*) 36, 56, 192

swallow/s 33, 191, 197

Symposium 93

synistēmi/synistamai 28, 89

synthesis ("body composition") 7, 14, 32, 118, 175, 179; *see also* animals, body parts

syntrophos 13

teleology 2, 6, 16, 55–97, 119, 191; *see also* finality

telos (end, completion) 7, 26–32, 35, 83, 87–88, 90, 94–96, 143; as life 32, 176, 217

temperance (*sophrosynē*) 55, 140–42, 148, 193

Theophrastus 13–14, 121, 146, 148

thought/intelligence (*dianoia*) 16–17, 59, 79, 110, 191

Timaeus 5–6, 12, 16–18, 23, 36, 55, 63

touch 1–3, 7, 8, 34, 64–67, 88, 90–91, 114, 116, 118–20, 124, 137–42, 146, 152, 179, 198, 215–16; as *prōtē aisthēsis* 118; sense organ of 15, 116–18, 120–21, 124

tykhē (chance) 6, 23–25, 83, 93, 218

wasps 2, 8, 183, 196, 198–200

water 24, 27, 52, 57, 80, 82–83, 85–87, 190–92, 196, 198, 213, 216

wisdom (*sophia*) 55, 161, 190–91, 201

wolf 186, 188

womb 79, 89–91

zoocentrism/zoocentric 3, 6–7, 22, 35, 67, 79, 82, 94, 219

zoology 6, 8, 13, 16–17, 29, 31, 34–35, 82, 113, 148, 175, 196, 199–200, 216, 218

zoophyta ("animal plants") 152